AF684931

THE BEHAVIOR of SANDWICH STRUCTURES of ISOTROPIC and COMPOSITE MATERIALS

HOW TO ORDER THIS BOOK

BY PHONE: 800-233-9936 or 717-291-5609, 8AM–5PM Eastern Time

BY FAX: 717-295-4538

BY MAIL: Order Department
Technomic Publishing Company, Inc.
851 New Holland Avenue, Box 3535
Lancaster, PA 17604, U.S.A.

BY CREDIT CARD: American Express, VISA, MasterCard

BY WWW SITE: http://www.techpub.com

PERMISSION TO PHOTOCOPY-POLICY STATEMENT

Authorization to photocopy items for internal or personal use, or the internal or personal use of specific clients, is granted by Technomic Publishing Co., Inc. provided that the base fee of US $3.00 per copy, plus US $ .25 per page is paid directly to Copyright Clearance Center, 222 Rosewood Drive, Danvers, MA 01923, USA. For those organizations that have been granted a photocopy license by CCC, a separate system of payment has been arranged. The fee code for users of the Transactional Reporting Service is 1-56676/99 $5.00 + $ .25.

The Behavior of SANDWICH STRUCTURES of ISOTROPIC and COMPOSITE MATERIALS

Jack R. Vinson, Ph.D., P.E.

The H. Fletcher Brown Professor of Mechanical and Aerospace Engineering
Department of Mechanical Engineering
Center for Composite Materials, and the College of Marine Studies
University of Delaware
Newark, Delaware

TECHNOMIC PUBLISHING CO., INC.
LANCASTER · BASEL

The Behavior of Sandwich Structures of Isotropic and Composite Materials

a TECHNOMIC® publication

Technomic Publishing Company, Inc.
851 New Holland Avenue, Box 3535
Lancaster, Pennsylvania 17604 U.S.A.

Copyright ©1999 by Technomic Publishing Company, Inc.
All rights reserved

No part of this publication may be reproduced, stored in a retrieval system, or transmitted, in any form or by any means, electronic, mechanical, photocopying, recording, or otherwise, without the prior written permission of the publisher.

Printed in the United States of America
10 9 8 7 6 5 4 3 2 1

Main entry under title:
The Behavior of Sandwich Structures of Isotropic and Composite Materials

A Technomic Publishing Company book
Bibliography: p.
Includes index p. 363

Library of Congress Catalog Card No. 99-60570
ISBN No. 1-56676-699-0

To my beautiful wife, Midge, for her love and encouragement, so necessary to writing this text, and to our wonderful grandchildren:

Kendra Vinson
Ryan Vinson
Lori Vinson
Dustin Davis
Scott Shore
Courtney Vinson
Devon Davis
Aaron Shore
Abby Shore
Tyler Vinson
Trudy Vinson
Tanner Vinson

PREFACE

The use of sandwich structures is growing very rapidly around the world. Its many advantages, the development of new materials, and the need for high-performance, low-weight structures insure that sandwich construction will continue to be in demand.

The present textbook emanates from class notes used in teaching sandwich structures over recent years. It is clear that a suitable textbook is needed both to train future engineers, as well as to provide additional insight and skills to practicing engineers.

One thing is emphasized herein: the equations describing the behavior of sandwich structures are compatible with the equations developed for composite material thin-walled structures, simply by employing the appropriate in-plane, flexural, and transverse shear stiffness quantities.

Hopefully, the text provides a satisfactory mixture of derivations; physical understanding; mathematical rigor; the means to analyze, design, and optimize various sandwich structures; meaningful results from previous research; and sufficient references for future study.

Chapter 1 provides a framework and a background for sandwich structures to give a perspective for the rest of the text. In Chapter 2 a complete description of the mathematical representation of anisotropic elasticity and the laminate theory for composite constructions is presented. The governing equations for flat sandwich panels are developed in Chapter 3, and these are simplified to the study of beams, columns, and rods in Chapter 4. Because energy methods are so valuable in finding solutions to real-life structural problems, Chapter 5 is devoted to the Theorem of Minimum Potential Energy and Reissner's Variational Theorem. Solutions of rectangular sandwich panels are given in Chapter 6 for many static load problems. Solutions to many problems with dynamic loads are the subject of Chapter 7. Chapter 8 deals with thermoelastic problems of sandwich structures, and for polymer matrix materials the hygrothermal considerations. Buckling of sandwich panels comprises Chapter 9, while Chapter 10 provides the means to obtain minimum weight optimized construction. Sandwich shell theory is treated in Chapter 11, primarily for cylindrical shells. The buckling of sandwich cylindrical shells is the subject of Chapter 12, and Chapter 13 provides the means to attain minimum weight sandwich cylindrical shells.

In Appendices 1 and 2, the properties of various honeycomb and solid core materials and face materials are given, while Appendix 3 lists ASTM Standard Test Methods for use.

Appreciation is hereby expressed to Ms. Donna L. Fritz, who typed the text and its many alterations and corrections, and to Dr. Andras Z. Szeri, who authorized my sabbatical leave to work on the textbook in the Spring of 1997 and authorized the typing assistance through February 15, 1998. Special thanks goes to Fred E. Shore for continuing to correct the text and assist me until the book's completion. Appreciation is also expressed to Eric Preissner and Alex Dee for proofreading the entire text and to Zhaohui Chen and Nihar Satapathy for proofing portions of the text.

As time goes by, the author will generate additional homework problems and their solutions, as well as examination problems and their solutions. These will be shared with professors who request them. The author would also appreciate receiving problems and solutions from the users of the text.

Unfortunately, no matter how hard one tries, undoubtedly, there are some errors in the text. Please inform the author of errors as they are found, and corrections will be provided to anyone upon request.

Best wishes toward the increased use of sandwich construction.

JACK R. VINSON

CHAPTER 1

Sandwich Structures: Origins, Advantages, and Uses

1.1 DESCRIPTION OF VARIOUS SANDWICH CONSTRUCTIONS

Everyone is familiar with a sandwich, because most of us grew up and continue eating them. Usually the faces are identical (as in two slices of store-bought and cut bread), but they can differ (analogous to a Kaiser roll). The inside (core) of a sandwich can include almost anything and can vary in thickness, density, and solidity. Occasionally, one can encounter a double-decker sandwich as well, as in many box materials used for shipping.

Structural sandwiches follow suit, as shown in Figure 1.1; most often there are two faces, identical in material and thickness, which primarily resist the in-plane and lateral (bending) loads. However, in special cases the faces may differ in either thickness or material or both, because perhaps one face is the primary load-carrying, low-temperature portion while the other face must withstand an elevated temperature, corrosive environment, etc. Assuming a uniform core, the former sandwich is regarded as a symmetric sandwich, the latter an asymmetric sandwich.

The core of a sandwich structure can be almost any material or architecture, but in general, cores fall into four types, as shown in Figure 1.1: (a) foam or solid core, (b) honeycomb core, (c) web core, and (d) a corrugated or truss core. Developments in new cores continue to be of primary interest, such as new cores developed by Christensen [1] and Czaplicki [2]. Czaplicki has introduced a cellular core structure formed by progressive corrugation of a single continuous sheet of material. Bitzer [3] recently described several new honeycomb cores: thermoplastic CECORE, Marine Nomex, Korex, and non-metallic Tube-core.

Foam or solid cores are relatively inexpensive and can consist of balsa wood, and an almost infinite selection of foam/plastic materials with a wide

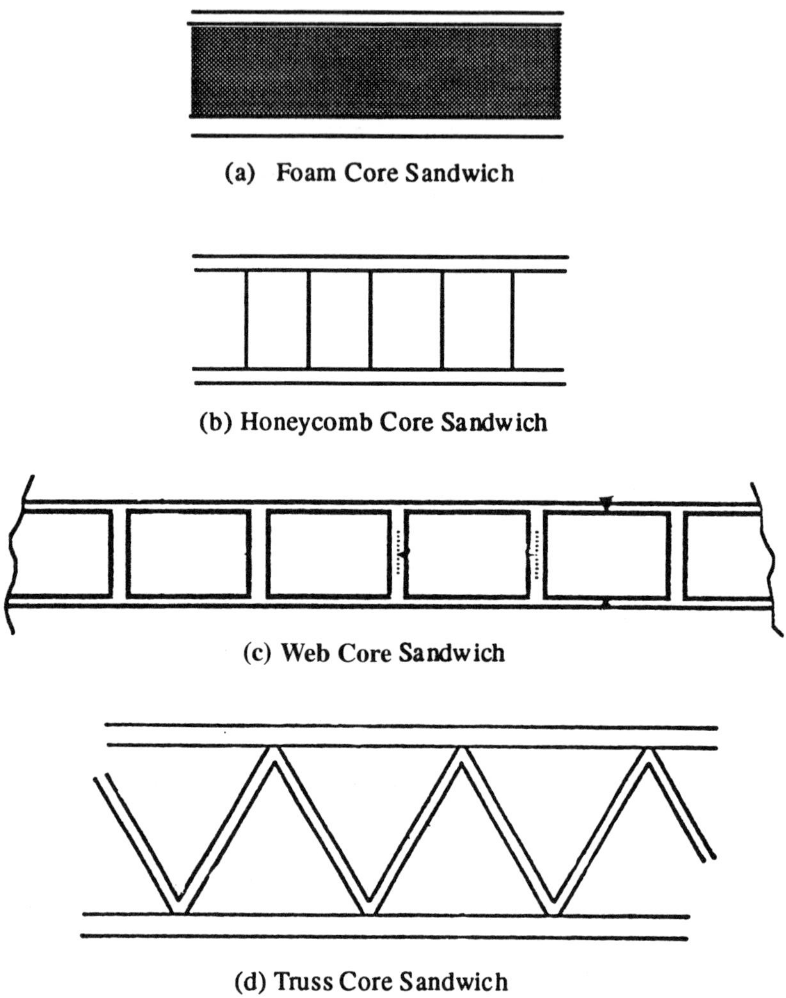

FIGURE 1.1. Types of sandwich construction.

variety of densities and shear moduli. Since World War II, honeycomb-core architectures have been widely used. The two most common types are the hexagonally-shaped cell structure (hexcell) and the square cell (egg-crate); see Figure 1.1(b). Web core construction is also used; see Figure 1.1(c), analogous to a group of I-beams with their flanges welded together. Truss or triangulated core construction is shown in Figure 1.1(d). In the web core and truss core constructions, the space in the core could be used for liquid storage or as a heat exchanger.

In all cases the primary loading, both in-plane and bending, are carried by the faces, while the core resists transverse shear loads (analogous to the web of an I-beam), and keeps the faces in place. In most foam-core and honeycomb-core sandwiches one can assume that all of the in-plane and bending loads are carried by the faces only. However, in web-core and truss-core construction, a portion of the in-plane and bending loads are also carried by the core elements.

The following foam cores are discussed in more detail by Davies [4].

(1) Polyurethane (PUR), a thermosetting material; widely used
(2) Polyisocyanurate (PIR), a thermosetting material
(3) Phenolic foam (PF), a thermosetting material; not yet widely used
(4) Polystyrene (expanded, EPS and extruded, XPS), a thermoplastic material

These foam core materials can be compared by a plot of core shear modulus as a function of density as in Figure 1.2.

As stated earlier, in addition to the four types of core shown in Figure 1.1, there are or can be many other core architectures. For instance, Sun, Hasebe, and Hua [5] have studied sandwich panels with a core that is composite vertical laminate-reinforced Rohacell foam, as shown in Figure 1.3. Thus, one could say that this is a foam-filled web core; see Figure 1.1(c). With the type of core shown in Figure 1.3, as with the web core and truss core of Figure 1.1, the extensional and flexural stiffness of the sandwich in the flute direction is greater than the honeycomb and foam core of Figure 1.1(a) and (b), because the core resists more than a negligible part of the load.

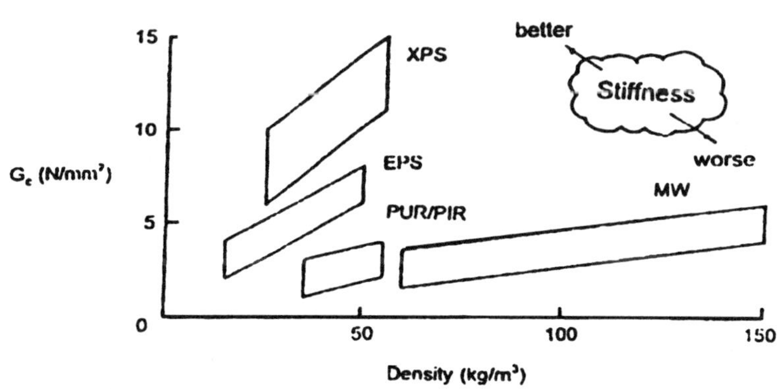

FIGURE 1.2. Shear stiffness of typical core materials as a function of density.

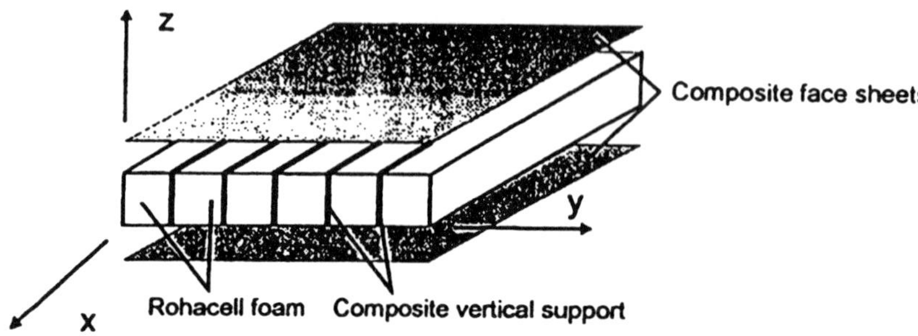

FIGURE 1.3. Sandwich construction with laminate-reinforced foam core.

1.2 ADVANTAGES OF SANDWICH CONSTRUCTION OVER CONVENTIONAL MONOCOQUE THIN-WALLED CONSTRUCTION

Sandwich construction is playing an increasingly important role in structures because of its exceptionally high flexural stiffness-to-weight ratio compared to monocoque and other architectures. As a result, sandwich construction results in lower lateral deformations, higher buckling resistance, and higher natural frequencies than do other constructions. Thus, for a given set of mechanical and environmental loads, sandwich construction often results in a lower structural weight than do other configurations.

Even with these advantages, it is important and desirable to develop means by which to optimize the sandwich construction in order to: (1) determine the absolute minimum weight for a given structural geometry, loading, and material system; (2) rationally compare one type of sandwich construction with others; (3) rationally compare the best sandwich construction with alternative structural configurations (monocoque, rib-reinforced, etc.); (4) rationally select the best face and core materials to minimize structural weight; (5) select the best stacking sequence for faces composed of laminated composite materials; and (6) rationally compare the optimum construction weight to weights required when there are some restrictions; i.e., the weight penalty due to restrictions of cost, minimum gage, manufacturing, material availability, etc.

In the following discussion only, foam-core and honeycomb-core sandwich constructions will be covered. It is interesting, at the outset, to compare certain structural parameters of sandwich construction with those of monocoque construction of the same approximate weight. In general, foam and honeycomb cores do not carry any appreciable in-plane or bending loads to which the structure is subjected. Their primary purpose is to insure the spacing between the faces and to carry the transverse shear loads to which the structure is subjected. Also, generally, the core is a very small percentage of the sandwich weight.

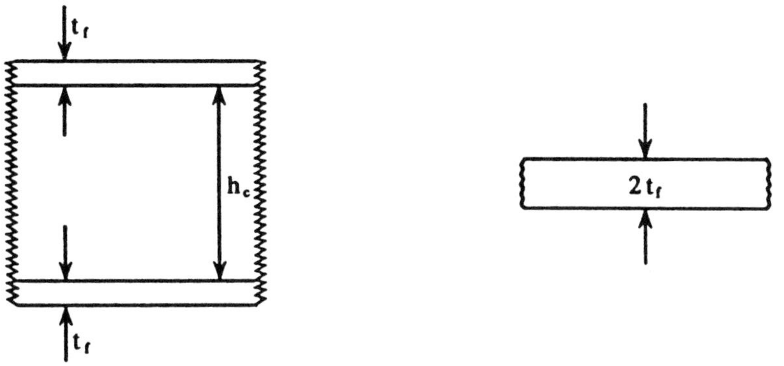

(a) Sandwich Construction **(b) Monocoque Construction**

FIGURE 1.4. Cross section of sandwich and monocoque construction. (Reprinted from Reference [53].)

It is also interesting to compare an isotropic sandwich construction with a monocoque (thin-walled) construction of approximately the same weight. In Figure 1.4, the sandwich construction employs two identical faces of thickness t_f, and a core depth of h_c. The monocoque construction on the right is a flat sheet construction of thickness $2t_f$, hence approximately the same weight as the sandwich construction using the same face material.

For an isotropic face material with a modulus of elasticity, E_f, the extensional stiffness per unit width, K, for both the sandwich and the monocoque construction is

$$K = 2E_f t_f/(1 - \nu_f^2) \tag{1.1}$$

Thus, for in-plane tensile and compressive (up to buckling) loads the two constructions have the same in-plane stiffness.

However, there is a marked difference in the flexural stiffness per unit width, D. For the panel construction of Figure 1.4(b) above, the flexural stiffness is

$$D_{\text{mon.}} = \frac{E_f(2t_f)^3}{12(1 - \nu_f^2)} = \frac{2E_f t_f^3}{3(1 - \nu_f^2)} \tag{1.2}$$

while the flexural stiffness for the isotropic foam or honeycomb-sandwich construction is given by

$$D_{\text{sand.}} = 2E_f\, t_f(h_c/2)^2/(1 - \nu_f^2) = \frac{E_f\, t_f h_c^2}{2(1 - \nu_f^2)} \tag{1.3}$$

where it is assumed the core does not contribute to the flexural stiffness and $t_f/h_c \ll 1$.

From the above, it is seen that the ratio of the flexural stiffness of foam- or honeycomb-core sandwich to a monocoque construction of approximately the same weight, using the same face material is

$$\frac{D_{\text{sand.}}}{D_{\text{mon.}}} = \frac{3}{4}\left(\frac{h_c}{t_f}\right)^2 \tag{1.4}$$

For example, if $t_f/h_c = 1/20$, then the flexural stiffness of the sandwich construction is 300 times the flexural stiffness of the monocoque construction. As a result, the use of sandwich construction, compared to monocoque construction employing the same materials, results in much lower lateral deflections, much higher *overall* buckling loads, and much higher flexural vibration natural frequencies than does the monocoque construction of nearly the same weight. However, the compressive face must be checked for face wrinkling or face dimpling, which is discussed later.

Now, looking at a comparison of stresses, consider the same sandwich construction and the monocoque construction to be subjected to an in-plane load of N, as a force per unit width of structure, and a bending moment per unit width of M.

For the in-plane loads, in each case the resulting in-plane face stress, σ_f, is given by

$$\sigma_f = \frac{N}{2t_f} \tag{1.5}$$

Thus, neither construction provides any advantage over the other regarding in-plane structural loads. However, for a bending moment, M, the monocoque construction results in maximum stresses at the top and bottom surface of

$$\sigma_{\text{mon.}} = \pm\frac{6M}{(2t_f)^2} = \pm\frac{3}{2}\frac{M}{t_f^2} \tag{1.6}$$

Similarly, for the bending moment M, the stresses in a sandwich face are

$$\sigma_{\text{sand.}} = \pm\frac{M}{t_f h_c} \tag{1.7}$$

Therefore, the ratio of the bending stress in a sandwich face to the maximum stress in a monocoque structure of approximately the same weight is

$$\frac{\sigma_{\text{sand.}}}{\sigma_{\text{mon.}}} = \frac{2t_f}{3h_c} \tag{1.8}$$

Therefore, for the example of a sandwich in which $t_f/h_c = 1/20$, the bending stress in a sandwich structure is 1/30 that of a monocoque construction of approximately the same weight.

1.3 ORIGINS OF SANDWICH CONSTRUCTION

Sandwich construction is relatively new, even though Noor, Burton, and Bert [6] report that the concept of sandwich construction goes back to Fairbairn in 1849 [7]. Feichtinger [8] states that during World War II, the concept of sandwich construction in the U.S. originated with faces of reinforced plastic and a lower density core. In 1943, Wright Patterson Air Force Base designed and fabricated the Vultee BT-15 fuselage using fiberglass-reinforced polyester as faces using both glass-fabric honeycomb and balsa-wood core [9].

The first research paper concerning sandwich construction was due to Marguerre [10], in 1944, and it dealt with in-plane compressive loads.

In 1948, Hoff [11] derived the differential equations and boundary conditions for the bending and buckling of sandwich plates using the principle of virtual displacements, but pursued only the buckling problem of panels under edgewise compression. In the same year, Libove and Batdorf [12] published a general small deflection theory for sandwich plates.

In 1949, Flügge [13] published a paper on the structural optimization of sandwich panels in which he presented nomograms for the solution of the following problems: (1) the geometric dimension and the core properties for a given compression load and minimum weight; (2) the geometrical dimension and core properties for a given weight and maximum compressive load; and (3) the ultimate strength of a given sandwich. In all cases studied, the materials were isotropic, and the following cases of failure were considered: overall buckling, crinkling (wrinkling), and the elastic limit of the faces. The unloaded edges were free or simply supported. Flügge continued his work and published another paper on the subject in 1952 [14].

In 1951, Bijlaard [15] approached the subject of sandwich optimization by considering plates with given weight per unit surface area, and computing the ratio of the elastic moduli of core and faces which lead to a maximum buckling load. He carried out the optimization for a given ratio between thickness of core and faces, as well as for a given total thickness of the sandwich plate. Only isotropic materials were considered. An abridgment of this publication appeared in the Proceedings of the 1st U.S. National Congress of Applied Mechanics in 1952 [16].

Ericksen of the U.S. Forest Products Laboratory (USFPL) issued a report in 1951 accounting for the effects of shear deformation on deflections of sandwich panels with isotropic cores and facings. The bending rigidities of the faces were also included. The development used a double Fourier series for representing the deflection, and results were presented in chart form including deflections of and strains in the panel. Explicit charts for stresses in the faces and the core were not included. Also presented were general expressions for strain components in sandwich panels with orthotropic faces and cores. In 1952, Eringen [17] obtained from the Theorem of Minimum Potential Energy four partial differential equations for the bending and buckling of rectangular sandwich plates with isotropic cores and faces under various loadings and edge conditions. However, the results presented included only information useful for buckling purposes. In 1952, March [18] of the USFPL studied the behavior of panels under lateral and compressive edge loads. The work was essentially for isotropic faces and cores and was limited to expressions for deflections, bending moments, and reactions. In 1956, Ericksen [19] again studied the deflections of panels by double Fourier series. In 1955, a USFPL report by March [20] took into account the effects of shear deformation on the deflection of panels of orthotropic faces and cores. A definitive study for use in design and optimization is the USFPL report by Raville [21] issued in 1955. Theoretical expressions for deflections and stresses in rectangular, simply supported panels are developed using double Fourier series to represent the deflections. Isotropic facings and orthotropic cores were considered and the results were presented in graphs. Military Handbook 23 [22], which reflects primarily the work of the USFPL, is used for design purposes.

In 1956, Gerard briefly discusses sandwich plate optimization in one chapter of his book, *Minimum Weight Analysis of Compression Structures* [23].

In 1957, Kaechele published a report [24] on the minimum-weight design of sandwich panels. He presented a method for determining the optimum configuration of flat, simply-supported sandwich panels under uniaxial compression, when the load, width, and stress-strain curves for the face material are given. He also invoked side conditions on the maximum allowable face stress (or strain) and on core strength or density. He applied the method to both hexagonal-cell and square-cell honeycomb-core panels, and discussed elevated temperature effects briefly.

Another series solution was presented by Cheng [25], which provided deflection, shear, and moment expressions.

In 1960, Heath published a paper on the correlation and extension of the existing theory of flat panels of sandwich construction subjected to lengthwise compression [26,27]. The second part of that paper is concerned with the optimum design of flat-sandwich panels.

The early theoretical work on the behavior of rectangular sandwich panels subjected to lateral loads was restricted to uniform loads and simply supported edge conditions.

During the early post-World-War-II period, the USFPL was the primary group in the development of analysis and design methods for sandwich structures. USFPL also led the efforts with the MIL-HDBK-23 publication and continuous updating of that document.

By the mid 1960s, efforts in sandwich research had spread widely. In 1966, Plantema published his famous, and the first, book [28] on sandwich structures. In 1969, this was followed by the book by H. G. Allen [29]. These books were the "bibles" for sandwich structures for many years.

Also in the mid 1960s, the Naval Air Engineering Center sponsored research to develop a fiberglass-sandwich construction to compete in weight with conventional aluminum construction for aircraft structures. Much of this effort was in minimum-weight optimization of the sandwich structures to enable the fiberglass sandwich to weigh less than the aluminum construction for the same load conditions [30–35]. A fairly complete bibliography (over 250 publications) describing all publications regarding sandwich construction before 1966 is provided in [30]. Since the 1960s, sandwich construction use has increased significantly, and publications regarding sandwich structures have blossomed; many of the most cogent of these are referenced in this text. A text containing many of N. J. Hoff's landmark papers on sandwich construction [36] was published in 1986.

To date, there have been four International Conferences on Sandwich Construction: the first in Stockholm in 1989 [37]; the second in Gainesville in 1992 [38]; the third in Southhampton in 1995 [39]; and the fourth in Stockholm in 1998.

In 1989, Ha [40] provided an overview of finite elements applied to sandwich plates. In 1991, Bert [41] provided a review of sandwich plate analysis.

Most recently, in 1995, a monograph by Zenkert [42] supplements much of the material contained in the Plantema and Allen books. Also, the review by Noor, Burton, and Bert in 1996 [6] provides over 800 references discussed in the review and another 559 references as a supplemental bibliography. At the time of this writing, ASME AD-Vol. 55 [43] is the most recent publication that includes numerous papers on sandwich structures.

1.4 USES OF SANDWICH CONSTRUCTION

Sandwich construction has been used primarily in the aircraft industry since the 1940s, with the development of the British Mosquito bomber, and later logically extended to missile and spacecraft structures.

TABLE 1.1. Use of Sandwich Construction in Boeing Aircraft.

Boeing Aircraft	Percent of Wetted Surface
707	8
727	18
737	26
747	36
757–767	46

In 1992, Bitzer [44] gave an excellent overview of the uses of honeycomb-core materials and applications. He points out that honeycomb-core sandwich construction is widely used on both commercial and military aircraft. Bitzer says that every two- (or more) engine aircraft in the west utilizes some honeycomb sandwich. He lists the quantity of honeycomb sandwich being used in various Boeing aircraft. In Table 1.1, "wetted surface" is defined as the airplane's surface that would be wet if the aircraft were submerged in water. In the Boeing 747, the fuselage cylindrical shell is primarily Nomex-honeycomb sandwich, and the floors, side panels, overhead bins, and ceiling are also of sandwich construction.

The Beech Starship uses Nomex honeycomb with graphite or Kevlar faces for the entire structure—the first all sandwich aircraft. A major portion of the space shuttle is a composite-faced honeycomb-core sandwich.

Europe leads the way in the use of sandwich constructions for lightweight railcars, while in the U.S. some of the rapid transit trains use honeycomb sandwich.

The U.S. Navy is using honeycomb-sandwich bulkheads to reduce the ship weight above the waterline. Sailboats, racing boats, and auto racing cars are all employing sandwich construction.

Sandwich construction is also used in snow skis, water skis, kayaks, canoes, pool tables, and platform tennis paddles.

Honeycomb-sandwich construction is also excellent for absorbing mechanical and sound energy. It has a high-crush strength-to-weight ratio. It can also be used to transmit heat or to be an insulative barrier. In the former, a metallic honeycomb is used plus natural convection; for the latter, a nonmetallic core is used with the cells filled with a foam. For a sound barrier, the honeycomb core is filled with a fiberglass batting, and a thin porous Tedlar skin can be used for the interior face. Also, honeycomb core has been used in direct fans, wind tunnels, air conditioners, heaters, grills, registers, etc.

Boat hulls became a logical use for fiberglass sandwich construction, particularly in pleasure craft where the foam core increases the chance of flotation in emergency situations. Ferry boats in the Scandinavian countries and the Pacific Rim use fiberglass-sandwich construction. Finally, the Royal Swedish Navy has been using fiberglass- and graphite-composite sandwich construction for their naval vessels for more than 20 years [45]. The hulls were designed to withstand underwater explosions and debris strikes as well as steel hull designs have done in the past. In addition, the fiberglass sandwich construction eliminated the attraction of the hulls to magnetic mines, as well as negating salt water corrosion. Furthermore, they state that the next generation of surface vessels will be stealth-optimized composite vessels using large amounts of GRP sandwich construction.

In 1995, Kimpara [46] reported the use of honeycomb-sandwich structures for an ultra-light-weight sailing dinghy, with a weight reduction as large as 70% of more conventional graphite-composite construction.

Until recently, locomotives have been made of steel or aluminum. Now, there is a change from a classical metal design using thick flat or cylindrically bent metal plates to spherical shapes, making composite production much more efficient. Since 1980, composite front cabs have been built for the XPT locomotives in Australia, the ETR 500 locomotives in Italy, the French TGV, and the Swiss locomotive 2000. The major considerations are the pressure waves occurring during the crossing of two high-speed trains in tunnels. Self-supporting sandwich construction was selected, and approximately 10^3 kg was saved per locomotive.

In 1995, Starlinger and Reif [47] reported that sandwich construction is now being used for roof panels and intermediate floor panels in such ground transportation vehicles as double-decker buses.

Kattan [48] states that corrugated-core sandwich panels have been used in light-weight structures, such as lightly loaded aircraft parts, for a long time, and only manufacturing difficulties (i.e., difficulties in rigidly joining faces and core) have prevented the widespread use of steel corrugated-core sandwich construction for shipbuilding. Kujala and Tuhkuri [49] investigated the use of steel-corrugated panels for superstructures in ships both analytically and experimentally. They found that the sandwich structures were 40–50% lighter than the conventional steel construction.

More recently, sandwich construction is being used increasingly in civil engineering infrastructure rehabilitation projects such as bridge decks. Karbhari [50] provides an overview of the use of composite-sandwich usage for the twenty-first century. Also, Woldesenbet [51,52] investigated the use of sandwich construction for low cost and emergency housing.

Davies [4] reports that sandwich-cladding panels composed of two metallic faces and a light-weight insulating core are finding ever-increasing use as wall and roof cladding for a wide variety of buildings. Here, the primary attraction is their outstanding thermal performance.

Also, boxes and packaging of all kinds comprise a multibillion dollar per year use of sandwich construction. This use involves low cost Kraft paper components and quite often triple-faced sandwich with truss cores.

1.5 PRESENT APPROACH TO ANALYSIS

In the early work up through the 1960s, including the books by Plantema [28] and Allen [29], a nomenclature and analytical presentation was developed that was unique to sandwich construction. While the structural mechanics and analytical solutions were correct, it required the user to become familiar with that special notation and approach.

Beginning with the 1970s, a new area of structural mechanics developed, dealing with laminated composite-material structures. Laminate theory developed, became established, and very widely used throughout the world. Thus, this analysis and design methodology is not only well-known and understood, but standardized computer codes have been developed to describe all types of structural laminates, including sandwich construction. Simply put, one lets one face be lamina 1, the core be lamina 2, and the other face be lamina 3. In this way, all sandwich structures can be analyzed using laminate theory. This approach is used herein because of its simplicity and generally wide acceptance throughout the world. One key advantage of this approach is that composite-material engineers/designers/analysts, who are already familiar with the composite laminate theory, can readily apply their knowledge to sandwich structures. Laminate theory is discussed in Chapter 2.

1.6 PROBLEMS

1.1 Look through the contents of three archival journals published during the last twelve months and list all of the papers dealing with sandwich construction using the same style as in the references above. Duplicate the list and share with all of your classmates. The result will be that each class member will have an up-to-date bibliography.

1.2 Look through at least three trade journals published during the last twelve months and list each different use of sandwich construction, and the reference in which it appears in the style of the references above. Likewise, share.

1.7 REFERENCES

1. Christensen, R. M., Patent Pending.

2. Czaplicki, R. M., "Cellular Core Structures Providing Gridlike Bearing Surfaces on Opposing

Parallel Planes of the core," in *Proceedings of the Second International Conference on Sandwich Construction,* Editors, D. Weissman-Berman and K-A. Olsson, EMAS Publications, United Kingdom, 1992, pp. 721–736.

3. Bitzer, T. N., "Recent Honeycomb Core Developments," in *Proceedings of the Second International Conference on Sandwich Construction,* Editors, D. Weissman-Berman and K-A. Olsson, EMAS Publications, United Kingdom, 1992, pp. 555–563.
4. Davies, J. M., "Design Criteria for Sandwich Panels for Building Construction," in *Proceedings of the ASME Aerospace Division, Structures and Materials Committee,* ASME-AD-Vol. 55, New York, 1997, pp. 273–284.
5. Sun, C. T., R. S. Hasebe and Y. Hua, "Properties of Sandwich Structures with Reinforced Core," in *Proceedings of the ASME Aerospace Division, Structures and Materials Committee,* ASME-AD-Vol. 55, New York, 1997, pp. 323–334.
6. Noor, A. K., W. Scott Burton and C. W. Bert, "Computational Models for Sandwich Panels and Shells," *Applied Mechanics Reviews,* Vol. 49, No. 3, March 1996, pp. 155–199.
7. Fairbairn, W., *An Account of the Construction of the Bretannia and Conway Tubular Bridges,* John Weale, London, 1849.
8. Feichtinger, K. A., "Test Methods and Performance of Structural Core Materials—1. Static Properties," *4th Annual ASM International/Engineering Society of Detroit—Advanced Composites Conference/Exposition,* September 13–15, 1988.
9. Rheinfrank, G. B. and W. A. Norman, "Molded Glass Fiber Sandwich Fuselage for BT-15 Airplane," Army Air Corps Technical Report No. 5159, November 8, 1944.
10. Marguerre, K., "The Optimum Buckling Load of a Flexibly Supported Plate Composed of Two Sheets Joined by a Light Weight Filler, When Under Longitudinal Compression," Deutsche Vierteljahrsschrist für Literalurwissenschaft und Giests Geschichte, D.V.L. (ZWB UM 1360/2) 28 October 1944, 11 p. 5 figs.
11. Hoff, N. J., "Bending and Buckling of Rectangular Sandwich Plates," N.A.C.A. Technical Note 2225, November 1950.
12. Libove, C. and S. B. Batdorf, "A General Small Deflection Theory for Sandwich Plates," NACA Report 899, Washington, D.C., 1948.
13. Flügge, W., "Determination of Optimum Dimension of Sandwich Panels," *La Recherche Aeronautique,* No. 7, Jan./Feb. 1949.
14. Flügge, W., "The Optimum Problem of the Sandwich Plate," *Journal of Applied Mechanics,* Vol. 19, No. 1, March 1952, pp. 104–108.
15. Bijlaard, P. P., "Analysis of Elastic and Plastic Stability of Sandwich Plates by Method of Split Rigidities," *Journal of the Aeronautical Sciences,* I May 1951; II, December 1951; III, July 1952.
16. Bijlaard, P. P., "On the Optimum Distribution of Material in Sandwich Plates Loaded in Their Plane," *Proceedings of the First U.S. National Congress of Applied Mechanics,* June 1952, ASME, New York, 1952, pp. 373–380.
17. Eringen, A. C., "Bending and Buckling of Rectangular Sandwich Plates," *Proceedings of the First U.S. National Congress of Applied Mechanics,* ASME, New York, 1952, pp. 381–390.
18. March, H. W., "Behavior of a Rectangular Sandwich Panel Under Uniform Lateral Load and Compressive Edge Load," Forest Products Laboratory Report 1834, September 1952.
19. Ericksen, W. S., "Supplement to Effects of Shear Deformation in the Core of a Flat Rectangular Sandwich Panel," Forest Products Laboratory Report 1583-C, March 1956.

20. March, H. W., "Effects of Shear Deformation in the Core of a Flat Rectangular Sandwich Panel: 1, Buckling Under Compressive End Load: 2, Deflection Under Uniform Transverse Load," Forest Products Laboratory Report 1583, August 1955.
21. Raville, M. E., "Deflection and Stresses in a Uniformly Loaded Simply Supported, Rectangular Sandwich Plate," Forest Products Laboratory Report 1847, 1955.
22. Anon., "Materials, Properties and Design Criteria Part II, Sandwich Construction for Aircraft," *Military Handbook 23* (ANC-23), Department of the Air Force Research and Development Command; Department of the Navy, Bureau of Aeronautics; and the Department of Commerce, Civil Aeronautics Administration, 2nd Edition, 1955.
23. Gerard, G., *Minimum Weight Analysis of Compressive Structures,* New York University Press, New York, 1956.
24. Kaechele, L. E., "Minimum Weight Design of Sandwich Panels," USAF. Project Rand Research Memorandum, RM 1895, AD-133011, March 1957.
25. Cheng, S., "On the Theory of Bending of Sandwich Plates," *Proceedings of the Fourth U.S. National Congress of Applied Mechanics,* ASME, New York, 1962.
26. Heath, W. G., "Sandwich Construction, Part I: The Strength of Flat Sandwich Panels," *Aircraft Engineering,* Vol. 32, July 1960, pp. 186–191.
27. Heath, W. G., "Sandwich Construction, Part 2: The Optimum Design of Flat Sandwich Panels," *Aircraft Engineering,* Vol. 32, August 1960, pp. 230–235.
28. Plantema, F. J., *Sandwich Construction: The Bending and Buckling of Sandwich Beams, Plates and Shells,* John Wiley and Sons, New York, 1966.
29. Allen, H. G., *Analysis and Design of Structural Sandwich Panels,* Pergamon Press, Oxford, 1969.
30. Vinson, J. R. and S. Shore, "Bibliography on Methods of Structural Optimization for Flat Sandwich Panels," Naval Air Engineering Center Report NAEC-ASC-1082, April 15, 1965.
31. Vinson, J. R. and S. Shore, "Methods of Structural Optimization for Flat Sandwich Panels," Naval Air Engineering Center Report, NAEC-ASC-1083, April 15, 1965.
32. Vinson, J. R. and S. Shore, "Design Procedures for the Structural Optimization of Flat Sandwich Panels," Naval Air Engineering Center Report NAEC-ASC-1084, April 15, 1965.
33. Vinson, J. R. and S. Shore, "Structural Optimization of Corrugated Core and Web Core Sandwich Panels Subjected to Uniaxial Compression," Naval Air Engineering Center Report NAEC-ASC-1109, May 1967.
34. Vinson, J. R. and S. Shore, "Structural Optimization of Flat Corrugated Core Sandwich Panels Under In-Plane Shear Loads and Combined Uniaxial Compression and In-Plane Shear Loads," Naval Air Engineering Center Report NAEC-ASC-1110, July 1967.
35. McCoy, T. T., J. R. Vinson and S. Shore, "A Method for Weight Optimization of Flat Truss Core Sandwich Panel Under Lateral Loads," Naval Air Engineering Center Report NAEC-ASC-1111, July 1967.
36. Hoff, N. J., *Monocoque, Sandwich and Composite Aerospace Structures,* Technomic Publishing Company, Lancaster, PA, 1986.
37. *Sandwich Constructions 1—Proceedings of the First International Conference on Sandwich Construction,* Stockholm, Sweden, June 19–21, 1989, Editors, K-A. Olsson and R. P. Reichard, EMAS Publications, United Kingdom, 1989.

38. *Sandwich Constructions 2—Proceedings of the Second International Conference on Sandwich Construction,* Gainesville, Florida, March 9–12, 1992, Editors, D. Weissman-Berman and K-A. Olsson, 1992, 2 Volumes, EMAS Publications, United Kingdom, 1992.

39. *Sandwich Construction 3—Proceedings of the Third International Conference on Sandwich Construction,* Southampton, Great Britain, Editor, H. G. Allen, EMAS Publications, United Kingdom, 1996, 2 Volumes.

40. Ha, K. H., "Finite Element and Sandwich Construction: A Critical Review," in *Sandwich Constructions I—Proceedings of the First International Conference on Sandwich Construction,* Editors, K-A Olsson and R. P. Reichard, EMAS Publications, United Kingdom, 1989, pp. 69–84.

41. Bert, C. W., Part I, *Shock and Vibration Digest,* Vol. 23, 1991, pp. 3–14; Part II, *Shock and Vibration Digest,* Vol. 23, 1991, pp. 19–21.

42. Zenkert, D., *An Introduction to Sandwich Construction,* EMAS Publications, West Midlands, United Kingdom, 1995.

43. Simitses, G. J., ed. "Analysis and Design Issues for Modern Aerospace Vehicles—1997," *Proceedings of the ASME Aerospace Division, Structures and Materials Committee,* ASME-AD-Vol. 55, New York, 1997.

44. Bitzer, T. N., "Honeycomb Materials and Applications," in *Sandwich Constructions 2—Proceedings of the Second International Conference on Sandwich Construction,* Editors, D. Weissman-Berman and K-A. Olsson, EMAS Publications, United Kingdom, 1992, pp. 681–691.

45. Lönnö, A. and S.-E. Hellbratt, "Use of Carbon Fibre in a 63M High Speed Vessel, YS2000, for the Swedish Navy," in *Sandwich Construction 3—Proceedings of the Third International Conference on Sandwich Construction,* Editor, H. G. Allen, EMAS Publications, United Kingdom, 1996, pp. 3–13.

46. Kimpara, I., K. Kageyama, T. Suzuki, I. Ohsawa and T. Takatoya, "Design and Evaluation of Advanced Composite Honeycomb Sandwich Structures for Marine Use," in *Sandwich Construction 3—Proceedings of the Third International Conference on Sandwich Construction,* Editor, H. G. Allen, EMAS Publications, United Kingdom, 1996, pp. 37–46.

47. Starlinger, A. and G. Reif, "Sandwich Design of Lightweight Bus Structures—Engineering and Cost Saving Aspects," in *Sandwich Construction 3—Proceedings of the Third International Conference on Sandwich Construction,* Editor, H. G. Allen, EMAS Publications, United Kingdom, 1996, pp. 103–115.

48. Kattan, M. R., "Steel Sandwich Construction for Ships—A Reality?" in *Sandwich Construction 3—Proceedings of the Third International Conference on Sandwich Construction,* Editor, H. G. Allen, EMAS Publications, United Kingdom, 1996, pp. 399–410.

49. Kujala, K. and J. Tuhkuri, "All-Steel Corrugated-Core Sandwich Panels for Ship Structures," in *Sandwich Construction 3—Proceedings of the Third International Conference on Sandwich Construction,* Editor, H. G. Allen, EMAS Publications, United Kingdom, 1996, pp. 411–422.

50. Karbhari, V. M., "Application of Composite Materials to the Renewal of Twenty-First Century Infrastructure," *Proceedings of the Eleventh International Conference on Composite Materials,* Gold Coast, July 1997, Australian Composite Structures Society, RMIT, Fishermens Bend, Melbourne, Australia, 1997.

51. Woldesenbet, E. and J. R. Vinson, "Sandwich Composite Structure for Low-Cost and Emergency Housing," in *Sandwich Construction 3—Proceedings of the Third International Confer-*

Before understanding the physical behavior of composite sandwich structures and before being able to quantitatively determine the stresses, strains, deformations, natural frequencies, and buckling loads in them, a clear understanding of anisotropic elasticity is mandatory. In general, isotropic: materials are mathematical approximations to the true situation. For instance, in polycrystalline metals, the structure is usually made up of numerous anisotropic grains, wherein macroscopic isotropy exists in a statistical sense only because the anisotropic individual grains are randomly oriented. However, the same materials could be macroscopically anisotropic due to cold working, forging, or spinning during a fabrication process. Other materials such as wood, human and animal bone, and all fiber reinforced materials are anisotropic.

Fiber-reinforced composite materials are uniquely useful because the use of long fibers results in a material which has a higher strength-to-density ratio and/or stiffness-to-density ratio than any other material system at moderate temperature, and there exists the opportunity to uniquely tailor the fiber orientations to a given geometry, applied load, and environmental system. For short fiber composites, used mainly in high production, low cost systems, the use of fibers makes the composites competitive and superior to the plastic and metal alternatives. Finally, the use of two or more fibers with one matrix is termed a hybrid composite, where one fiber is stronger or stiffer while the other fiber is less expensive but desirable for less critical locations in an overall structural component. In other uses the laminae of stronger and stiffer (but more brittle) fibers are protected from impact by outer plies of a tougher fiber composite. Through the use of composite materials, the engineer is not merely a materials selector, but he is also a materials designer.

There are several texts and reference books dealing with sandwich structures, such as Plantema [57], Allen [58], and Zenkert [59]. For the small deflection, linear elastic analysis of such sandwich structures, the equilibrium equations, strain-displacement relations, and compatibility equations remain the same whether the structure is composed of an isotropic material or an anisotropic composite material. However, it is very necessary to drastically alter the stress-strain relations, also called the constitutive relations, to account for the anisotropy of the composite materials system.

A quantitative understanding of the virtues of using composite materials in a structure is found through systematically deriving the anisotropic elasticity tensor matrix, as will be done in Section 2.2.

2.2 DERIVATION OF THE ANISOTROPIC ELASTIC STIFFNESS AND COMPLIANCE MATRICES

Consider an elastic solid body of any general shape, and assume it is composed of an infinity of *material points* within it. In order to deal with a

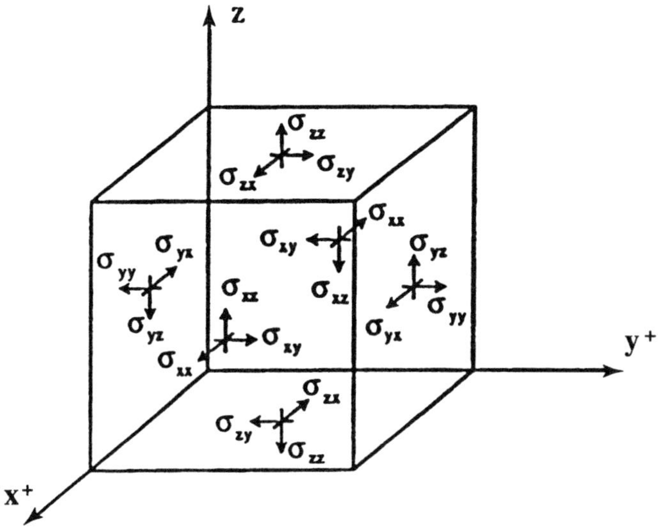

FIGURE 2.1. Positive stresses on a control element of an elastic body. (Reprinted from Reference [4].)

continuum, one also assumes that the material points are infinitely large compared to the molecular lattice spacing of the particular material. If one assigns a Cartesian reference frame to the elastic body shown in Figure 2.1, one then calls this rectangular parallelepiped material point a *control element* or control volume of dimension dx, dy, and dz in a Cartesian coordinate system.

On the surface of the control element there can exist both normal stresses (those perpendicular to the plane of the face) and shear stresses (those parallel to the plane of the face). On any one face the three mutually orthogonal stress components comprise a vector, which is called a *surface traction.*

It is important to note the sign convention and the meaning of the subscripts of these surface stresses. For a stress component on a face whose outward normal is in the direction of a positive axis, the stress component is positive when it is in the direction of a positive axis. Also, when a stress component is on a face whose outward normal is in the direction of a negative axis, the stress component is positive when it is in the direction of a negative axis. This can be seen clearly in Figure 2.1.

The first subscript of any stress component on any face of the control element signifies the axis to which the outward normal of the face is parallel the second subscript refers to the axis to which that stress component is parallel. Again, see Figure 2.1.

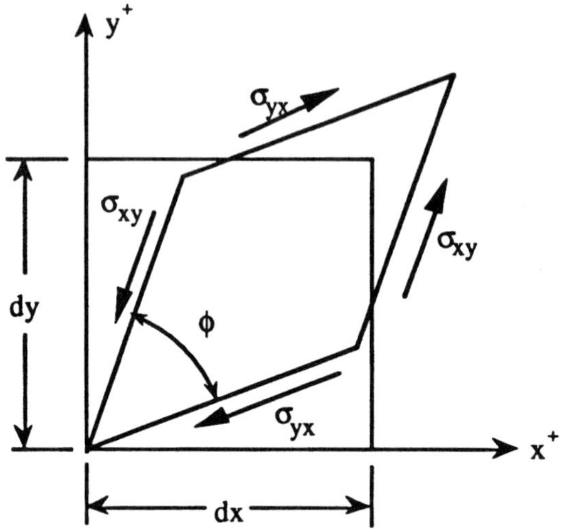

FIGURE 2.2. Shearing of a control element.

The strains occurring in an elastic body have the same subscripts as the stress components, but are of two types. Dilatational or extensional strains are denoted by ε_{ii}, where $i = x, y, z$, and are a measure of the change in dimension of the control volume in the subscripted direction due to normal stresses, σ_{ii}, acting on the control volume. Shear strains ε_{ij} $(i \neq j)$ are proportional to the change in angles of the control volume from 90°, changing the rectangular control volume into a parallelogram due to the shear stresses, σ_{ij}, $i \neq j$. For example, looking at the control volume x-y plane shown in Figure 2.2, shear stresses σ_{xy} and σ_{yx} cause the square control element with 90° corner angles to become a parallelogram with the corner angle ϕ as shown. Here, the change in angle γ_{xy} is

$$\gamma_{xy} = \frac{\pi}{2} - \phi$$

The shear strain ε_{xy}, a tensor quantity, is defined by

$$\varepsilon_{xy} = \gamma_{xy}/2$$

Similarly, $\varepsilon_{xz} = \gamma_{xz}/2$, and $\varepsilon_{yz} = \gamma_{yz}/2$.

Having defined all of the elastic stress and strain tensor components, the stress-strain relations are now used to derive the anisotropic stiffness and compliance matrices.

The following derivation of the stress-strain relations for an anisotropic material parallels the derivation of Sokolnikoff [1], Vinson, and Chou [2], and Vinson and Sierakowski [3], and Vinson [4]. Although the derivation is very formal mathematically to the reader who is primarily interested with the end result, the systematic derivation does provide confidence in the extended use of the results for everyone.

From knowledge of the basic strength of materials [5], both stress, σ_{ij}, and strain ε_{ij}, are second order tensor quantities, where in three dimensional space they have $3^2 = 9$ components. They are equated by means of the fourth order elasticity tensor, C_{ijkl}, which therefore has $3^4 = 81$ components, with the resulting constitutive equation:

$$\sigma_{ij} = C_{ijkl}\varepsilon_{kl} \tag{2.1}$$

where i, j, k and l assume values of 1, 2, 3 or x, y, z in a Cartesian coordinate system. Fortunately, there is no actual material which has 81 elastic constants. Both the stress and strain tensors are symmetric, i.e., $\sigma_{ij} = \sigma_{ji}$ and $\varepsilon_{kl} = \varepsilon_{lk}$, and therefore the following shorthand notation may be used:

$$\begin{array}{llll} \sigma_{11} = \sigma_1 & \sigma_{23} = \sigma_4 & \varepsilon_{11} = \varepsilon_1 & 2\varepsilon_{23} = \varepsilon_4 \\ \sigma_{22} = \sigma_2 & \sigma_{31} = \sigma_5 & \varepsilon_{22} = \varepsilon_2 & 2\varepsilon_{31} = \varepsilon_5 \\ \sigma_{33} = \sigma_3 & \sigma_{12} = \sigma_6 & \varepsilon_{33} = \varepsilon_3 & 2\varepsilon_{12} = \varepsilon_6 \end{array} \tag{2.2}$$

At the outset it is noted that ε_4, ε_5, and ε_6, which are quantities widely used in composite analyses, are not tensor quantities and therefore do not transform from one set of axes to another by affine transformation relationships. Care must also be taken regarding whether or not to use the factor of "two" when using shear strain relations. Using Equation (2.2), Equation (2.1) can be written:

$$\sigma_i = C_{ij}\,\varepsilon_j \qquad (i,j = 1 \ldots 6) \tag{2.3}$$

It should be noted that the contracted C_{ij} quantities are also not tensor quantities, and therefore cannot be transformed as such.

Hence, by the symmetry in the stress and strain tensors the elasticity tensor immediately reduces to the 36 components shown in Equation (2.3). In addition, if a strain energy density function, W, exists [2–5], i.e.,

$$W = \frac{1}{2}\sigma_{ij}\varepsilon_{ij}$$

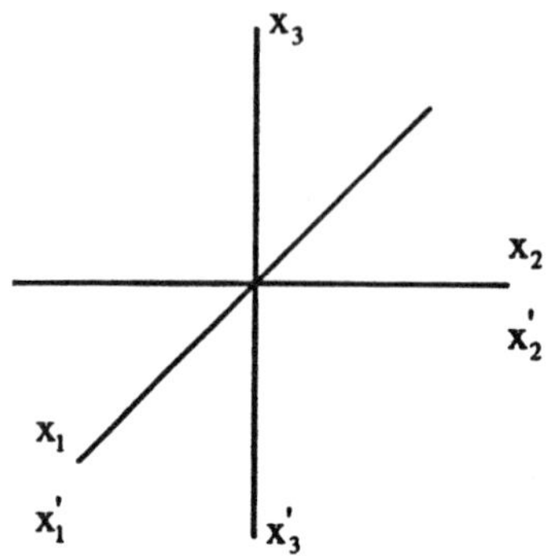

Directional Cosines

	X_1	X_2	X_3
X_1'	1	0	0
X_2'	0	1	0
X_3'	0	0	−1

FIGURE 2.3. One plane of symmetry. (Reprinted from Reference [4].)

in such a way that

$$\frac{\partial W}{\varepsilon_{ij}} = C_{ijkl}\varepsilon_{kl} = \sigma_{ij} \tag{2.4}$$

then the independent components of C_{ijkl} are reduced to 21 elastic constants, since $C_{ijkl} = C_{klij}$ and now it can be written $C_{ij} = C_{ji}$.

Next, to simplify the general mathematical anisotropy to the cases of very practical importance, consider the Cartesian coordinate system only. (However, the results are applicable to any of the 12 curvilinear orthogonal coordinate systems, some of which are spherical, cylindrical, elliptical, etc.)

First, consider an elastic body whose properties are symmetric with respect to the $X_1 - X_2$ plane. The resulting symmetry can be expressed by the fact that the C_{ij}'s discussed above must be invariant under the transformation $x_1 = x_1'$, $x_2 = x_2'$ and $x_3 = -x_3'$, shown in Figure 2.3.

Also shown in the chart in Figure 2.3 are the direction cosines, t_{ij}, associated with this transformation. The stresses and strains of the primed coordinate system are related to those of the original (unprimed) coordinate system by the well-known relationships:

$$\sigma'_{\alpha\beta} = t_{\alpha i}t_{\beta j}\sigma_{ij} \quad \text{and} \quad \varepsilon'_{\alpha\beta} = t_{\alpha i}t_{\beta j}\varepsilon_{ij}$$

Therefore, for $i = 1, 2, 3, 6$, $\sigma_i' = \sigma_i$ and $\varepsilon_i' = \varepsilon_i$, i.e., $\sigma_{11}' = t_{11}t_{11}\sigma_{11} = \sigma_{11}$. However, from the direction cosines, $\varepsilon_{23}' = -\varepsilon_{23}$ or $\varepsilon_4' = -\varepsilon_4$, and $\sigma_4' =$

$-\sigma_4$; likewise $\varepsilon'_{31} = -\varepsilon_{31}$, hence $\varepsilon'_5 = -\varepsilon_5$ and $\sigma'_5 = -\sigma_5$. For example, $\sigma'_{23} = \sigma'_4 = t_{22}t_{33}\sigma_{23} = (1)(-1)\sigma_{23} = -\sigma_{23} = -\sigma_4$.

If one looks in detail at Equation (2.3) then,

$$\sigma'_4 = C_{41}\varepsilon'_1 + C_{42}\varepsilon'_2 + C_{43}\varepsilon'_3 + C_{44}\varepsilon'_4 + C_{45}\varepsilon'_5 + C_{46}\varepsilon'_6$$

$$\sigma_4 = C_{41}\varepsilon_1 + C_{42}\varepsilon_2 + C_{43}\varepsilon_3 + C_{44}\varepsilon_4 + C_{45}\varepsilon_5 + C_{46}\varepsilon_6$$

It is clearly seen from these two equations that $C_{41} = C_{42} = C_{43} = C_{46} = 0$. From similar examinations of the other two symmetries, it is seen also that $C_{25} = C_{35} = C_{64} = C_{65} = 0$, $C_{51} = C_{52} = C_{53} = C_{56} = 0$, and $C_{14} = C_{15} = C_{16} = C_{24} = C_{34} = 0$.

So, for a material having only one plane of symmetry the number of elastic constants is now reduced to 13. Note that from a realistic engineering point of view this would still require 13 independent physical tests (at each temperature and humidity condition)—an almost impossible task both in manpower and budget.

Now, materials which have three mutually-orthogonal planes of elastic symmetry are called "orthotropic" (a shortened term for orthogonally anisotropic). In that case, other terms in the elasticity matrix are also zero, namely

$$C_{16} = C_{26} = C_{36} = C_{45} = 0$$

Therefore, the elasticity tensor for orthotropic materials is shown below, remembering that $C_{ij} = C_{ji}$,

$$C_{ij} = \begin{bmatrix} C_{11} & C_{12} & C_{13} & 0 & 0 & 0 \\ C_{21} & C_{22} & C_{23} & 0 & 0 & 0 \\ C_{31} & C_{32} & C_{33} & 0 & 0 & 0 \\ 0 & 0 & 0 & C_{44} & 0 & 0 \\ 0 & 0 & 0 & 0 & C_{55} & 0 \\ 0 & 0 & 0 & 0 & 0 & C_{66} \end{bmatrix} \tag{2.5}$$

So, for orthotropic elastic bodies, such as most composite materials in three dimensional configuration, there are nine elastic constants.

Hence, with Equations (2.5) and (2.3), the explicit strain relations for an orthotropic, three dimensional material are: $\sigma_i = C_{ij}\varepsilon_j$ $(i, j = 1, 2, \ldots, 6)$ or more explicitly,

$$
\begin{aligned}
\sigma_1 &= C_{11}\varepsilon_1 + C_{12}\varepsilon_2 + C_{13}\varepsilon_3 \\
\sigma_2 &= C_{21}\varepsilon_1 + C_{22}\varepsilon_2 + C_{23}\varepsilon_3 \\
\sigma_3 &= C_{31}\varepsilon_1 + C_{32}\varepsilon_2 + C_{33}\varepsilon_3 \\
\sigma_4 &= \sigma_{23} = C_{44}\varepsilon_4 = 2C_{44}\varepsilon_{23} \\
\sigma_5 &= \sigma_{31} = C_{55}\varepsilon_5 = 2C_{55}\varepsilon_{31} \\
\sigma_6 &= \sigma_{12} = C_{66}\varepsilon_6 = 2C_{66}\varepsilon_{12}
\end{aligned}
\tag{2.6}
$$

It should be noted that in the latter three relationships, which involve shear relations, the factor of two is present because of the widely used definitions of ε_4, ε_5, and ε_6.

If Equation (2.6) is inverted, then, through standard matrix transformation:

$$
\begin{aligned}
\varepsilon_1 &= a_{11}\sigma_1 + a_{12}\sigma_2 + a_{13}\sigma_3 \\
\varepsilon_2 &= a_{21}\sigma_1 + a_{22}\sigma_2 + a_{23}\sigma_3 \\
\varepsilon_3 &= a_{31}\sigma_1 + a_{32}\sigma_2 + a_{33}\sigma_3 \\
\varepsilon_4 &= 2\varepsilon_{23} = a_{44}\sigma_{23} = a_{44}\sigma_4 \\
\varepsilon_5 &= 2\varepsilon_{31} = a_{55}\sigma_{31} = a_{55}\sigma_5 \\
\varepsilon_6 &= 2\varepsilon_{12} = a_{66}\sigma_{12} = a_{66}\sigma_6
\end{aligned}
\tag{2.7}
$$

The a_{ij} matrix, called the compliance matrix, is the transpose of the cofactor matrix of the C_{ij}'s divided by the determinant of the C_{ij} matrix and each term is defined as

$$
a_{ij} = \frac{[CoC_{ij}]^T}{|C_{ij}|} \tag{2.8}
$$

Again, the a_{ij} quantities are not tensors, and cannot be transformed as such. In fact, factors of 1, 2, and 4 appear in various terms when relating the tensor

TABLE 2.1. Summary of the Number of Elastic Coefficients Involved for Certain Classes of Materials.

Class of Material	Number of Nonzero Coefficients	Number of Independent Coefficients
Three-dimensional case		
General anisotropic	36	21
One-plane of symmetry	20	13
Two-planes of symmetry	12	9
Transversely isotropic	12	5
Isotropic	12	2
Two-dimensional case		
General anisotropic	9	6
One-plane of symmetry	9	6
Two-planes of symmetry	5	4
Transversely isotropic	5	4
Isotropic	5	2

compliance quantities a_{ijkl} and the contracted compliance quantities a_{ij}. It can easily be shown that $a_{ij} = a_{ji}$ and that

$$\varepsilon_i = a_{ij}\sigma_j \qquad (\text{where } i, j = 1, 2, \ldots, 6) \tag{2.9}$$

Table 2.1 is useful for listing the number of elastic coefficients present in both two- and three-dimensional elastic bodies.

2.3 THE PHYSICAL MEANING OF THE COMPONENTS OF THE ORTHOTROPIC ELASTICITY TENSOR

So far, the components of both the stiffness matrix, C_{ij}, and the compliance matrix, a_{ij}, are mathematical symbols to relate stresses and strains. By performing simple tensile and shear tests, all of the components above can be related to physical or mechanical properties.

Consider a simple, standard tensile test in the x_1 direction. The resulting stress and strain tensors are

$$\sigma_{ij} = \begin{bmatrix} \sigma_{11} & 0 & 0 \\ 0 & 0 & 0 \\ 0 & 0 & 0 \end{bmatrix}, \qquad \varepsilon_{ij} = \begin{bmatrix} \varepsilon_{11} & 0 & 0 \\ 0 & -\nu_{12}\varepsilon_{11} & 0 \\ 0 & 0 & -\nu_{13}\varepsilon_{11} \end{bmatrix} \tag{2.10}$$

where the Poisson's ratio, ν_{ij}, is very carefully defined as the negative of the ratio of the strain in the x_j direction to the strain in the x_i direction due to an applied stress in the x_i direction. In other words, it is seen above that $\varepsilon_{22} = -\nu_{12}\varepsilon_{11}$ or $\nu_{12} = -\varepsilon_{22}/\varepsilon_{11}$.

Also, the constant of proportionality between stress and strain is noted to be E_i, the modulus of elasticity in the x_i direction. Thus,

$$\varepsilon_1 = a_{11}\sigma_1 = \frac{\sigma_1}{E_1}$$

$$\varepsilon_2 = a_{21}\sigma_1 = -\nu^{12}\varepsilon_1 = -\frac{\nu_{12}\sigma_1}{E_1}$$

$$\varepsilon_3 = a_{31}\sigma_1 = -\nu_{13}\varepsilon_1 = -\frac{\nu_{13}\sigma_1}{E_1}$$

Therefore

$$a_{11} = 1/E_1, \quad a_{21} = -\nu_{12}/E_1, \quad a_{31} = -\nu_{13}/E_1 \tag{2.11}$$

For a simple tensile test in the x_2 direction, it is found that

$$a_{12} = -\nu_{21}/E_2, \quad a_{22} = 1/E_2, \quad a_{32} = -\nu_{23}/E_2 \tag{2.12}$$

Likewise, a tensile test in the x_3 direction yields

$$a_{13} = -\nu_{31}/E_3, \quad a_{23} = -\nu_{32}/E_3, \quad a_{33} = 1/E_3 \tag{2.13}$$

From the fact that $a_{ij} = a_{ji}$, then

$$\frac{\nu_{ij}}{E_i} = \frac{\nu_{ji}}{E_j} \qquad (i, j = 1, 2, 3) \tag{2.14}$$

Equation (2.14) is most valuable and widely used in the analysis of all composite material bodies.

Next, consider a hypothetical simple shear test as shown in Figure 2.4. In this case the stress, strain, and displacement tensor components are

$$\sigma_{ij} = \begin{bmatrix} 0 & \sigma_{12} & 0 \\ \sigma_{21} & 0 & 0 \\ 0 & 0 & 0 \end{bmatrix}, \ \varepsilon_{ij} = \begin{bmatrix} 0 & \varepsilon_{12} & 0 \\ \varepsilon_{21} & 0 & 0 \\ 0 & 0 & 0 \end{bmatrix}, \ u_{i,j} = \begin{bmatrix} 0 & 0 & 0 \\ \sigma_{21}/G_{21} & 0 & 0 \\ 0 & 0 & 0 \end{bmatrix}$$

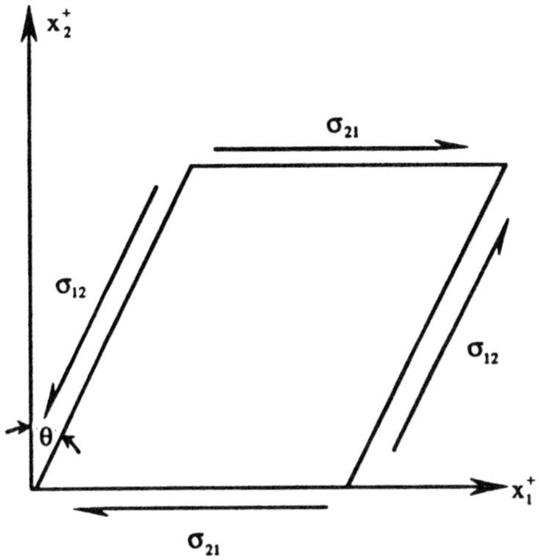

FIGURE 2.4. Shear stresses and strains. (Reprinted from Reference [4].)

In the above, u_i is the displacement and $u_{i,j} = (\partial u_i)/(\partial x_j)$. From elementary strength of materials the constant of proportionality between the shear stress σ_{21} and the angle θ is G_{21}, the shear modulus in the x_1–x_2 plane.

From the theory of elasticity

$$\varepsilon_{12} = \frac{1}{2}(u_{1,2} + u_{2,1}) = \frac{\sigma_{21}}{2G_{21}} = \frac{\tan\theta}{2} \tag{2.15}$$

From Equation (2.7), $\varepsilon_6 = a_{66}\sigma_6$, or

$$\varepsilon_{12} = \frac{a_{66}\sigma_{21}}{2} = \frac{\sigma_{21}}{2G_{21}}$$

Hence,

$$a_{66} = \frac{1}{G_{21}} = \frac{1}{G_{12}} \tag{2.16}$$

Similarily,

$$a_{44} = \frac{1}{G_{23}} \quad \text{and} \quad a_{55} = \frac{1}{G_{13}} \tag{2.17}$$

Thus, all a_{ij} components have now been related to mechanical properties, and we see that to characterize a three-dimensional orthotropic body, nine physical quantities—hence nine separate tests—are needed [i.e., E_1, E_2, E_3, G_{12}, G_{23}, G_{31}, ν_{12}, ν_{13}, ν_{21}, ν_{23}, ν_{31}, and ν_{32}, and the utilization of Equation (2.14)]. The standardized tests to conduct to obtain these anisotropic elastic constants are given in ASTM tests, and are described in the text by Carlsson and Pipes [6]. For convenience, the compliance matrix is given explicitly as

$$a_{ij} = \begin{bmatrix} \frac{1}{E_1} & -\frac{\nu_{21}}{E_2} & -\frac{\nu_{31}}{E_3} & 0 & 0 & 0 \\ -\frac{\nu_{12}}{E_1} & \frac{1}{E_2} & -\frac{\nu_{32}}{E_3} & 0 & 0 & 0 \\ -\frac{\nu_{13}}{E_1} & -\frac{\nu_{23}}{E_2} & \frac{1}{E_3} & 0 & 0 & 0 \\ 0 & 0 & 0 & \frac{1}{G_{23}} & 0 & 0 \\ 0 & 0 & 0 & 0 & \frac{1}{G_{13}} & 0 \\ 0 & 0 & 0 & 0 & 0 & \frac{1}{G_{12}} \end{bmatrix} \tag{2.18}$$

2.4 METHODS TO OBTAIN COMPOSITE ELASTIC PROPERTIES FROM FIBER AND MATRIX PROPERTIES

There are several sets of equations for obtaining the composite elastic properties from those of the fiber and matrix materials. These include those of Halpin and Tsai [7], Hashin [8], and Christensen [9]. In 1980, Hahn [10] codified certain results for fibers of circular cross section which are randomly distributed in a plane normal to the unidirectionally oriented fibers. For that case the composite is macroscopically transversely isotropic, that is $(\ )_{12} = (\ )_{13}$, $(\ )_{22} = (\ )_{33}$ and $(\ )_{55} = (\ )_{66}$, where in the parentheses the quantity could be E, G, or ν; hence, the elastic properties involve only five independent constants, namely $(\ )_{11}$, $(\ )_{22}$, $(\ )_{12}$, $(\ )_{23}$, and $(\ )_{66}$.

For several of the elastic constants, Hahn [10] states that they all have the same functional form:

$$P = \frac{(P_f V_f + \eta P_m V_m)}{(V_f + \eta V_m)} \tag{2.19}$$

TABLE 2.2. Determination of Composite Properties From Fiber and Matrix Properties

Elastic Constant	P	P_f	P_m	η
E_{11}	E_{11}	E_{11f}	E_m	1
ν_{12}	ν_{12}	ν_{12f}	ν_m	1
G_{12}	$1/G_{12}$	$1/G_{12f}$	$1/G_m$	η_6
G_{23}	$1/G_{23}$	$1/G_{23f}$	$1/G_m$	η_4
K_T	$1/K_T$	$1/K_f$	$1/K_m$	η_K

where for the elastic constant P, the P_f, P_m and η are given in Table 2.2, and where V_f and V_m are the volume fractions of the fibers and matrix respectively (and whose sum equals unity).

The expressions for E_{11} and ν_{12} are called the Rule of Mixtures. In the above K_T is the plane strain bulk modulus, $K_f = [E_f/2(1 - \nu_f)]$ and $K_m = [E_m/2(1 - \nu_m)]$. Also, the η's are given as follows:

$$\eta_6 = \frac{1 + G_m/G_{12f}}{2}, \quad \eta_4 = \frac{3 - 4\eta_m + G_m/G_{23f}}{4(1 - \nu_m)}, \quad \eta_K = \frac{1 + G_m/K_f}{2(1 - \nu_m)}$$

The shear modulus of the matrix material, G_m, if isotropic, is given by $G_m = E_m/2(1 + \nu_m)$.

The transverse moduli of the composite, $E_{22} = E_{33}$, are found from the following equation:

$$E_{22} = E_{33} = \frac{4K_T G_{23}}{K_T + mG_{23}} \tag{2.20}$$

where

$$m = 1 + \frac{4K_T\nu_{22}^2}{E_{11}}$$

The equations above have been written specifically for composites reinforced with anisotropic fibers such as graphite and aramid (Kevlar) fibers. If the fibers are isotropic, the fiber properties involve E_f, G_f, and ν_f, where $G_f = \dfrac{E_f}{2(1 + \nu_f)}$. In that case also η_K becomes

$$\eta_K = \frac{1 + (1 - 2\nu_f)G_m/G_f}{2(1 - \nu_m)} \tag{2.21}$$

Hahn notes that for most polymeric matrix structural composites, $G_m/G_f <$ 0.05. If that is the case then the η parameters are approximately

$$\eta_6 \approx 0.5; \quad \eta_4 = \frac{3 - 4\nu_m}{4(1 - \nu_m)}; \quad \eta_K = \frac{1}{2(1 - \nu_m)} \tag{2.22}$$

Finally, noting that $\nu_m = 0.35$ for most epoxies, then $\eta_4 = 0.62$ and $\eta_K =$ 0.77. Also, the Poisson's ratio, ν_{23}, can be written as

$$\nu_{23} = \nu_{f12}\,V_f + \nu_m\,(1 - V_f)\left[\frac{1 + \nu_m - \nu_{12}\left(\dfrac{E_m}{E_{11}}\right)}{1 - \nu_m^2 + \nu_m\nu_{12}\left(\dfrac{E_m}{E_{11}}\right)}\right] \tag{2.23}$$

The above equations along with Equation (2.14) provide the engineer with the resources to estimate the elastic constants for a composite material if the constituent properties and volume fractions are known. In a few instances the weight fraction of the fiber, W_f, is known. In that case the volume fraction is obtained from the following equation, where W_m is the weight fraction of the matrix, and ρ_m and ρ_f are the respective densities:

$$V_f = \frac{\rho_m W_f}{\rho_m W_f + \rho_f W_m} \tag{2.24}$$

Chou [11] provides the most comprehensive treatment for determining the composite elastic constants for short-fiber composites, hybrid composites, textile composites, and very flexible composites.

2.5 THERMAL AND HYGROTHERMAL CONSIDERATIONS

In the previous two sections, the elastic relations developed pertain only to an anisotropic elastic body at one temperature, that temperature being the "stress-free" temperature; i.e., the temperature at which the body is considered to be free of stress if it is under no mechanical static or dynamic loadings.

However, in both metallic and composite structures,, changes in temperature are commonplace both during fabrication and during structural usage. Changes in temperature result in two effects that are very important. First, most materials expand when heated and contract when cooled, and in most cases this expansion is proportional to the temperature change. If, for instance, one had a long

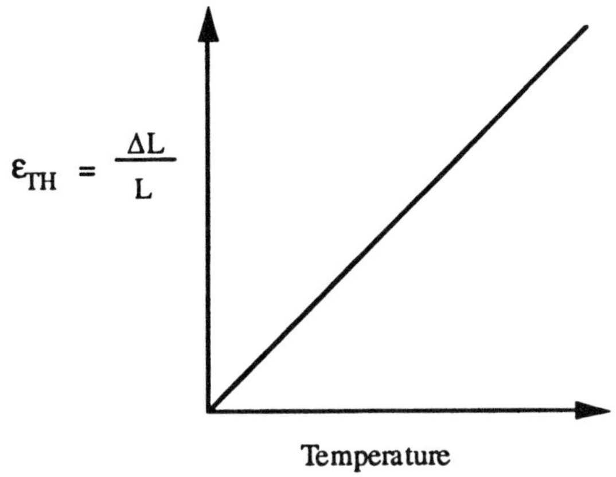

FIGURE 2.5. Change in length of a bar or rod as a function of temperature. (Reprinted from Reference [4].)

thin bar of a given material, then with change in temperature, the ratio of the change in length of the bar, ΔL, to the original length L, is related to the temperature of the bar T, as shown in Figure 2.5. Mathematically, this can be written as

$$\varepsilon_{\text{thermal}} = \frac{\Delta L}{L} = \alpha \Delta T \tag{2.25}$$

where α is the coefficient of thermal expansion, i.e., the proportionality constant between the "thermal" strain ($\Delta L/L$) and the change in temperature, ΔT, from some reference temperature at which there are no thermal stresses or thermal strains.

The second major effect of temperature change relates to stiffness and strength. Most materials become softer, more ductile, and weaker as they are heated. Typical plots of ultimate strength, yield stress, and modulus of elasticity as functions of temperature are shown in Figure 2.6. In performing a stress analysis, determining the natural frequencies, or finding the buckling load of a heated or cooled structure one must use the strengths and the moduli of elasticity of the material at the temperature at which the structure is expected to perform.

In an orthotropic material, such as a composite, there will or could be three different coefficients of thermal expansion, and three different thermal strains, one in each of the orthogonal directions comprising the orthotropic material [Equation (2.25) would then have subscripts of 1, 2, and 3 on both the strains and the coefficients of thermal expansion].

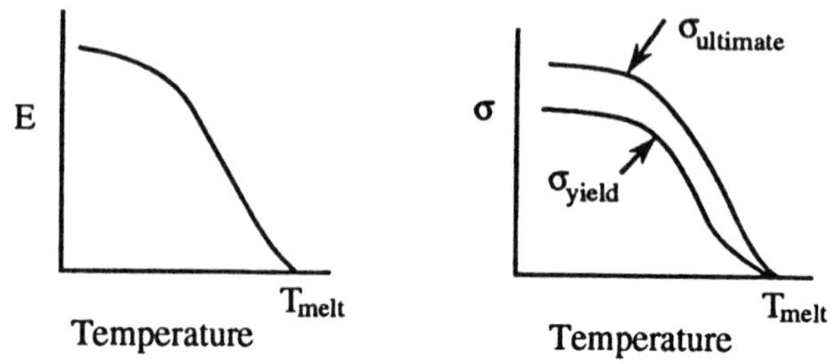

FIGURE 2.6. Modulus of elasticity and strengths as functions of temperature. (Reprinted from Reference [4].)

During the mid-seventies another physical phenomenon associated with polymer matrix composites was recognized as important. It was found that the combination of high temperature and high humidity caused a doubly deleterious effect on the structural performance of these composites. Engineers and material scientists became very concerned about these effects, and considerable research effort was expended in studying this new phenomenon. Conferences [12] were held which discussed the problem, and both short- and long-range research plans were proposed. The twofold problem involves the fact that the combination of high temperature and high humidity results in the entrapment of moisture in the polymer matrix, with attendant weight increase ($\leq 2\%$) and more importantly, a swelling of the matrix. It was realized [13] that the ingestion of moisture varied linearly with the swelling so that in fact

$$\varepsilon_{\text{hygrothermal}} = \frac{\Delta L}{L} = \beta \Delta m \tag{2.26}$$

where Δm is the increase from zero moisture measured in percentage weight increase, and β is the coefficient of hygrothermal expansion, analogous to the coefficient of thermal expansion, depicted in Equation (2.25). This analogy is very important because one can see that the hygrothermal effects are entirely analogous mathematically to the thermal effect. Therefore, if one has the solutions to a thermoelastic problem, merely substituting $\beta\Delta m$ for or adding it to the $\alpha\Delta T$ terms provides the hygrothermal solution. The reader, however, should be aware that ambiguities exist in the literature regarding the coefficient of hygrothermal expansion. The test methods to obtain values of the coefficient of hygrothermal expansion β are given in Reference [14].

The second effect (i.e., the reduction of strength and stiffness) is also similar to the thermal effect. This is shown qualitatively in Figure 2.7. Dry polymers have properties that are usually rather constant until a particular temperature is reached, traditionally called by polymer chemists the "glass transition tem-

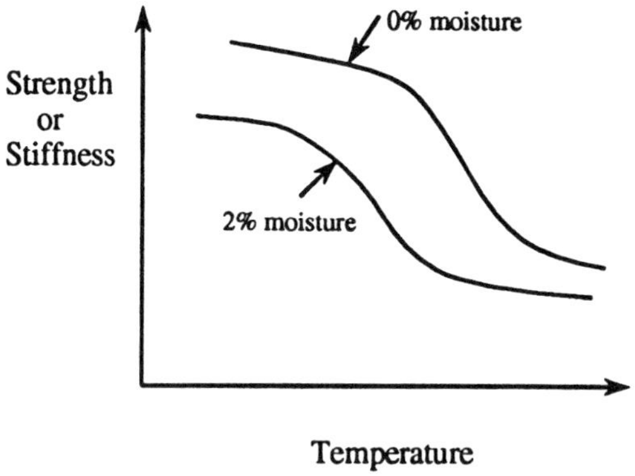

FIGURE 2.7. Mechanical properties as a function of temperature and moisture absorption. (Reprinted from Reference [4].)

perature," above which both strength and stiffness deteriorate rapidly. If the same polymer is saturated with moisture, not only are the mechanical properties degraded at any one temperature, but the glass transition temperature for that polymer is significantly lower.

As a quantitative example, Figure 2.8 clearly shows the diminution in tensile and shear strength due to the long-term hygrothermal environment. Short-time ASTM tensile and shear tests were performed on random-mat glass/polyester resin specimens. It is clearly seen that there is a significant reduction

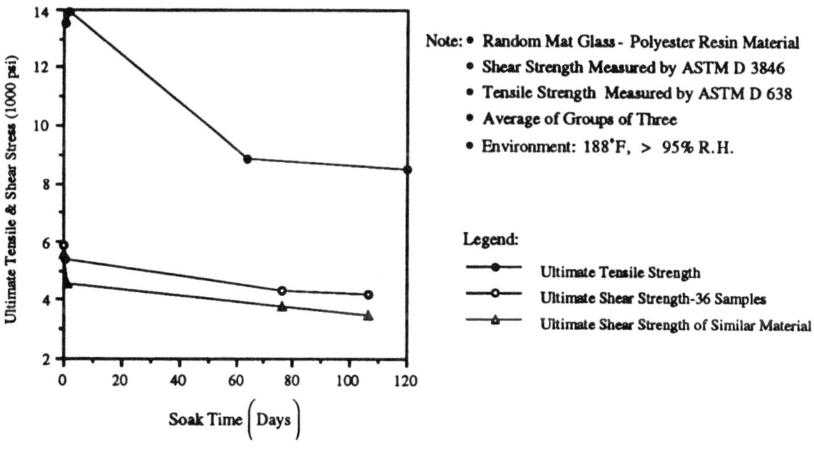

FIGURE 2.8. Strength as a function of soak time in a hot, wet environment. (Reprinted from Reference [4].)

in tensile strength, and a 29.3% and a 37.1% reduction in ultimate shear strength of these materials over a 100⁺ day soak period. If these effects are not accounted for in design analysis, catastrophic failures will occur.

Thus, for modern polymer matrix composites one must include not only the thermal effects but also the hygrothermal effects, or the structure could be considerably underdesigned, resulting in potential failure. To deal with the real world of polymer composites, Equation (2.9) must therefore be modified to read

$$\varepsilon_i = a_{ij}\sigma_j + \alpha_i \Delta T + \beta_i \Delta_m \qquad (i, j = 1, 2, 3) \tag{2.27}$$

$$\varepsilon_i = a_{ij}\sigma_j \qquad (i, j = 4, 5, 6) \tag{2.28}$$

where in each equation $j = 1\text{–}6$. Two types of equations are shown above because both thermal effects and hygrothermal effects are dilatational only; that is, they cause an expansion or contraction, but do not affect the shear stresses or strains. This is important to remember.

Although the thermal effect and moisture effect are analogous, one major difference is the timescale involved. A structure subjected to a change in temperature environment would require minutes or at most hours to come to equilibrium at the new temperature; the same structure would require weeks, months, or even years to come to moisture equilibrium (saturation) if that dry structure were placed in a 95–100% relative humidity environment. Figure 2.9

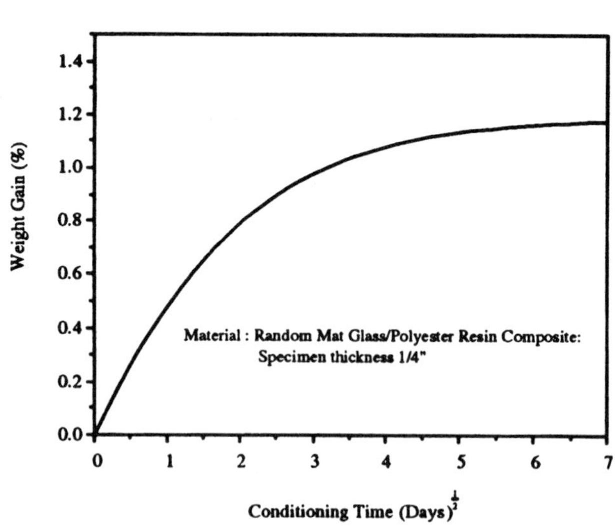

FIGURE 2.9. Moisture absorption as a function of soak time. (Reprinted from Reference [4].)

illustrates the point. A 1/4″ thick random-mat glass polyester matrix material requires 49 days of soak time at 188°F and 95% relative humidity to become saturated.

Recently, Woldesenbet [15] soaked a large number of IM7/8551-7 graphite-epoxy unidirectional test pieces, some in room-temperature water, to saturation. The test pieces were 1/4″ diameter by 3/8″ long cylinders. At room temperature, the time required to reach saturation was 55 weeks; other test pieces were soaked at an elevated temperature to reduce soak time.

Foam cores may also be affected by hygrothermal environments. Den Haan [16] recently reported on a new foam core, a polyetherimide-based Airex R82.80 foam, which he claims is excellent in wet environments and for damage tolerance.

2.6 TIME-TEMPERATURE EFFECTS ON COMPOSITE MATERIALS

In addition to the effects of temperature and moisture on the short-time properties discussed above, if a structure is maintained under a constant load for a period of time, then creep and viscoelastic effects can become very important in the design and analysis of that structure. The subject of creep is discussed in numerous materials science and strength of materials texts and will not be described here in detail.

Creep and viscoelasticity can become significant in any material above certain temperatures, but can be particularly important in polymer matrix materials whose operating temperatures must be kept below maximum temperatures of 250°F, 350°F, or in some cases 600°F for short periods of time.

From a structural mechanics point of view, almost all of the viscoelastic effects occur in the polymer matrix, while little or no creep occurs in the fibers. Thus, the study of creep in the polymeric materials that comprise the matrix provides the data necessary to study creep in composites. Jurf [17] experimentally studied the effects of temperature and moisture (hygrothermal effects) on various epoxy materials (FM 73M and FM 300M adhesives). He established that at least for some epoxy materials it is possible to construct a master creep curve using a temperature shift factor, and established the fact that a moisture-shift factor can also be employed. The importance of this is that by these experimentally determined temperature and moisture-shift factors, for the shear modulus of the epoxy, the results of short-time creep tests can be used for a multitude of time/temperature/moisture combinations over the lifetime and environment of a structure comprised of that material. Wilson [18,19] studied the effects of viscoelasticity on the buckling of columns and rectangular plates and found that significant reductions of the buckling loads can occur.

Wilson found that for the materials he studied, the buckling load diminished over the first 400 hours, then stabilized at a constant value. However, that value may be a small fraction of the elastic buckling load if the composite properties in the load direction were matrix-dominated properties (described later in the text). Wilson also established that for the problems studied it was quite satisfactory to bypass the complexities of a full-scale viscoelastic analysis using the Correspondence Principle and Laplace transformations. He suggested the use of appropriate short-time stiffness properties of the composite experimentally determined with specimens that have been held at the temperature until the time for which the structural calculations are being made.

Chevalier [20] has studied effects of both creep and fatigue on various sandwich core materials including end-grain balsa (EGB), hexagonal honeycomb made of aluminum and aramid fiber-paper (NOMEX), and PVC. He notes that there is little data available, and that data is badly needed.

Gellhorn and Starlinger [21] studied creep of foams used in sandwich structures, and emphasize the importance of including creep considerations in the design of long-term loaded sandwich structures. Shenoi, Clark, Allen, and Hicks [22] have also studied the effects of creep on foam-core sandwich structures, primarily concentrating on sandwich beams. Feichtinger and Martensen [23] recently compared the creep properties of various core materials including aluminum and Nomex honeycomb cores, PVC foam cores, and balsa wood. Caprino, Teti, and Messa [24] also investigated the creep behavior of various foam-core sandwich structures. Caprino, Teti, and Messa [24] point out that while aluminum and Nomex honeycomb cores are satisfactory for heavily loaded structures, their high cost makes them less acceptable for mass production structural components. They state that the reasonable compromise between cost and mechanical properties rests with polymeric foam cores. In particular, they praise PVC because of its high stiffness and strength combined with good thermal insulation and low water absorption.

2.7. HIGH STRAIN RATE EFFECTS ON MATERIAL PROPERTIES

Another consideration in the analysis of all sandwich structures is the effect of high-strain rate on the strength and stiffness properties of the materials used. Most materials have significantly different strengths, moduli, and strains to failure at high strain rates compared to static values. However, even in the fall of 1992, most if not all of the major finite element codes involving 10^5–10^6 elements using 10^1–10^2 hours of computer time to describe underwater and other explosion effects on structures, still utilized static material properties. Such high strain rate properties of materials are sorely needed. Some dynamic properties have been found, and test techniques established. For more information see Lindholm [25], Daniel, La Bedz, and Liber [26], Nicholas [27], Zukas [28], Sierakowski [29,30], and Feichtinger [31].

Gellhorn and Reif [32] studied the high-strain rate effects on the compressive- and shear properties of three different crosslinked PVC foam and one linear PVC foam for compressive-strain rates up to 500/s and shear strain rates up to 200/s in order to ascertain the viscoelastic behavior as a function of strain rate. They found that the linear PVC was the only material which had no important loss in strain to failure. At the highest strain rate the linear PVC foam was 34% higher in compressive strength and 34% higher in strain to failure than the cross-linked PVC foams. Feichtinger [31] presented shear strength and stiffness data for strain rates up to 3.75/s.

Gellhorn and Reif [32] state that a boat traveling at 10 knots combined with a panel length of 1.5 meters results in a shear strain rate of 6.9/s. They also conclude that high strain rate material property data cannot be extrapolated from static data. Materials that exhibit high stiffness and strength at low strain rates may fail at high strain rates because they turn brittle, while other materials such as linear PVC that has modest stiffness and strength at low strain rates have superior properties at higher strain rates.

Recent research on the high strain rate effects on numerous composite materials at strain rates up to 1600 sec^{-1}, has resulted in over 20 publications involving those by Choe, Finch, and Vinson [33], Frey, Vinson, and Prewo [34], Woldesenbet and Vinson [35], Greenfield [36], Dee and Vinson [37] and Preissner, Woldesenbet and Vinson [38].

Over the range of strain rates tested up to 1600/sec, for the various composite materials tested, it was found that with high strain rates compared to static values, the yield stresses may increase by factors up to 3.6, the yield strains may change by factors of 3.1, strains to failure may change by factors up to 4.7, moduli of elasticity may change by factors up to 2.4, elastic strain energy densities may change by factors up to 6, while strain energy densities to failure may change by factors up to 8.1. Thus, the use of static material properties to analyze and design structures subjected to impact, explosions, crashes, or other dynamic loads is not recommended.

2.8 LAMINAE OF COMPOSITE MATERIALS

Almost all practical composite material structures are thin in the thickness direction because their superior properties permit the use of thin-walled structures. Most polymeric matrix composites are made in the form of a uniaxial set of fibers surrounded by a polymeric matrix in the form of a tape several inches wide termed "prepreg." The basic element in most long fiber composite structures is a lamina of fiber plus matrix, all fibers oriented in one direction,

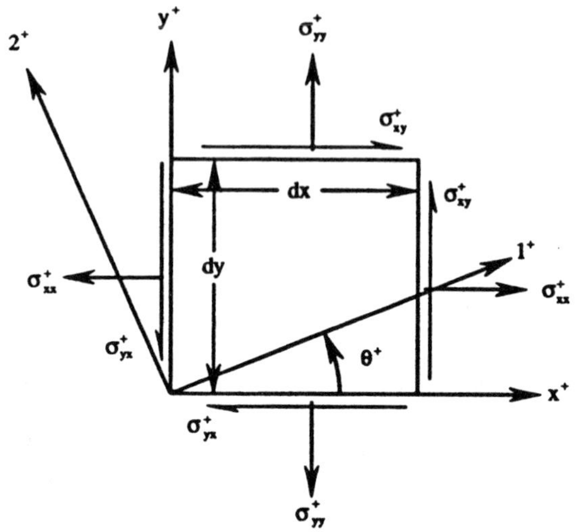

FIGURE 2.10. Laminae coordinate systems. (Reprinted from Reference [61].)

made by laying the prepreg tape of a certain length side by side. In Section 2.9, the stacking of various laminae to form a superior structure, termed a laminate, will be discussed. For sandwich structures, the faces may be made of "prepregs." In that case these faces may comprise several laminae forming a laminate.

To describe this, consider a small element of a lamina of constant thickness h, wherein the principal material axes are labeled 1 and 2, that is the 1 direction is parallel to the fibers, the 2-direction is normal to them, and consider that the beam, plate, or shell geometric axes are x and y as depicted in Figure 2.10.

This element has the positive directions of all stresses shown in a consistent manner [1–4,25,26]. If one performs a force equilibrium study to relate σ_x, σ_y, and σ_{xy} to σ_1, σ_2, and σ_{12}, it is exactly analogous to the Mohr's circle analysis in basic strength of materials with the result that, in matrix form,

$$\begin{Bmatrix} \sigma_1 \\ \sigma_2 \\ \sigma_6 \end{Bmatrix} = [T]_{CL} \begin{Bmatrix} \sigma_x \\ \sigma_y \\ \sigma_{xy} \end{Bmatrix} \tag{2.29}$$

where

$$[T]_{CL} = \begin{bmatrix} m^2 & n^2 & +2mn \\ n^2 & m^2 & -2mn \\ -mn & mn & (m^2 - n^2) \end{bmatrix} \tag{2.30}$$

where here $m = \cos\theta$, $n = \sin\theta$, and θ is defined positive as shown in Figure 2.10, and where the subscripts CL refers to the classical two-dimensional case only, that is, in the 1-2 plane or the x-y plane only. Analogously, a strain relationship also follows for the classical isothermal case

$$\begin{Bmatrix} \varepsilon_1 \\ \varepsilon_2 \\ \varepsilon_{12} \end{Bmatrix} = [T]_{CL} \begin{Bmatrix} \varepsilon_x \\ \varepsilon_y \\ \varepsilon_{xy} \end{Bmatrix} \tag{2.31}$$

However, these classical two-dimensional relationships must be modified to treat a composite material to include thermal effects, hygrothermal effects, and the effects of transverse-shear deformation treated in detail elsewhere [2–4,6,7]. The effects of transverse-shear deformation, shown through the inclusion of the $\sigma_4 - \varepsilon_4$ and $\sigma_5 - \varepsilon_5$ relations in Equations (2.32) and (2.34), must be included in composite materials, because in the fiber direction the composite has many of the mechanical properties of the fiber itself (strong and stiff), while in the thickness direction the fibers are basically ineffective and the shear properties are dominated by the weaker matrix material. Similarly, because quite often the matrix material has much higher coefficients of thermal and hygrothermal expansion (α and β), thickening and thinning of the lamina cannot be ignored in some cases. Hence, without undue derivation, the Equations (2.29) through (2.31) are modified to be:

$$\begin{Bmatrix} \sigma_1 \\ \sigma_2 \\ \sigma_3 \\ \sigma_4 \\ \sigma_5 \\ \sigma_6 \end{Bmatrix} = [T] \begin{Bmatrix} \sigma_x \\ \sigma_y \\ \sigma_z \\ \sigma_{yz} \\ \sigma_{xz} \\ \sigma_{xy} \end{Bmatrix} \quad \text{and} \quad \begin{Bmatrix} \varepsilon_1 \\ \varepsilon_2 \\ \varepsilon_3 \\ \dfrac{\varepsilon_4}{2} \\ \dfrac{\varepsilon_5}{2} \\ \dfrac{\varepsilon_6}{2} \end{Bmatrix} = [T] \begin{Bmatrix} \varepsilon_x \\ \varepsilon_y \\ \varepsilon_z \\ \varepsilon_{yz} \\ \varepsilon_{xz} \\ \varepsilon_{xy} \end{Bmatrix} \tag{2.32}$$

wherein

$$[T] = \begin{bmatrix} m^2 & n^2 & 0 & 0 & 0 & 2mn \\ n^2 & m^2 & 0 & 0 & 0 & -2mn \\ 0 & 0 & 1 & 0 & 0 & 0 \\ 0 & 0 & 0 & m & -n & 0 \\ 0 & 0 & 0 & n & m & 0 \\ -mn & mn & 0 & 0 & 0 & (m^2 - n^2) \end{bmatrix} \tag{2.33}$$

Please note the factor of $\frac{1}{2}$ seen in the strain expressions, because of the way ε_4, ε_5, and ε_6 are defined in Equation (2.2).

For completeness, the reverse transformations are given.

$$\begin{Bmatrix} \sigma_x \\ \sigma_y \\ \sigma_z \\ \sigma_{yz} \\ \sigma_{xz} \\ \sigma_{xy} \end{Bmatrix} = [T]^{-1} \begin{Bmatrix} \sigma_1 \\ \sigma_2 \\ \sigma_3 \\ \sigma_4 \\ \sigma_5 \\ \sigma_6 \end{Bmatrix} \text{ and } \begin{Bmatrix} \varepsilon_x \\ \varepsilon_y \\ \varepsilon_z \\ \varepsilon_{yz} \\ \varepsilon_{xz} \\ \varepsilon_{xy} \end{Bmatrix} = [T]^{-1} \begin{Bmatrix} \varepsilon_1 \\ \varepsilon_2 \\ \varepsilon_3 \\ \frac{\varepsilon_4}{2} \\ \frac{\varepsilon_5}{2} \\ \frac{\varepsilon_6}{2} \end{Bmatrix} = [T]^{-1} \begin{Bmatrix} \varepsilon_{11} \\ \varepsilon_{22} \\ \varepsilon_{33} \\ \varepsilon_{23} \\ \varepsilon_{13} \\ \varepsilon_{12} \end{Bmatrix} \tag{2.34}$$

where[1]

$$\begin{bmatrix} m^2 & n^2 & 0 & 0 & 0 & -2mn \\ n^2 & m^2 & 0 & 0 & 0 & 2mn \\ 0 & 0 & 1 & 0 & 0 & 0 \\ 0 & 0 & 0 & m & n & 0 \\ 0 & 0 & 0 & -n & m & 0 \\ mn & -mn & 0 & 0 & 0 & (m^2 - n^2) \end{bmatrix} \tag{2.35}$$

Again, please note that the transformations can be made only with *tensor* strains. Hence, from Equation (2.2), it is necessary to divide ε_4, ε_5, and ε_6 by two.

If one systematically uses these expressions, and utilizes Hooke's Law relating stress and strain, and includes the thermal and hygrothermal effects, one can produce the following overall general equations for a lamina of a fiber-reinforced composite material in terms of the principal material directions (1, 2, 3); see Equations (2.5)–(2.8).

$$\begin{Bmatrix} \sigma_1 \\ \sigma_2 \\ \sigma_3 \\ \sigma_4 \\ \sigma_5 \\ \sigma_6 \end{Bmatrix} = \begin{bmatrix} Q_{11} & Q_{12} & Q_{13} & 0 & 0 & 0 \\ Q_{12} & Q_{22} & Q_{23} & 0 & 0 & 0 \\ Q_{13} & Q_{23} & Q_{33} & 0 & 0 & 0 \\ 0 & 0 & 0 & Q_{44} & 0 & 0 \\ 0 & 0 & 0 & 0 & Q_{55} & 0 \\ 0 & 0 & 0 & 0 & 0 & Q_{66} \end{bmatrix} \begin{Bmatrix} \varepsilon_1 - \alpha_1 \Delta T - \beta_1 \Delta m \\ \varepsilon_2 - \alpha_2 \Delta T - \beta_2 \Delta m \\ \varepsilon_3 - \alpha_3 \Delta T - \beta_3 \Delta m \\ 2\varepsilon_{23} \\ 2\varepsilon_{31} \\ 2\varepsilon_{12} \end{Bmatrix} \tag{2.36}$$

In the above, the Q_{ij} quantities are used for the stiffness-matrix quantities obtained directly from Equations (2.5) through (2.18). One should also remem-

[1] $[T]^{-1}$ can be found by replacing θ by $(-\theta)$ in $[T]$.

ber that $\varepsilon_{23} = \left(\frac{1}{2G_{23}}\right)\sigma_4$, $\varepsilon_{31} = \left(\frac{1}{2G_{31}}\right)\sigma_5$, and $\varepsilon_{12} = \left(\frac{1}{2G_{12}}\right)\sigma_6$, hence the coefficients of "two" appearing with the tensor-shear strains ε_{23}, ε_{31}, and ε_{12} above. Using the notation of Sloan [39],

$$\begin{aligned}
&Q_{11} = E_{11}(1 - \nu_{23}\nu_{32})/\Delta, \quad Q_{22} = E_{22}(1 - \nu_{31}\nu_{13})/\Delta \\
&Q_{33} = E_{33}(1 - \nu_{12}\nu_{21})/\Delta, \quad Q_{44} = G_{23}, \quad Q_{55} = G_{13}, \quad Q_{66} = G_{12} \\
&Q_{12} = (\nu_{21} + \nu_{31}\nu_{23})E_{11}/\Delta = (\nu_{12} + \nu_{32}\nu_{13})E_{22}/\Delta \\
&Q_{13} = (\nu_{31} + \nu_{21}\nu_{32})E_{11}/\Delta = (\nu_{13} + \nu_{12}\nu_{23})E_{33}/\Delta \\
&Q_{23} = (\nu_{32} + \nu_{12}\nu_{31})E_{22}/\Delta = (\nu_{23} + \nu_{21}\nu_{13})E_{33}/\Delta \\
&\Delta = 1 - \nu_{12}\nu_{21} - \nu_{23}\nu_{32} - \nu_{31}\nu_{13} - 2\nu_{21}\nu_{32}\nu_{13}
\end{aligned} \tag{2.37}$$

Incidentally in the above expressions, if the lamina is transversely isotropic, i.e., same properties in both the 2- and 3-direction, then $\nu_{12} = \nu_{13}$, $G_{12} = G_{13}$, $E_{22} = E_{33}$ with resulting simplification.

For preliminary calculation in design or where great accuracy is not needed, simpler forms [2,3] for some of the expressions in Equation (2.37) can be used, as shown below:

$$\begin{aligned}
&Q_{11} = E_{11}/(1 - \nu_{12}\nu_{21}), \; Q_{22} = E_{22}/(1 - \nu_{12}\nu_{21}) \\
&Q_{12} = Q_{21} = \nu_{21}E_{11}/(1 - \nu_{12}\nu_{21}) = \nu_{12}E_{22}/(1 - \nu_{12}\nu_{21}) \\
&Q_{66} = G_{12}
\end{aligned} \tag{2.38}$$

If these simpler forms are used, then one would use the classical form of the constitutive relations instead of Equation (2.36), neglecting transverse shear deformation and transverse normal stress, namely

$$\begin{Bmatrix} \sigma_1 \\ \sigma_2 \\ \sigma_6 \end{Bmatrix} = \begin{bmatrix} Q_{11} & Q_{12} & 0 \\ Q_{12} & Q_{22} & 0 \\ 0 & 0 & Q_{66} \end{bmatrix} \begin{Bmatrix} \varepsilon_1 - \alpha_1\Delta T - \beta_1\Delta m \\ \varepsilon_2 - \alpha_2\Delta T - \beta_2\Delta m \\ 2\varepsilon_{12} \end{Bmatrix} \tag{2.39}$$

where one should remember also that $2\varepsilon_{12} = \varepsilon_6$, hence the appearance of two before ε_{12}. As stated above for many cases it is sufficient to use Equations (2.38) and (2.39) rather than Equations (2.36) and (2.37) for faster and easier calculation.

Now, to relate these relationships to the x-y-z coordinate system, one utilizes Equations (2.34) through (2.36). The result is

$$\begin{Bmatrix} \sigma_x \\ \sigma_y \\ \sigma_z \\ \sigma_{yz} \\ \sigma_{xz} \\ \sigma_{xy} \end{Bmatrix} = \begin{bmatrix} \overline{Q}_{11} & \overline{Q}_{12} & \overline{Q}_{13} & 0 & 0 & \overline{Q}_{16} \\ \overline{Q}_{12} & \overline{Q}_{22} & \overline{Q}_{23} & 0 & 0 & \overline{Q}_{26} \\ \overline{Q}_{13} & \overline{Q}_{23} & \overline{Q}_{33} & 0 & 0 & \overline{Q}_{36} \\ 0 & 0 & 0 & \overline{Q}_{44} & \overline{Q}_{45} & 0 \\ 0 & 0 & 0 & \overline{Q}_{45} & \overline{Q}_{55} & 0 \\ \overline{Q}_{16} & \overline{Q}_{26} & \overline{Q}_{36} & 0 & 0 & \overline{Q}_{66} \end{bmatrix} \begin{Bmatrix} \varepsilon_x - \alpha_x \Delta T - \beta_x \Delta m \\ \varepsilon_y - \alpha_y \Delta T - \beta_y \Delta m \\ \varepsilon_z - \alpha_z \Delta T - \beta_z \Delta m \\ 2\varepsilon_{yz} \\ 2\varepsilon_{xz} \\ 2(\varepsilon_{xy} - \alpha_{xy}\Delta T - \beta_{xy}\Delta m) \end{Bmatrix} \tag{2.40}$$

where $[\overline{Q}] = [T]^{-1}[Q][T]$, or more explicitly,

$$\begin{aligned}
\overline{Q}_{11} &= Q_{11}m^4 + 2(Q_{12} + 2Q_{66})m^2n^2 + Q_{22}n^4 \\
\overline{Q}_{12} &= (Q_{11} + Q_{22} - 4Q_{66})m^2n^2 + Q_{12}(m^4 + n^4) \\
\overline{Q}_{13} &= Q_{12}m^2 + Q_{23}n^2 \\
\overline{Q}_{16} &= -mn^3Q_{22} + m^3nQ_{11} - mn(m^2 - n^2)(Q_{12} + 2Q_{66}) \\
\overline{Q}_{22} &= Q_{11}n^4 + 2(Q_{12} + 2Q_{66})m^2n^2 + Q_{22}m^4 \\
\overline{Q}_{23} &= n^2Q_{13} + m^2Q_{23} \\
\overline{Q}_{33} &= Q_{33} \\
\overline{Q}_{26} &= -m^3nQ_{22} + mn^3Q_{11} + mn(m^2 - n^2)\,(Q_{12} + 2Q_{66}) \\
\overline{Q}_{36} &= (Q_{13} - Q_{23})mn \\
\overline{Q}_{44} &= Q_{44}m^2 + Q_{55}n^2 \\
\overline{Q}_{45} &= (Q_{55} - Q_{44})mn \\
\overline{Q}_{55} &= Q_{55}m^2 + Q_{44}n^2 \\
\overline{Q}_{66} &= (Q_{11} + Q_{22} - 2Q_{12})m^2n^2 + Q_{66}(m^2 - n^2)^2 \\
\alpha_x &= \alpha_1 m^2 + \alpha_2 n^2 \qquad \beta_x = \beta_1 m^2 + \beta_2 n^2 \\
\alpha_y &= \alpha_2 m^2 + \alpha_1 n^2 \qquad \beta_y = \beta_2 m^2 + \beta_1 n^2 \\
\alpha_z &= \alpha_3 \qquad \beta_z = \beta_3 \\
\alpha_{xy} &= (\alpha_1 - \alpha_2)mn \qquad \beta_{xy} = (\beta_1 - \beta_2)mn
\end{aligned} \tag{2.41}$$

It should be remembered that although the coefficients of both thermal and hygrothermal expansion are purely dilatational in the material coordinate system 1-2, rotation into the structural coordinate system x-y, results in an α_{xy} and a β_{xy}.

Again, for preliminary design purposes or for approximate but usually accurate calculations one can use the simpler classical form of

$$\begin{Bmatrix} \sigma_x \\ \sigma_y \\ \sigma_{xy} \end{Bmatrix} = \begin{bmatrix} \overline{Q}_{11} & \overline{Q}_{12} & \overline{Q}_{16} \\ \overline{Q}_{12} & \overline{Q}_{22} & \overline{Q}_{26} \\ \overline{Q}_{16} & \overline{Q}_{26} & \overline{Q}_{66} \end{bmatrix} \begin{Bmatrix} \varepsilon_x - \alpha_x \Delta T - \beta_x \Delta m \\ \varepsilon_y - \alpha_y \Delta T - \beta_y \Delta m \\ 2(\varepsilon_{xy} - \alpha_{xy} \Delta T - \beta_{xy} \Delta m) \end{Bmatrix} \tag{2.42}$$

where the $\overline{Q}_{ij}$ are defined in Equation (2.41), but one can use the $\overline{Q}_{ij}$ of Equation (2.38) instead of Equation (2.37) for consistency with the simpler expressions above.

One interesting variation of the above classical quantities of Equation (2.42) resulted when Tsai and Pagano [40] rewrote many of the above quantities in a manner that is very useful in comparing various material systems in the design of a composite structure:

$$\begin{aligned}
\overline{Q}_{11} &= U_1 + U_2 \cos(2\theta) + U_3 \cos(4\theta) \\
\overline{Q}_{22} &= U_1 - U_2 \cos(2\theta) + U_3 \cos(4\theta) \\
\overline{Q}_{12} &= U_4 - U_3 \cos(4\theta) \\
\overline{Q}_{66} &= U_5 - U_3 \cos(4\theta) \\
\overline{Q}_{16} &= +\frac{1}{2} U_2 \sin(2\theta) + U_3 \sin(4\theta) \\
\overline{Q}_{26} &= +\frac{1}{2} U_2 \sin(2\theta) - U_3 \sin(4\theta)
\end{aligned} \tag{2.43}$$

where

$$U_1 = \frac{1}{8}(3Q_{11} + 3Q_{22} + 2Q_{12} + 4Q_{66})$$

$$U_2 = \frac{1}{2}(Q_{11} - Q_{22})$$

$$U_3 = \frac{1}{8}(Q_{11} + Q_{22} - 2Q_{12} - 4Q_{66})$$

$$U_4 = \frac{1}{8}(Q_{11} + Q_{22} + 6Q_{12} - 4Q_{66})$$

$$U_5 = \frac{1}{8}(Q_{11} + Q_{22} - 2Q_{12} + 4Q_{66})$$

In the above, the U_i quantities are invariant with respect to axis rotation and therefore are truly composite lamina properties. In any case Equations (2.43) and (2.41) can be used to check numerical accuracy of the results.

At this point, given a lamina of a unidirectional composite of known elastic properties, if used in a plate or panel, (as well as a beam, ring, or shell, discussed later) with the 1-2 material axis at an angle θ from the plate or panel x-y axes, all stiffness quantities Q_{ij} and $\overline{Q}_{ij}$ can be determined relating stresses and strains in either coordinate system. It is also worthwhile to mention that if the $\overline{Q}_{ij}$ quantities are calculated by both Equations (2.41) and (2.43), some obvious mistakes can be avoided.

Keep in mind that in sandwich construction each face may itself be composed of many laminae, each oriented in a different direction to attain the best macroscopic properties. Also, in sandwich construction one can consider the two faces and the core to be the three laminae comprising the overall sandwich laminate, as stated earlier in Chapter 1.

2.9. LAMINATE ANALYSIS

The generalized constitutive equations for one lamina of a composite material were formulated in Section 2.8. Many structures of composite materials, including sandwich structures, are composed of numerous laminae which are bonded and/or cured together. In fact, over and above the superior properties in strength and stiffness that composites possess, the ability to optimize the orientation of each lamina in order to tailor the laminate material properties for a given structural size and set of loadings, is one of the major advantages that composites have over more conventional structures. Up to this point, the concentration has been on the stress-strain or constitutive relations. Now the other three sets of equations comprising the equations of elasticity will be considered: the strain-displacement relations, the equilibrium equations, and the compatibility equations. As stated above, for sandwich structures, each face can be considered as one or more laminae, and the core can be considered as a lamina as well, regardless of the magnitude of the core depth (h_c).

Consider a laminate composed of N laminae. For the kth lamina of the laminate, Equation (2.40) can be written

$$\begin{Bmatrix} \sigma_x \\ \sigma_y \\ \sigma_z \\ \sigma_{yz} \\ \sigma_{xz} \\ \sigma_{xy} \end{Bmatrix}_k = [\overline{Q}]_k \begin{Bmatrix} \varepsilon_x - \alpha_x \Delta T - \beta_x \Delta m \\ \varepsilon_y - \alpha_y \Delta T - \beta_y \Delta m \\ \varepsilon_z - \alpha_z \Delta T - \beta_z \Delta m \\ 2\varepsilon_{yz} \\ 2\varepsilon_{xz} \\ 2(\varepsilon_{xy} - \alpha_{xy} \Delta T - \beta_{xy} \Delta m) \end{Bmatrix}_k \tag{2.44}$$

where all matrices must have the subscript k due to the orientation of the particular lamina with respect to the structural x-y coordinates and its unique $[\overline{Q}]_k$, $\{\alpha_i\}_k$, and $\{\beta_i\}_k$.

For any elastic body the strain-displacement equations, i.e., those kinematic relations describing the functional relations between the elastic strains in the body and its displacements, are given by

$$\varepsilon_{ij} = \frac{1}{2}(u_{i,j} + u_{j,i}) \tag{2.45}$$

where $i, j = x, y, z$ in a Cartesian coordinate frame, and the comma denotes partial differentiation with respect to the coordinate denoted by the symbol after the comma. Explicitly, the relations are

$$\begin{aligned} &\varepsilon_x = \frac{\partial u}{\partial x}, \ \varepsilon_y = \frac{\partial v}{\partial y}, \ \varepsilon_z = \frac{\partial w}{\partial z} \\ &\varepsilon_{xz} = \frac{1}{2}\left(\frac{\partial u}{\partial z} + \frac{\partial w}{\partial x}\right), \ \varepsilon_{yz} = \frac{1}{2}\left(\frac{\partial v}{\partial z} + \frac{\partial w}{\partial y}\right) \\ &\varepsilon_{xy} = \frac{1}{2}\left(\frac{\partial u}{\partial y} + \frac{\partial v}{\partial x}\right) \end{aligned} \tag{2.46}$$

In the above u, v, and w are the displacements in the structural x, y, and z directions respectively. In the linear elastic beam, plate and shell theory, it is assumed that a lineal element extending through the thickness of a thin plate and perpendicular to the middle surface (that is, the x-y plane in Figure 2.11) prior to loading, upon the load application of load undergoes at most a translation and a rotation with respect to the original coordinate system. Based on that one assumption, the functional form of the displacements for the laminated and/or sandwich plate are

$$\begin{aligned} u(x, y, z) &= u_0(x, y) + z\overline{\alpha}(x, y) \\ v(x, y, z) &= v_0(x, y) + z\overline{\beta}(x, y) \\ w(x, y, z) &= w(x, y) \end{aligned} \tag{2.47}$$

where u_0, v_0, and w are the middle surface displacements; i.e., the translations of the lineal element, and the second terms in the first two equations are the

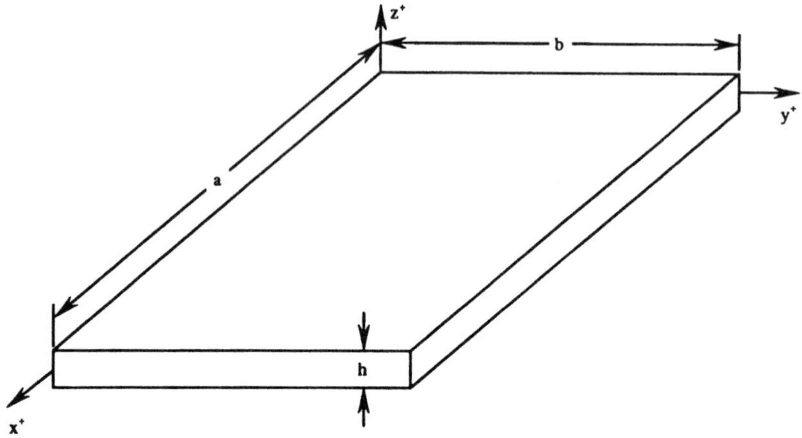

FIGURE 2.11. Typical rectangular plate. (Reprinted from Reference [4].)

rotations of a lineal element through the thickness. In classical beam and plate theory, $\overline{\alpha}$ and $\overline{\beta}$ are the negative of the first derivative of the lateral displacement with respect to the x and y coordinates respectively (i.e., $\overline{\alpha} = -(\partial w/\partial x)$ and $\overline{\beta} = -(\partial w/\partial y)$, the negative of the slope), but if transverse shear deformation is nonzero, $\overline{\alpha}$ and $\overline{\beta}$ are unknown dependent variables which must be solved for; this will be discussed later. Also, in classical theory, it is assumed that the lineal element across the thickness of the beam, plate, or shell cannot extend nor shrink because at most it undergoes a translation and rotation, hence $w = w(x, y)$ only. Sloan [39] has shown that for many practical composites, plate thickening (where the lateral deflection is a function of z) is unimportant and can be neglected; hence it is not included here, but can be found in detail in [39]. For sandwich plates, core thickening or thinning may be important, however, and that too will be considered later.

Substituting Equation (2.47) into Equation (2.45) results in:

$$\varepsilon_x = \frac{\partial u_0}{\partial_x} + z\frac{\partial \overline{\alpha}}{\partial x}, \ \varepsilon_y = \frac{\partial v_0}{\partial y} + z\frac{\partial \overline{\beta}}{\partial y}, \ \varepsilon z = 0$$

$$\varepsilon_{xz} = \frac{1}{2}\left(\overline{\alpha} + \frac{\partial w}{\partial x}\right), \ \varepsilon_{yz} = \frac{1}{2}\left(\overline{\beta} + \frac{\partial w}{\partial y}\right) \tag{2.48}$$

$$\varepsilon_{xy} = \frac{1}{2}\left(\frac{\partial u_0}{\partial y} + \frac{\partial v_0}{\partial x}\right) + \frac{z}{2}\left(\frac{\partial \overline{\alpha}}{\partial y} + \frac{\partial \overline{\beta}}{\partial x}\right).$$

The mid-surface strains can be written as

$$\varepsilon_{x_0} = \frac{\partial u_0}{\partial x}, \ \varepsilon_{y_0} = \frac{\partial v_0}{\partial y}, \ \varepsilon_{xy_0} = \frac{1}{2}\left(\frac{\partial u_0}{\partial y} + \frac{\partial v_0}{\partial x}\right) \tag{2.49}$$

The curvatures can be written as

$$\kappa_x = \frac{\partial \overline{\alpha}}{\partial x}, \ \kappa_y = \frac{\partial \overline{\beta}}{\partial y}, \ \kappa_{xy} = \frac{1}{2}\left(\frac{\partial \overline{\alpha}}{\partial y} + \frac{\partial \overline{\beta}}{\partial x}\right) \tag{2.50}$$

The theory assumes displacements and strains are employed to describe the strains and displacements of a laminated or sandwich structure of composite materials, because all of the individual laminae are bonded together; therefore, the same assumptions are made regarding the lineal element through the laminate thickness. Thus, a continuity of strains and displacements occurs regardless of the orientation of individual plies or laminae.

Substituting Equation (2.48) into Equation (2.44) results in (2.51); wherein, because we have assumed $\varepsilon_z = 0$ (because plate thickening is neglected), σ_z for a thin-walled structure of composite material is usually negligible and will not be considered further.

$$\begin{Bmatrix} \sigma_x \\ \sigma_y \\ \sigma_{yz} \\ \sigma_{xz} \\ \sigma_{xy} \end{Bmatrix}_k = [\overline{Q}]_k \begin{Bmatrix} \varepsilon_{x_0} + z\kappa_x - \alpha_x \Delta T - \beta_x \Delta m \\ \varepsilon_{y_0} + z\kappa_y - \alpha_y \Delta T - \beta_y \Delta m \\ 2\varepsilon_{yz} \\ 2\varepsilon_{xz} \\ 2(\varepsilon_{xy_0} + \kappa_{xy} z - \alpha_{xy} \Delta T - \beta_{xy} \Delta m) \end{Bmatrix}_k \tag{2.51}$$

Note that without the hygrothermal terms, the strain-curvature matrix at the right in Equation (2.51) would suffice for the entire laminate independent of orientation, because the displacements and strains are continuous over the thickness of the laminate. In that case the subscript k on that matrix would not be needed. However, even though there is continuity of the midsurface strains and curvatures across the thickness of the laminate, the stresses are discontinuous across the laminate thickness because of the various orientations of each lamina; hence, the subscript k in the stress matrix above. It is seen from Equation (2.51) that if all quantities on the right-hand side are known, one can easily calculate each stress component in each lamina comprising the laminate.

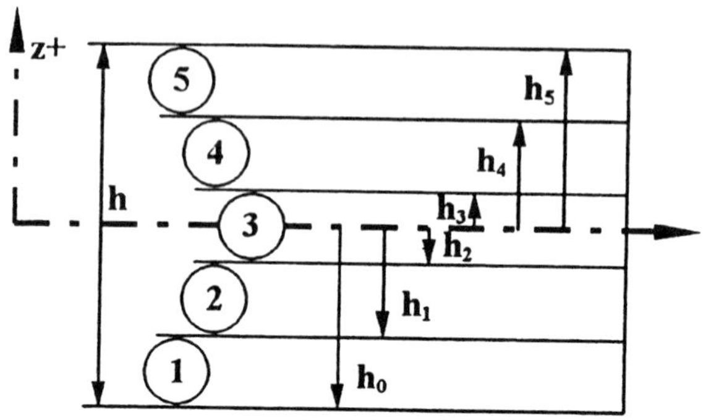

FIGURE 2.12. Nomenclature for the stacking sequence.

Consider a laminated plate or panel of thickness h as shown in Figure 2.12. It is seen that h_k is the *vectorial* distance from the panel midplane, $z = 0$, to the upper surface of the kth lamina, i.e., *any dimension below the midsurface is a negative dimension* and any dimension above the midsurface is positive. For example, consider a laminate 0.52 mm (0.020″) thick, composed of four equally thick laminae, each being 0.13 mm (0.005″) thick. Then $h_0 = -0.26$ mm ($-0.010''$), $h_1 = -0.13$ mm ($-0.005''$), $h_2 = 0$, $h_3 = 0.13$ mm (0.005″), and $h_4 = 0.26$ mm (0.010″).

As in classical beam, plate, and shell theory [2–5,41,42], one defines and uses stress resultants (N), stress couples (M), and transverse shear resultants (Q) per unit width for the overall structure regardless of the number and the orientation of the laminae; hence:

$$\begin{Bmatrix} N_x \\ N_y \\ N_{xy} \\ Q_x \\ Q_y \end{Bmatrix} = \int_{-h/2}^{+h/2} \begin{Bmatrix} \sigma_x \\ \sigma_y \\ \sigma_{xy} \\ \sigma_{xz} \\ \sigma_{yz} \end{Bmatrix} dz, \quad \begin{Bmatrix} M_x \\ M_y \\ M_{xy} \end{Bmatrix} = \int_{-h/2}^{+h/2} \begin{Bmatrix} \sigma_x \\ \sigma_y \\ \sigma_{xy} \end{Bmatrix} z dz \tag{2.52}$$

It must be emphasized that all of the above quantities are force per unit width and couple per unit width, because in plate and shell structures these quantities vary in both the x and y directions.

On the plate shown in Figure 2.13, the positive directions of all the resultants, and stress couples are shown, consistent with the definitions of the quantities given in Equation (2.52).

For a laminated or a sandwich plate, the stress components can be integrated

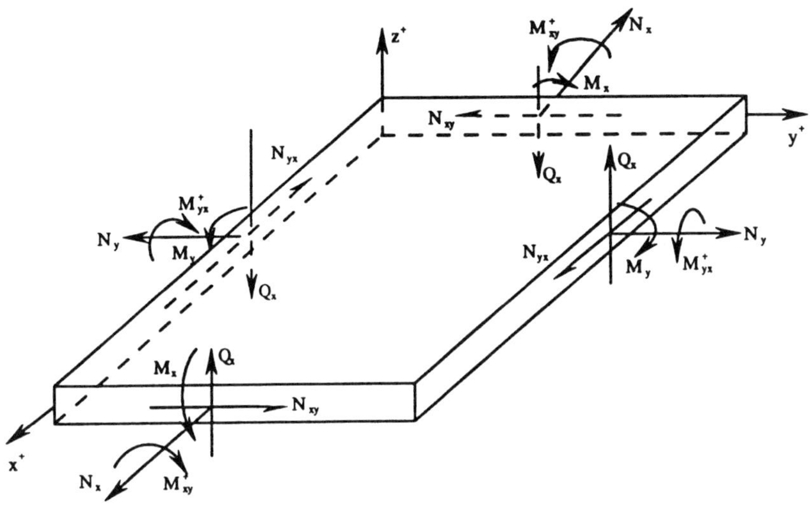

FIGURE 2.13 Positive directions for stress resultants and stress couples for a plate. (Reprinted from Reference [4].)

across each lamina, but must then be added together as follows: Employing Equations (2.44) and (2.48) through (2.50):

$$\begin{bmatrix} N_x \\ N_y \\ N_{xy} \end{bmatrix} = \sum_{k=1}^{N} \int_{h_{k-1}}^{h_k} \begin{bmatrix} \sigma_x \\ \sigma_y \\ \sigma_{xy} \end{bmatrix} dz$$

$$\begin{Bmatrix} N_x \\ N_y \\ N_{xy} \end{Bmatrix} = \sum_{k=1}^{N} \left(\int_{h_{k-1}}^{h_k} [\overline{Q}]_k \begin{Bmatrix} \varepsilon_{x_0} \\ \varepsilon_{y_0} \\ \varepsilon_{xy_0} \end{Bmatrix} dz + \int_{h_{k-1}}^{k} [\overline{Q}]_k \begin{Bmatrix} \kappa_x \\ \kappa_y \\ \kappa_{xy} \end{Bmatrix} zdz \right.$$

$$\left. - \int_{h_{k-1}}^{h_k} [\overline{Q}]_k \begin{Bmatrix} \alpha_x \\ \alpha_y \\ \alpha_{xy} \end{Bmatrix}_k \Delta T dz - \int_{h_{k-1}}^{h_k} [\overline{Q}]_k \begin{Bmatrix} \beta_x \\ \beta_y \\ \beta_{xy} \end{Bmatrix}_k \Delta m dz \right) \quad (2.53)$$

where here only the pertinent portions of the $[\overline{Q}]_k$ matrix are used.

Since the derivatives of the midsurface displacements (u_0 and v_0) and the rotations; ($\overline{\alpha}$ and $\overline{\beta}$) and the $\overline{Q}$'s are not functions of z, Equation (2.53) can be rewritten as

$$\begin{Bmatrix} N_x \\ N_y \\ N_{xy} \end{Bmatrix} = \sum_{k=1}^{N} \left([\overline{Q}]_k \begin{Bmatrix} \varepsilon_{x_0} \\ \varepsilon_{y_0} \\ \varepsilon_{xy_0} \end{Bmatrix}_k \int_{h_{k-1}}^{h_k} dz + [\overline{Q}]_k \begin{Bmatrix} \kappa_x \\ \kappa_y \\ \kappa_{xy} \end{Bmatrix}_k \int_{h_{k-1}}^{h_k} zdz \right.$$

$$\left. - \int_{h_{k-1}}^{h_k} [\overline{Q}]_k \begin{Bmatrix} \alpha_x \\ \alpha_y \\ \alpha_{xy} \end{Bmatrix}_k \Delta T dz - \int_{h_{k-1}}^{h_k} [\overline{Q}]_k \begin{Bmatrix} \beta_x \\ \beta_y \\ \beta_{xy} \end{Bmatrix}_k \Delta m dz \right) \tag{2.54}$$

Finally, Equation (2.54) can be written succinctly as

$$\{N\} = [A]\{\varepsilon_0\} + [B]\{\kappa\} - \{N\}^T - \{N\}^m \tag{2.55}$$

where it is shown later in Equation (2.64) that a factor of two is necessary in some terms, and where

$$A_{ij} = \sum_{k=1}^{N} (\overline{Q}_{ij})_k (h_k - h_{k-1}) \qquad [i,j = 1, 2, 6] \tag{2.56}$$

$$B_{ij} = \frac{1}{2} \sum_{k=1}^{N} (\overline{Q}_{ij})_k (h_k^2 - h_{k-1}^2) \qquad [i,j = 1, 2, 6] \tag{2.57}$$

$$N_{ij}^T = \sum_{k=1}^{N} \int_{h_{k-1}}^{h_k} (\overline{Q}_{ij})_k (\alpha_{ij})_k \Delta T dz \qquad [i,j = 1, 2, 6] \tag{2.58}$$

$$N_{ij}^m = \sum_{k=1}^{N} \int_{h_{k-1}}^{h_k} (\overline{Q}_{ij})_k (\beta_{ij})_k \Delta m dz \qquad [i, j = 1, 2, 6] \tag{2.59}$$

where it is obvious from Equation (2.54) how the $[\alpha_{ij}]_k$ and $[\beta_{ij}]_k$ matrices are defined.

From Equation (2.55), it is seen that the in-plane stress resultants for a laminated thin-walled structure are not only functions of the midplane strains ($\varepsilon_{x_0} = \partial u_0/\partial x$, etc.) as they are in a homogeneous beam, plate, or shell, but they can also be functions of the curvatures and twists ($\kappa_x = \partial\overline{\alpha}/\partial x$, etc.) as well. Also, it is seen that in-plane forces can cause curvatures or twisting deformations.

Similar to the above, but multiplying Equation (2.51) through by z first before integrating, as in Equation (2.52), the following can be found:

$$\{M\} = [B]\{\varepsilon_0\} + [D]\{\kappa\} - [M]^T - [M]^m \tag{2.60}$$

where it will be shown in Equation (2.63) that factors of 2 are necessary in some terms, and where

$$D_{ij} = \frac{1}{3}\sum_{k=1}^{N} (\overline{Q}_{ij})_k\,[h_k^3 - h_{k-1}^3] \qquad [i,j = 1, 2, 6] \tag{2.61}$$

$$M_{ij}^T = \sum_{k=1}^{N} \int_{h_{k-1}}^{h_k} (\overline{Q}_{ij})_k\,(\alpha_{ij})_k\,\Delta Tzda \qquad [i,j = 1, 2, 6] \tag{2.62}$$

$$M_{ij}^m = \sum_{k=1}^{N} \int_{h_{k-1}}^{h_k} (\overline{Q}_{ij})_k\,(\beta_{ij})_k\,\Delta mzdz \qquad [i,j = 1, 2, 6] \tag{2.63}$$

where $i, j = 1, 2, 6$.

Finally, the results of Equations (2.55) and (2.60) can be combined as follows:

$$\begin{Bmatrix} N_x \\ N_y \\ N_{xy} \\ --- \\ M_x \\ M_y \\ M_{xy} \end{Bmatrix} = \begin{bmatrix} A_{11}A_{12}A_{16} & | & B_{11}B_{12}B_{16} \\ A_{12}A_{22}A_{26} & | & B_{12}B_{22}B_{26} \\ A_{16}A_{26}A_{66} & | & B_{16}B_{26}B_{66} \\ \text{-- -- --} & & \text{-- -- --} \\ B_{11}B_{12}B_{16} & | & D_{11}D_{12}D_{16} \\ B_{12}B_{22}B_{26} & | & D_{12}D_{22}D_{26} \\ B_{16}B_{26}B_{66} & | & D_{16}D_{26}D_{66} \end{bmatrix} \begin{Bmatrix} \varepsilon_{x0} \\ \varepsilon_{y0} \\ 2\varepsilon_{xy0} \\ --- \\ \kappa_x \\ \kappa_y \\ 2\kappa_{xy} \end{Bmatrix} - \begin{Bmatrix} N_x^T \\ N_y^T \\ N_{xy}^T \\ --- \\ M_x^T \\ M_y^T \\ M_{xy}^T \end{Bmatrix} - \begin{Bmatrix} N_x^m \\ N_y^m \\ N_{xy}^m \\ --- \\ M_x^m \\ M_y^m \\ M_{xy}^m \end{Bmatrix} \tag{2.64}$$

It is seen that the $[A]$ matrix is the extensional stiffness matrix relating the in-plane stress resultants (N's) to the midsurface strains (ε_0's) and the $[D]$ matrix is the flexural stiffness matrix relating the stress couples (M's) to the curvatures (κ's). Since the $[B]$ matrix relates M's to ε_0's and N's to κ's, it is called the bending-stretching coupling matrix. It should be noted that a laminated structure can have bending-stretching coupling even if all laminae are isotropic; for example, a laminate composed of one lamina of steel and another of polyester. In fact, only when the structure is exactly symmetric about its middle surface are all of the B_{ij} components equal to zero, and this requires symmetry in laminae properties, orientation, and location from the middle surface.

It is seen that stretching-shearing coupling occurs when A_{16} and A_{26} are nonzero. Twisting-stretching coupling as well as bending-shearing coupling occurs when the B_{16} and B_{26} terms are nonzero, and bending-twisting coupling comes from nonzero values of the D_{16} and D_{26} terms. Usually the $(\ )_{16}$ and $(\ )_{26}$ terms are avoided by proper stacking sequences, but there could be some structural applications where these effects could be used to advantage, such as in aeroelastic tailoring.

Examples of these effects in several cross-ply laminates (i.e., combinations of 0° and 90° plies), angle ply laminates (combinations of $+\theta$ and $-\theta$ plies), and unidirectional laminates (all 0° plies) are involved in problems at the end of this chapter. It is seen in Equation (2.56) that $(h_k - h_{k-1})$ is always positive, and from Equation (2.61) $(h_k^3 - h_{k-1}^3)$ is always positive; hence, in all cross-ply laminates, all $(\ )_{16}$ and $(\ )_{26}$ terms are zero.

The inclusion of transverse shear deformation effects on the structural behavior, results in an improved theory as follows. To determine the transverse shear resultants Q_x and Q_y, defined in Equation (2.52), it is assumed that the transverse shear stresses are distributed parabolically across the laminate thickness. In spite of the discontinuities at the interface between laminae, a continuous function $f(z)$ is used as a weighting function by some authors, which includes a factor of 5/4 so that the shear factor calculated for a layered orthotropic shell wall is consistent with the established shear factor from the previous work of Reissner [43] and Mindlin [44] for the homogeneous case:

$$f(z) = \frac{5}{4}\left[1 - \left(\frac{z}{h/2}\right)^2\right] \tag{2.65}$$

Then from Equations (2.37), (2.40), (2.48), (2.52), and (2.65)

$$\sigma_{xz_k} = 2\overline{Q}_{55_k}\varepsilon_{xz} + 2\overline{Q}_{45_k}\varepsilon_{yz} \qquad \sigma_{yz_k} = 2\overline{Q}_{45_k}\varepsilon_{xz} + 2\overline{Q}_{44_k}\varepsilon_{yz}$$

where $\overline{Q}_{44} = Q_{44}m^2 + Q_{55}n^2$, $\overline{Q}_{45} = (Q_{55} - Q_{44})mn$, $\overline{Q}_{55} = Q_{55}m^2 + Q_{44}n^2$, $Q_{44} = G_{23}$, and $Q_{55} = G_{13}$.

Hence, the transverse shear resultants Q_x and Q_y are as follows:

$$Q_x = 2(A_{55}\varepsilon_{xz} + A_{45}\varepsilon_{yz}) \tag{2.66}$$

$$Q_y = 2(A_{45}\varepsilon_{xz} + A_{44}\varepsilon_{yz}) \tag{2.67}$$

where,

$$A_{ij} = \frac{5}{4}\sum_{k=1}^{N} (\overline{Q}_{ij})_k \left[h_k - h_{k-1} - \frac{4}{3}(h_k^3 - h_{k-1}^3)\frac{1}{h^2}\right] \tag{2.68}$$

where h is the total thickness of laminate or sandwich plate and $i, j = 4, 5$ only. Some authors use other weighting functions. As an example if one considers the laminate to be only one lamina, $-\frac{h}{2} \leq z \leq \frac{h}{2}$, then $A_{55} = \left(\frac{5}{6}\right) Q_{55}h$, illustrating clearly the weighting factor introduced by Reissner of 5/6 [43].

At this point the stress-strain relations, or constitutive relations of Equation (2.64) can be combined with the appropriate stress equations of equilibrium, and the strain-displacement relations to form an appropriate beam, plate, or shell theory including thermal and hygrothermal effects as well as transverse shear deformation.

2.10 [*A*], [*B*], AND [*D*] STIFFNESS MATRICES FOR A MIDPLANE SYMMETRIC SANDWICH STRUCTURE

Consider a cross-section of a sandwich structure shown in Figure 2.14. To illustrate how the same methods used to determine the stiffness matrix quantities for a laminated structure given in Figure 2.12 can be used to obtain the stiffness quantities for a sandwich plate, simply define lamina 1 as the lower face, lamina 2 as the core, and lamina 3 as the top face.

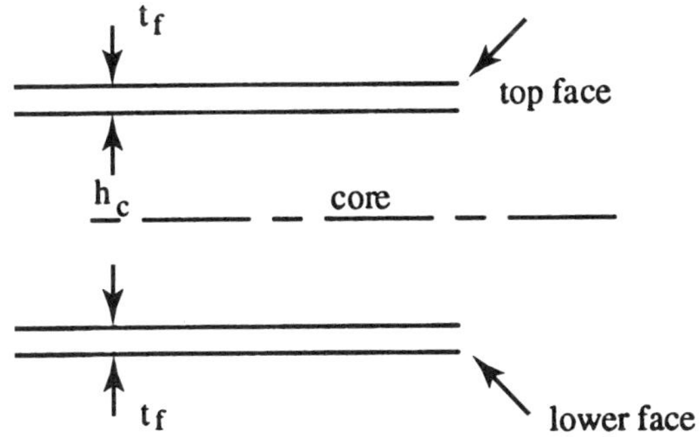

FIGURE 2.14. Cross section of a symmetric sandwich structure. (Reprinted from Reference [4].)

Therefore, in this example because the materials are assumed to be isotropic $\overline{Q}_{ij} = Q_{ij}$, where face quantities use subscript f, core quantities use subscript c,

$$A_{ij} = \sum_{k=1}^{3} (\overline{Q}_{ij})_k(h_k - h_{k-1}) = \sum_{k=1}^{3} (Q_{ij})_k(h_k - h_{k-1})$$

$$= (Q_{ij})_f\left[-\frac{h_c}{2} - \left(-\frac{h_c}{2} - t_f\right)\right] + [Q_{ij}]_c\left[\left(\frac{h_c}{2}\right) - \left(-\frac{h_c}{2}\right)\right]$$

$$+ (Q_{ij})_f\left[\left(\frac{h_c}{2} + t_f\right) - \frac{h_c}{\mathrm{c}}\right]$$

$$= (Q_{ij})_f(t_f) + (Q_{ij})_c(h_c) + (Q_{ij})_f(t_f) = (Q_{ij})_f\,(2t_f) + (Q_{ij})_c(h_c) \quad (2.69)$$

For i and j = 1 or 2, to calculate the in-plane stiffness terms,

$$(Q_{ij})_f = \frac{E_f}{(1 - \nu_f^2)} \quad \text{and} \quad (Q_{ij})_c = \frac{E_c}{(1 - \nu_c^2)}$$

So

$$A_{11} = A_{22} = \frac{E_f}{(1 - \nu_f^2)}(2t_f) + \frac{E_c}{(1 - \nu_c^2)} \quad (2.70)$$

and if E_c is negligible compared to E_f, then the in-plane stiffness per unit width is the very familiar sandwich expression

$$A_{11} = A_{22} = \frac{2E_f t_f}{(1 - \nu_f^2)} \quad (2.71)$$

Similarly

$$D_{ij} = \frac{1}{3}\sum_{k=1}^{3} (\overline{Q}_{ij})_k(h_k^3 - h_{k-1}^3) = \frac{1}{3}(Q_{ij})_f\left[\left(-\frac{h_c}{2}\right)^3 - \left(-\frac{h_c}{2} - t_f\right)^3\right]$$

$$+ \frac{1}{3}(Q_{ij})_c\left[\left(\frac{h_c}{2}\right)^3 - \left(-\frac{h_c}{2}\right)^3\right] + \frac{1}{3}(Q_{ij})_f\left[\left(\frac{h_c}{2} + t_f\right)^3 - \left(\frac{h_c}{2}\right)^3\right] \quad (2.72)$$

For i and $j = 1$ or 2, the resulting flexural stiffness quantities are seen to be

$$D_1 = D_2 = \frac{1}{3}\left\{\frac{2E_f}{(1 - \nu_f^2)}\left[\left(\frac{3}{4}\right) h_c^2 t_f + \frac{3}{2} h_c t_f^2 + t_f^3\right] + \frac{E_c}{(1 - \nu_c^2)}\frac{h_c^3}{4}\right\}$$

Neglecting higher powers of t_f and factoring out the first terms gives

$$D_1 = D_2 = \frac{1}{2}\frac{E_f h_c^2 t_f}{(1 - \nu_f^2)}\left[1 + \frac{1}{6}\frac{E_c(1 - \nu_f^2)h_c}{E_f(1 - \nu_c^2)t_f}\right] \qquad (2.73)$$

Now for many sandwich structures both $\frac{E_c}{E_f} \ll 1$, and $\frac{t_f}{h_c} \ll 1$, such that $\frac{E_c h_c}{E_f t_f} \ll 1$; therefore the second term in the bracket of Equation (2.73) is very small compared to unity. In that case

$$D_1 = D_2 = \frac{1}{2}\frac{E_f h_c^2 t_f}{1 - \nu_f^2} \qquad (2.74)$$

which is the very familiar expression for the flexural stiffness for an isotropic sandwich structure.

Likewise all $[A]$, $[B]$, and $[D]$ matrix quantities can be derived from the laminate analysis for a sandwich structure, isotropic or anisotropic, midplane symmetric or asymmetric. Quite often the resulting expressions can be simplified as shown above. For example, consider a midsurface symmetric rectangular sandwich panel, wherein both faces are made of a unidirectional continuous fiber composite with the material 1 axis coincident with the panel's x-axis etc., and where the core is assumed to contribute nothing to the stiffness matrix. In that case

$$A_{ij} = (\overline{Q}_{ij})_f 2t_f = (Q_{ij})_f 2t_f \qquad (i, j = 1, 2, 6)$$

$$B_{ij} = 0$$

$$D_{ij} = \frac{(\overline{Q}_{ij})_f t_f h_c^2}{2} = \frac{(Q_{ij})_f t_f h_c^2}{2} \qquad (i, j = 1, 2, 6)$$

or

$$A_{ij} = \begin{bmatrix} \dfrac{2E_{11}t_f}{(1 - \nu_{12}\nu_{21})} & \dfrac{2\nu_{21}E_{11}t_f}{(1 - \nu_{12}\nu_{21})} & 0 \\ \dfrac{2\nu_{21}E_{11}t_f}{(1 - \nu_{12}\nu_{21})} & \dfrac{2E_{22}t_f}{(1 - \nu_{12}\nu_{21})} & 0 \\ 0 & 0 & 2G_{12}t_f \end{bmatrix} \tag{2.75}$$

$$D_{ij} = \begin{bmatrix} \dfrac{E_{11}h_c^2t_f}{2(1 - \nu_{12}\nu_{21})} & \dfrac{\nu_{21}E_{11}h_c^2t_f}{2(1 - \nu_{12}\nu_{21})} & 0 \\ \dfrac{\nu_{21}E_{11}h_c^2t_f}{2(1 - \nu_{12}\nu_{21})} & \dfrac{E_{22}h_c^2t_f}{2(1 - \nu_{12}\nu_{21})} & 0 \\ 0 & 0 & \dfrac{G_{12}h_c^2t_f}{2} \end{bmatrix} \tag{2.76}$$

It should be noted that for this sandwich

$$(D_{ij}) = (A_{ij})h_c^2/4$$

For the transverse shear quantities A_{44} and A_{55}, following the same procedures for the isotropic sandwich

$$A_{44} = A_{55} = G_c h_c + 2t_f G_f \tag{2.77}$$

and when in many cases $t_f/h_c \ll 1$, then

$$A_{44} = A_{55} = G_c h_c \tag{2.78}$$

In this expression, it is obvious once again that the core acts as the web in an I-beam in that the core (not the faces) provides the transverse shear stiffness for the structure.

In all of the above, it is assumed that the sandwich core is very weak compared to the face materials; that the core does not contribute significantly to the $[A]$, $[B]$, or $[D]$ stiffness matrix quantities; and that the thickness, h, of the sandwich panel remains constant. This is true in many cases, and almost all of the cases treated in this textbook.

However, in some cases the loads are such that the core material is compressed significantly, and then the question arises as to how does the core behave. According to Sikarski and Mercado [45], most of the common core materials behave linearly in shear initially, and then behave nonlinearly:

elastic–perfectly plastic for PVC foams, and bilinear for end-grain balsa, as examples in the extremes. Sikarskie and Mercado then analyze sandwich beams under four-point bending and sandwich plates showing the growth of damage and behavior with the nonlinear core materials.

Many modern sandwich structures involve foam cores that are compressible, and under some loadings the upper and lower faces undergo differing deformation patterns. This occurs particularly under localized lateral loads, and can lead to premature failure of the structure. Frostig [46] and numerous colleagues including Baruch [47], Patel [48], Shenhar [49], and Thomsen [50] have studied these problems extensively regarding buckling, vibrations, and delamination. Frostig [46] emphasizes that a stiffener edge support always causes stress concentrations that affect the faces as well as the skin-core interfaces with any type of loading.

2.11 PIEZOELECTRIC EFFECTS

A piezoelectric material is one of several intelligent materials (others include shape memory alloys, electro-rheological fluids, etc.) used to develop adaptive structures. Piezoelectric materials are those that produce strains if an electrical potential is applied, or if strained produce an electric potential. At present, piezoelectric materials are in the form of thin layers of polymer or ceramic, and can be integrated into or onto structures for both sensing deformations on one hand, or controlling structural deformations and damping vibrations on the other. Most work in this area has been performed since 1985, but it is developing rapidly. In 1990, Lee and Moon [51] treated the beams and plates with thin piezoelectric PVDF layers. The linear vibrations of piezoelectric laminates were also treated in Reference [51]. Yu [52] has also presented the nonlinear equations for a sandwich plate with a thin piezoelectric face layer.

Basically the piezoelectric effect can be described by

$$\varepsilon_{ij}^{P} = d_{kij}E_k' \ (i,j,k = 1,2,3) \tag{2.79}$$

where ε_{ij}^{P} is the usual strain tensor, caused by a piezoelectric effect, E_k' is the electrical field intensity vector, and d_{kij} is the piezoelectric tensor. It is of rank three, and hence has 27 (3^3) components in three-dimensional space. It is easily seen that it is symmetric with respect to i and j, because the strain tensor ε_{ij} is symmetric. It is the only third order tensor having any application to structural mechanics.

It has been found, see Leibowitz and Vinson [53], Larson [54], and Newill [55], that the piezoelectric effect is analogous to the thermal and the hygrothermal effect. This means the total strain equals the sum of the isothermal strain, plus any thermal strain, plus any moisture (hygrothermal) strain, plus the strains caused by the piezoelectric effect, and therefore a piezoelectric stress resultant and a piezoelectric stress couple can be defined exactly analogous to thermal and moisture stress resultants and couples. These are

$$N_{ij}^{P} = \sum_{k=1}^{N} (\overline{Q}_{ij})_k \int_{h_{k-1}}^{h_k} d_{kij}\, E_k dz \qquad (2.80)$$

$$M_{ij}^{P} = \sum_{k=1}^{N} (\overline{Q}_{ij})_k \int_{h_{k-1}}^{h_k} d_{kij}\, E_k \zeta dz \qquad (2.81)$$

and they can appear as another column matrix on the right-hand side of Equation (2.63). This analogy is extremely important because it enables the analyst and designer to utilize all available thermoelastic and hygrothermal, solutions to solve sandwich structure (as well as other) problems involving piezoelectric materials. Other recent references for study are papers by Hilton, Yi, and Vinson, e.g., [56].

2.12 PROBLEMS

2.1 Consider a foam-core sandwich panel composed of two identical faces and a foam core. Each face is a cross-ply laminate of [0°, 90°, 90°, 0°] construction, with each ply being 0.25 mm (0.01 in.) thick. The core is made of Klegecell foam of density 16.02 kg/m^3 (1.0 lb/ft^3) and shear modulus G_c of 68.95 MPa (10,000 psi), $\nu_c = 0.4$, and is 25.4 mm (1 in.) thick. The stiffness matrix quantities for the boron/epoxy faces are

$$Q_{11} = 2.43 \times 10^5 \text{ MPa } (35.32 \times 10^6 \text{ psi})$$

$$Q_{22} = 2.43 \times 10^4 \text{ MPa } (3.532 \times 10^6 \text{ psi})$$

$$Q_{12} = 7.30 \times 10^3 \text{ MPa } (1.06 \times 10^6 \text{ psi})$$

$$Q_{66} = 1.034 \times 10^4 \text{ MPa } (1.5 \times 10^6 \text{ psi})$$

and the boron/epoxy density is $\rho_f = 2.0$ gkm^3 (0.0721 lb/in^3).

a. Using Equations (2.56), (2.57), and (2.61), what are the [A] and [D] stiffness matrices for this sandwich panel?

b. Using the simpler equations analogous to Equations (2.71) and (2.74), in which the core contribution is ignored, what are the $[A]$ and $[D]$ stiffness matrices for this sandwich panel?

c. What is the largest percentage difference between (a) and (b) above of any component in either matrix?

d. If the panel is subjected to in-plane tension in the x-direction (i.e., the 0° direction) such that the faces and core are equally strained, what percentage of the load is carried by the faces, and what percentage is carried by the core, using the results of (a) above.

2.2 Consider a midplane symmetric sandwich panel, wherein the faces are made of a unidirectional Kevlar/49 epoxy composite with the 1-axis coincident with the x-axis. The properties are given in Table 2.3, and $\nu_{12} = 0.34$. The core is assumed to contribute nothing to the extensional or bending stiffness. This sandwich panel has faces 1 mm (0.04 inches) thick, and the core thickness is 25.4mm (1 inch).

a. Provide the values of the $[A]$ matrix and the $[D]$ matrix in (2.64), using the methods indicated in section 2.10.

b. How does this panel compare with that of Problem 2.1 which geometrically is the same?

TABLE 2.3. Unidirectional Properties.

	Elastic Moduli			Ultimate Strength			
Material	Axial (E_{11})	Transverse (E_{22})	Shear (G_{12})	Axial Tens. (σ_{11})	Trans. Tens. (σ_{22})	Shear (τ_{12})	Density
High strength	20	1.0	0.65	220	6	14	0.057
GR/epoxy	(138)	(6.9)	(4.5)	(1517)	(41)	(97)	(1.57)
High modulus	32	1.0	0.7	175	5	10	0.058
GR/epoxy	(221)	(6.9)	(4.8)	(1206)	(34)	(69)	(1.60)
Ultra high	44	1.0	0.95	110	4	7	0.061
modulus	(303)	(6.9)	(6.6)	(758)	(28)	(48)	(1.68)
GR/epoxy							
Kevlar 49/	12.5	0.8	0.3	220	4	6	0.050
epoxy	(86)	(5.5)	(2.1)	(1517)	(28)	(41)	(1.38)
S glass	8	1.0	0.5	260	6	10	0.073
epoxy	(55)	(14)	(3.4)	(1793)	(41)	(69)	(2.00)
Steel	30	30	11.5	60	60	35	0.284
	(207)	(207)	(79)	(414)	(414)	(241)	(7.83)
Aluminum	10.5	10.5	3.8	42	42	28	0.098
6061-T6	(72)	(72)	(26)	(290)	(290)	(193)	(2.70)

Moduli in Msi (GPa); stress in Ksi (MPa); density in lb/in^3 (g/cm^3).

TABLE 2.4. Quasi-Isotropic Composites Properties.

Material	Configuration	V_f (%)	(E) (10^6 psi)	ρ (lb/in³)	σ_{max} 10^3 psi
C/epoxy	Cross ply	58	12.04	0.0555	55.1
B/epoxy	Cross ply	60	15.37	0.0721	55.1
SiC/Ti-6A1-4V	Woven	35	27.0	0.1492	245
2024-T3 Alum	Isotropic	N/A	10.88	0.0973	50.0
Ti	Isotropic	N/A	16.5	0.160	128.0
E-glass/epoxy	Cross ply	57	3.12	0.0710	82.0
Welton steel	Isotropic	N/A	30.45	0.283	126.9

2.3 Consider a midplane asymmetric sandwich panel, wherein the two faces differ in material such that the upper and lower material properties are subscripted U and L. Assume $t_f \ll h_c$, that the core material contributes nothing to the stiffness discussed here, and that both faces are specially orthotropic [i.e., all $(\ )_{16}$ and $(\ )_{26}$ quantities are zero]. Determine the A_{ij}, B_{ij}, and D_{ij} stiffness components in Equation (2.64) in terms of the material properties and sandwich dimensions analogous to problem 2.2 above.

2.4 In selecting face materials, quite often one selects a material which has the highest specific strength, defined here as σ_{11}/ρ, or the highest specific stiffness defined here as E_{11}/ρ. In Tables 2.3 and 2.4,

a. Which material has the highest specific strength? Which has the lowest?
b. Which material has the highest specific stiffness? Which has the lowest?

2.5 The elastic properties of a unidirectional Kevlar 49/epoxy composite are given in Table 2.3. A cross-ply laminate consists of 4 plies such that the stacking sequence is [0°, 90°, 90°, 0°]. Each ply is 0.0055 inches thick.

a. Determine each element of the [A], [B], and [D] matrices.
b. What if any couplings exist in this cross-ply configuration?
c. Concerning in-plane loads only, is the plate stiffer in the x direction, the y direction, or are they the same?
d. Concerning plate bending, is the plate stiffer in the x direction, the y direction, or are they equally stiff in bending?

2.6 Given an angle ply construction of five plies of the same unidirectional composite material oriented as $[+\theta°/-\theta°/+\theta°/-\theta°/+\theta°]$, with each ply of equal thickness, which elements of the [A], [B], and [D] matrices given by Equations (2.56), (2.57), and (2.61) will be equal to zero?

2.13 REFERENCES

1. Sokolnikoff, I. S., *The Mathematical Theory of Elasticity, McGraw-Hill Book Company, Inc., New York, Second Edition, 1956.*
2. Vinson, J. R. and T-W. Chou, *Composite Materials and Their Use in Structures*. Applied Science Publishers, London, 1975.
3. Vinson, J. R. and R. L. Sierakowski, *The Behavior of Structures Composed of Composite Materials,* Martinus-Nijhoff Publishers (now Kluwer Academic Publishers) Dordrecht, The Netherlands, 1986.
4. Vinson, J. R., *The Behavior of Shells Composed of Isotropic and Composite Materials,* Kluwer Academic Publisher, Dordrecht, The Netherlands, 1993.
5. Shames, I. H., *Introduction to Solid Mechanics*. New York: Prentice-Hall, Inc., 1975.
6. Carlsson, L. A. and R. B. Pipes, *Experimental Characterization of Advanced Composite Materials,* Prentice-Hall Publishing Co., Inc., Englewood Cliffs, NJ, Second Edition, 1996.
7. Halpin, J. C. and S. W. Tsai, "Environmental Factors in Composite Materials Design," Air Force Materials Laboratory Technical Report 67-423, 1967.
8. Hashin, Z., "Theory of Fiber Reinforced Materials," National Aeronautics and Space Administration Contractors Report 1974 (1972).
9. Christensen, R. M., *Mechanics of Composite Materials*. New York: John Wiley and Sons, Inc., 1979.
10. Hahn, H. T., "Simplified Formulas for Elastic Moduli of Unidirectional Continuous Fiber Composites," *Composites Technology Review,* Vol. 2, No. 3, 1980, pp. 5–7.
11. Chou, T. W., *Microstructural Design of Fiber Composites,* Cambridge University Press, 1992.
12. Vinson, J. R., W. J. Walker, R. B. Pipes, and O. R. Ulrich, "The Effects of Relative Humidity and Elevated Temperatures on Composite Structures," Air Force Office of Scientific Research Technical Report 77-0030, February 1977.
13. Pipes, R. B., J. R. Vinson, and T. W. Chou, "On the Hygrothermal Response of Laminated Composite Systems," *Journal of Composite Materials,* April 1976, pp. 130–148.
14. ASTM Standard D5229-92. Test Method for Moisture Absorption Properties and Equilibrium Conditioning of Polymer Matrix Composite Materials.
15. Woldesenbet, E., "High Strain Rate Properties of Composites," Ph.D. Dissertation, Department of Mechanical Engineering, University of Delaware, January 1998.
16. Den Haan, L., "A New Foam Core for Demanding Sandwich Structures," in *Sandwich Construction 3—Proceedings of the Third International Conference on Sandwich Construction,* Southampton, Great Britain, Editor, H. G. Allen, EMAS Publications, United Kingdom, 1996, pp. 513–524.
17. Jurf, R. A. and J. R. Vinson, "Effect of Moisture on the Static and Viscoelastic Shear Properties of Epoxy Adhesives," *Journal of Materials Science,* Vol. 20, 1985, pp. 2979–2989.
18. Wilson, D. W. and J. R. Vinson, "Viscoelastic Effects on Buckling of Laminated Plates Subjected to Hygrothermal Conditions," ASME Publications AD-03, "Advances in Aerospace Structures and Materials," 1982.
19. Wilson, D. W. and J. R. Vinson, "Viscoelastic Effects on the Buckling Response of Laminated

Columns," ASTM Special Publications, 864, "Recent Advances in Composites in the United States and Japan," 1985.

20. Chevalier, J. L., "Creep and Fatigue Properties of End-Grain Balsa and Other Typical Sandwich Cores," in *Sandwich Constructions 2—Proceedings of the Second International Conference on Sandwich Construction,* Gainesville, Florida, Editors, D. Weissman-Berman and K-A. Olsson, EMAS Publications, United Kingdom, 1992, pp. 519–539.
21. Gellhorn, E. v., and A. Starlinger, "Air Efficient Extender Procedure for Creep Parameters of Structural Foams," in *Sandwich Constructions 2—Proceedings of the Second International Conference on Sandwich Construction,* Gainesville, Florida, Editors, D. Weissman-Berman and K-A. Olsson, EMAS Publications, United Kingdom, 1992, pp. 779–788.
22. Shenoi, R. A., S. D. Clark, H. G. Allen and I. A. Hicks, "Steady State Creep Behavior of FRP Foam Cored Sandwich Beams," in *Sandwich Construction 3—Proceedings of the Third International Conference on Sandwich Construction,* Southampton, Great Britain, Editor, H. G. Allen, EMAS Publications, United Kingdom, 1996, pp. 789–799.
23. Feichtinger, K. A. and S. Martensen, "A Comparison of the Creep Properties of Aluminum and Nomex Honeycombs, PVCS and Balsa," in *Sandwich Construction 3—Proceedings of the Third International Conference on Sandwich Construction,* Southampton, Great Britain, Editor, H. G. Allen, EMAS Publications, United Kingdom, 1996, pp. 801–812.
24. Caprino, G., R. Teti and M. Messa, "Long Term Behavior of PVC Foam Cores for Structural Sandwich Construction," in *Sandwich Construction 3—Proceedings of the Third International Conference on Sandwich Construction,* Southampton, Great Britain, Editor, H. G. Allen, EMAS Publications, United Kingdom, 1996, pp. 813–824.
25. Lindholm, V. S., "Some Experiments with the Split Hopkinson Pressure Bar," *Journal of Mechanics and Physics of Solids,* Vol. 12, 1964, pp. 317–335.
26. Daniel, I. M., R. H. La Bedz, and T. Liber, "New Method for Testing Composites at Very High Strain Rates," *Experimental Mechanics,* Vol. 2, February 1981, pp. 71–77.
27. Nicholas, T., "Tensile Testing of Materials at High Rates of Strain," *Experimental Mechanics,* May 1981, pp. 177–185.
28. Zukas, J. A., "Impact Dynamics," John Wiley and Sons, New York, 1982.
29. Sierakowski, R. L. and S. K. Chaturvedi, *Dynamic Loading and Characterization of Fiber-Reinforced Composites,* Wiley Interscience, 1997.
30. Sierakowski, R. L, "Strain Rate Effects in Composites," *Applied Mechanics Reviews,* forthcoming.
31. Feichtinger, K. A., "Test Methods and Performance of Structural Core Materials - IIA Strain Rate Dependence of Shear Properties," *5th Autumn INERN Conference, Composite Materials and Sandwich Structures,* Lorient, October 1990.
32. Gellhorn, E. v. and G. Reif, "Thick Dynamic-Dynamic Test Data for the Design of Dynamically Loaded Structures," in *Sandwich Constructions 2—Proceedings of the Second International Conference on Sandwich Construction,* Gainesville, Florida, Editors, D. Weissman-Berman and K-A. Olsson, EMAS Publications, United Kingdom, 1992, pp. 541–557.
33. Choe, G. H., W. W. .Finch, Jr., and J. R. Vinson, "Compressive Testing of Composite Materials at High Strain Rates," *Proceedings of the Fourth Japan-U.S. Conference on Composite Materials,* Technomic Publishing Company, Lancaster, PA, pp. 82–91, 1988.
34. Frey, T. J., J. R. Vinson, and K. M. .Prewo, K. M., "High Strain Rate Mechanical Properties of Carbon/Glass and Graphite/Epoxy Composite Materials," *Proceedings of the 8th International*

Conference on Composite Materials, Honolulu, July 1991, EMAS Publications, United Kingdom, 1991.

35. Woldesenbet, E. and J. R. Vinson, "Effect of Moisture on High Strain Rate Properties of Graphite/Epoxy Composites," *Proceedings of the Twelfth American Society for Composites Technical Conference,* Dearborn, October 1997, Technomic Publishing Co., Inc., Lancaster, PA, 1997.
36. Greenfield, R. J., "Dynamic Tensile Properties of Woven Glass Fabric Composites," Master's Thesis, Mechanical Engineering, University of Delaware, December 1997.
37. Dee, A., and J. R. Vinson, "The Effect of High Strain Rate on the Compressive Mechanical Properties of Unidirectional Glass/Epoxy (3M Scotchply 1003)," *Proceedings of the Twelfth American Society for Composites Technical Conference,* Dearborn, October 1997, Technomic Publishing Co., Inc., Lancaster, PA, 1997.
38. Preissner, E., E. Woldesenbet and J. R. Vinson, "An Examination of the High Strain Rate Compressive Stress Strain Behavior of a KEVLAR/Epoxy Composite," *Proceedings of the Twelfth American Society for Composites Technical Conference,* Dearborn, October 1997, Technomic Publishing Co., Inc., Lancaster, PA, 1997.
39. Sloan, J. G., "The Behavior of Rectangular Composite Material Plates Under Lateral and Hygrothermal Loads," MMAE Thesis, University of Delaware, 1979 (also, AFOSR-TR-78-1477, July 1978).
40. Tsai, S. W., and N. J. Pagano, "Invariant Properties of Composite Materials," *Composite Materials Workshop,* Tsai, S. W. et al., eds., Technomic Publishing Co., Inc., Lancaster, PA, 1968, pp. 233–253.
41. Vinson, J. R., *Structural Mechanics: The Behavior of Plates and Shells.* New York: Wiley-Interscience, John Wiley and Sons, Inc., 1974.
42. Vinson, J. R., *The Behavior of Thin Walled Structures: Beams, Plates and Shells,* Kluwer Academic Publishers, Dordrecht, 1989.
43. Reissner, E., "On a Variational Theorem in Elasticity," *J. Math. Phys.,* Vol. 29, 1950, p. 90.
44. Mindlin, R. D., "Influence of Rotatory Inertia and Shear on Flexural Motions of Isotropic Elastic Plates," *Journal of Applied Mechanics,* 73, 1951.
45. Sikarski, D. L. and L. L. Mercado, "On the Response of a Sandwich Panel with a Bilinear Core," *Analysis and Design Issues for Modern Aerospace Vehicles,* Proceedings of the 1997 ASME IMECE, Nov. 16–21, 1997, Dallas, TX, ASME Aerospace Division AO-55, 4, 1997, pp. 365–377.
46. Frostig, Y., "Bending of Curved Sandwich Panels with Transersely Flexible Cores - Closed-Form High-Order Theory," *Analysis and Design Issues for Modern Aerospace Vehicles,* Proceedings of the 1997 ASME IMECE, Nov. 16–21, 1997, Dallas, TX, ASME Aerospace Division AO-55, 4, 1997, pp. 335–354.
47. Frostig, Y. and M. Baruch, "Localized Load Effects in High-Order Bending of Sandwich Panels with Transversely Flexible Core," *Journal of the ASCE, Engineering Mechanics Division,* Vol. 122, No. 11, 1996, p. 1069.
48. Patel, D., and Y. Frostig, "High Order Bending of Piecewise Uniform Sandwich Beams with a Tapered Transition Zone and a Transversly Flexible Core," *Composite Structures,* Vol. 31, 1995, pp. 151–162.
49. Frostig, Y. and I. Shenhar, "High-Order Bending of Sandwich Beams with a Transversely

Flexible Core and Unsymmetrical Laminated Composite Skins," *Composites Engineering,* Vol. 5, No. 4, 1995, pp. 405–414.

50. Thomsen, O. T. and Y. Frostig, "Localized Bending Effects in Sandwich Panels: Photoelastic Investigation Versus High-Order Sandwich Theory Results," *Composite Structures,* Vol. 37, No. 1, January 1997, pp. 97–108.
51. Lee, C. K. and F. C. Moon, "Modal Sensors/Actuators," *Journal of Applied Mechanics,* Vol. 57, 1990, pp. 434–441.
52. Yu, Y. Y., "Dynamics for Large Deflections of a Sandwich Plate with Thin Piezoelectric Face Layers," *Analysis and Design Issues for Modern Aerospace Vehicles,* ASME AD-Vol. 55, Ed. G. S. Simitses, pp. 285–292.
53. Leibowitz, M. and J. R. Vinson, "Intelligent Composites: Design and Analysis of Composite Material Structures Involving Piezoelectric Material Layers. Part A—Basic Formulation," Center for Composite Materials Technical Report 91-54, University of Delaware, November 1991.
54. Larson, P. H., "The Use of Piezoelectric Materials in Creating Adaptive Shell Structures," Ph.D. Dissertation, Mechanical Engineering, University of Delaware, 1993.
55. Newill, J. F., "Composite Sandwich Structures Incorporating Piezoelectric Materials," Ph.D. Dissertation, Mechanical Engineering, University of Delaware, 1996.
56. Hilton, H. H., S. Yi, and J. R. Vinson, "Probabalistic Structural Integrity of Piezoelectric Viscoelastic Composite Structures," *Proceedings of the 39th AIAA/ASME/ASCE/AHS/ASC SDM Conference, Vol. 2,* AIAA Paper 98-1848, AIAA Reston, VA, April 1998.
57. Plantema, F. J., *Sandwich Construction: The Bending and Buckling of Sandwich Beams, Plates and Shells,* John Wiley and Sons, New York, 1966.
58. Allen, H. G., *Analysis and Design of Structural Sandwich Panels,* Pergamon Press, Oxford, 1969.
59. Zenkert, D., *An Introduction to Sandwich Construction,* EMAS Publications, West Midlands, United Kingdom, 1995.
60. Vinson, J. R., "On the Analysis and Optimization of Sandwich Structures," Workshop on Advanced Materials Sandwich Structures, Capri, October 20–23, 1991.
61. Vinson, J.R. and Flaggs, D. L., Hygrothermal Effects on the Buckling of Composite Plates," *Fibre Science and Technology,* Vol. 11, No. 5, 1978, pp. 353–365.

CHAPTER 3

Derivation of the Governing Equations for Sandwich Plates (Panels)

3.1 INTRODUCTION

In this chapter, the plate (panel) theory will be discussed and used to develop the governing differential equations for a variety of theories, loadings, and environmental conditions. This provides the physical and mathematical basis for subsequent chapters that provide solutions useful for design and analysis.

In Chapter 2, the governing constitutive equations were developed in detail, describing the relationships between integrated stress resultants (N_x, N_y, N_{xy}), integrated stress couples (M_x, M_y, M_{xy}), in-plane midsurface strains (ε_x^0, ε_y^0, ε_{xy}^0), and the curvatures (κ_x, κ_y, κ_{xy}), as seen in Equation (2.64). These will be utilized with the strain-displacement relations of Equation (2.46), the displacement forms of Equation (2.47), and the equilibrium equations to be developed in Section 3.2 to provide structural theories for thin-walled bodies, which is the form in which composite materials are most generally employed. Beams, rods, and columns are discussed in Chapter 4. The use of energy methods for solving structural problems is discussed in Chapter 5. Shells will be the subject of Chapter 11. However, to study any of the structural equations it is necessary to first develop suitable equilibrium equations.

3.2 PLATE EQUILIBRIUM EQUATIONS

The integrated stress resultants (N), shear resultants (Q), and stress couples (M), with appropriate subscripts, are defined by Equation (2.52), and their positive directions are shown in Figure 2.13, for a rectangular plate, defined as a body of length a in the x-direction, width b in the y-direction, and thickness h in the z-direction, where $h \ll b$, $h \ll a$, i.e., a thin plate. It is now necessary to derive the equilibrium relations that exist for this plate structure. Keep in mind that this derivation for a laminated composite plate applies as well to a sandwich plate or panel.

In mathematically modeling solid materials, including the laminates of Chapter 2, a continuum theory is generally employed. In doing so, a representative material point within the elastic solid or lamina is selected as being typical of all material points in the body or lamina. The material point is assumed to be infinitely smaller than any dimension of the structure containing it, but infinitely larger than the size of the molecular lattice spacing of the material comprising it. Moreover, the material point is given a convenient shape; in a Cartesian reference frame that convenient shape is a small cube of dimensions *dx*, *dy*, and *dz*, as shown in Figure 3.1.

This cubic material point of dimensions *dx*, *dy*, and *dz* is termed a control element. The positive values of all stresses acting on each surface of the control element are shown in Figure 3.1, along with how they vary from one surface to another, using the positive sign convention consistent with most scientific literature and consistent with Figure 2.1. Details of the nomenclature

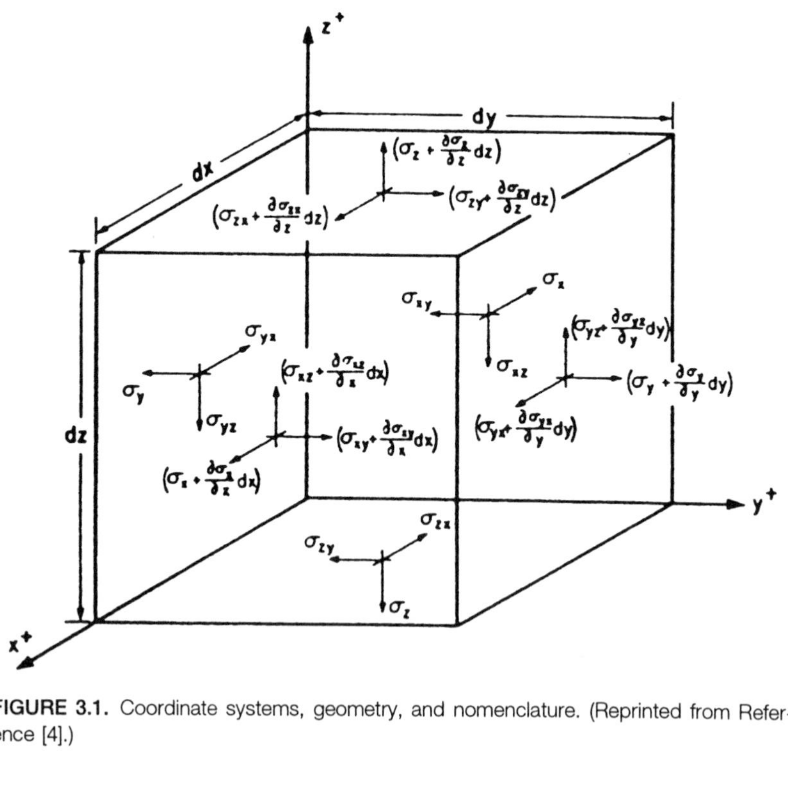

FIGURE 3.1. Coordinate systems, geometry, and nomenclature. (Reprinted from Reference [4].)

can be found in any text on solid mechanics, including References [1,2]. In addition to the surface stresses acting on the control element shown in Figure 3.1, body-force components F_x, F_y, and F_z can also act on the body. These force components are proportional to the control-element volume (i.e., its mass), such as gravitational, magnetic, or centrifugal forces.

A force balance can now be made in the x, y, and z directions resulting in three equations of equilibrium. For instance, a force balance on the x-direction would yield

$$\left(\sigma_x + \frac{\partial \sigma_x}{\partial x} dx\right) dydz + \left(\sigma_{yx} + \frac{\partial \sigma_{yx}}{\partial y} dy\right) dxdz + \left(\sigma_{zx} + \frac{\partial \sigma_{zx}}{\partial z} dz\right) dxdy$$

$$-\sigma_x dydz - \sigma_{yx} dxdz - \sigma_{zx} dydx + F_x dxdydz = 0$$

Canceling terms and dividing the remaining terms by the volume ($V = dxdydz$) results in the following:

$$\frac{\partial \sigma_x}{\partial x} + \frac{\partial \sigma_{yx}}{\partial_y} + \frac{\partial \sigma_{zx}}{\partial z} + F_x = 0 \tag{3.1}$$

Similarly, equilibrium in the y and z directions yields

$$\frac{\partial \sigma_{xy}}{\partial x} + \frac{\partial \sigma_y}{\partial y} + \frac{\partial \sigma_{zy}}{\partial z} + F_y = 0 \tag{3.2}$$

$$\frac{\partial \sigma_{xz}}{\partial x} + \frac{\partial \sigma_{yz}}{\partial y} + \frac{\partial \sigma_z}{\partial z} + F_z = 0 \tag{3.3}$$

These three equations comprise the equilibrium equations for three-dimensional elasticity. However, for beam, plate, and shell theory, whether involving composite materials or not, one must integrate the stresses across the thickness of the thin-walled structures to obtain solutions.

Recalling the definitions of Chapter 2, the stress resultants and stress couples defined previously are

$$\begin{Bmatrix} N_x \\ N_y \\ N_{xy} \\ Q_x \\ Q_y \end{Bmatrix} = \int_{-h/2}^{+h/2} \begin{Bmatrix} \sigma_x \\ \sigma_y \\ \sigma_{xy} \\ \sigma_{xz} \\ \sigma_{yz} \end{Bmatrix} dz = \sum_{k=1}^{N} \int_{h_{k-1}}^{h_k} \begin{Bmatrix} \sigma_x \\ \sigma_y \\ \sigma_{xy} \\ \sigma_{xz} \\ \sigma_{yz} \end{Bmatrix}_k dz_k \tag{3.4}$$

$$\begin{Bmatrix} M_x \\ M_y \\ M_{xy} \end{Bmatrix} = \int_{-h/2}^{h/2} \begin{Bmatrix} \sigma_x \\ \sigma_y \\ \sigma_{xy} \end{Bmatrix} z dz = \sum_{k=1}^{N} \int_{h_{k-1}}^{h_k} \begin{Bmatrix} \sigma_x \\ \sigma_y \\ \sigma_{xy} \end{Bmatrix}_k z_k dz_k \tag{3.5}$$

The first form of each is applicable to a single-layer plate, while the second form is necessary for a laminated plate due to the stress discontinuities associated with different materials and/or differing orientations in the various plies. The second form is, therefore, necessary for a sandwich plate, because of the different face and core material properties.

Turning now to Equation (3.1), neglecting the body force term, F_x, for simplicity of example, integrating term by term across each ply, and summing across the plate provides

$$\sum_{k=1}^{N} \int_{h_{k-1}}^{h_k} \frac{\partial \sigma_{x_k}}{\partial x} dz + \sum_{k=1}^{N} \int_{h_{k-1}}^{h_k} \frac{\partial \sigma_{yx_k}}{\partial y} dz + \sum_{k=1}^{N} \int_{h_{k-1}}^{h_k} \frac{\partial \sigma_{zx_k}}{\partial z} dz = 0 \tag{3.6}$$

In the first two terms integration and differentiation can be interchanged, hence:

$$\frac{\partial}{\partial x}\left[\sum_{k=1}^{N} \int_{h_{k-1}}^{h_k} \sigma_{x_k} dz\right] + \frac{\partial}{\partial y}\left[\sum_{k=1}^{N} \int_{h_{k-1}}^{h_k} \sigma_{yx_k} dz\right] + \sum_{k=1}^{N} \sigma_{zx} \Big|_{h_{k-1}}^{h_k} = 0 \tag{3.7}$$

In the first two terms N_x and N_{yx} appear explicitly as the bracketed quantities. In the third term it is clear that between all plies and between the face and core materials of a sandwich panel, the interlaminar shear stresses cancel each other out, and one can define externally applied surface shear stresses on the top ($z = h_N$) and bottom ($z = h_0$) surfaces as shown below (see Figure 2.1):

$$\sigma_{zx}(h_N) \equiv \tau_{1x} \quad \text{and} \quad \sigma_{zx}(h_0) \equiv \tau_{2x} \tag{3.8}$$

Equation (3.7) can therefore be written as

$$\frac{\partial N_x}{\partial x} + \frac{\partial N_{yx}}{\partial y} + \tau_{1x} - \tau_{2x} = 0 \tag{3.9}$$

Similarly, integrating the equilibrium equation in the y-direction provides

$$\frac{\partial N_{xy}}{\partial x} + \frac{\partial N_y}{\partial y} + \tau_{1y} - \tau_{2y} = 0 \tag{3.10}$$

where

$$\sigma_{zy}(h_N) \equiv \tau_{1y} \quad \text{and} \quad \sigma_{zy}(h_0) \equiv \tau_{2y} \tag{3.11}$$

Likewise equilibrium in the z-direction upon integration and summing provides

$$\frac{\partial Q_x}{\partial x} + \frac{\partial Q_y}{\partial y} + p_1 - p_2 = 0 \tag{3.12}$$

where

$$\sigma_z(h_N) \equiv p_1 \quad \text{and} \quad \sigma_z(h_0) \equiv p_2 \tag{3.13}$$

In addition to the integrated force equilibrium equations above, two equations of moment equilibrium are also needed, one for the x-direction and one for the y-direction. Multiplying Equation (3.1) through by zdz, integrating across each ply, and summing across all laminae results in the following:

$$\sum_{k=1}^{N} \int_{h_{k-1}}^{h_k} \frac{\partial \sigma_{x_k}}{\partial x} zdz + \sum_{k=1}^{N} \int_{h_{k-1}}^{h_k} \frac{\partial \sigma_{yx_k}}{\partial y} zdz + \sum_{k=1}^{N} \int_{h_{k-1}}^{h_k} \frac{\partial \sigma_{zx_k}}{\partial z} zdz = 0$$

Again, in the first two terms integration and summation can be interchanged with differentiation with the result that the first two terms become $(\partial M_x/\partial x) + (\partial M_{xy}/\partial y)$. However, the third term must be integrated by parts as follows:

$$\sum_{k=1}^{N} \int_{h_{k-1}}^{h_k} \frac{\partial \sigma_{zx_k}}{\partial z} zdz = \sum_{k=1}^{N} \left\{ z\sigma_{zx} \Big|_{h_{k-1}}^{h_k} - \int_{h_{k-1}}^{h_k} \sigma_{zx} dz \right\}$$

Here the last term is seen to be $-Q_x$. Again in the first term on the right, clearly the moments of all the interlaminar stresses between plies cancel each other out, and the only nonzero terms are the moments of the applied surface shear stresses; hence, that term becomes

$$h_N \, \tau_{1x} - h_0 \, \tau_{2x} = \frac{h}{2} [\tau_{1x} + \tau_{2x}]$$

Using the former expression, the equation of equilibrium of moments in the x-direction is

$$\frac{\partial M_x}{\partial x} + \frac{\partial M_{xy}}{\partial y} - Q_x + \frac{h}{2}[\tau_{1x} + \tau_{2x}] = 0 \tag{3.14}$$

Similarly in the y-direction the moment equilibrium equation is

$$\frac{\partial M_{xy}}{\partial x} + \frac{\partial M_y}{\partial y} - Q_y + \frac{h}{2}[\tau_{1y} + \tau_{2y}] = 0 \tag{3.15}$$

where all the terms are defined above. Thus, there are five equilibrium equations for a rectangular plate, regardless of what material or materials are utilized in the plate: (3.9), (3.10), (3.12), (3.14), and (3.15).

3.3 THE BENDING OF COMPOSITE MATERIAL LAMINATED AND/OR SANDWICH PLATES: CLASSICAL THEORY

Consider a plate composed of a laminated composite material that is mid-plane symmetric, i.e., $B_{ij} = 0$, and has no other coupling terms $(\ )_{16} = (\ )_{26} = 0$, no surface shear stresses, and no hygrothermal effects. The plate equilibrium equations for the bending of the plate subjected to lateral loads given by Equations (3.14), (3.15), and (3.12) become

$$\frac{\partial M_x}{\partial x} + \frac{\partial M_{xy}}{\partial y} - Q_x = 0 \tag{3.16}$$

$$\frac{\partial M_{xy}}{\partial x} + \frac{\partial M_y}{\partial y} - Q_y = 0 \tag{3.17}$$

$$\frac{\partial Q_x}{\partial x} + \frac{\partial Q_y}{\partial y} + p(x,y) = 0 \tag{3.18}$$

where

$$p(x,y) = p_1(x,y) - p_2(x,y)$$

These three equations apply also to a sandwich plate. Solving for Q_x and Q_y, Equations (3.16) and (3.17) can be substituted into Equation (3.18) with the result that

$$\frac{\partial^2 M_x}{\partial x^2} + 2\frac{\partial^2 M_{xy}}{\partial x \partial y} + \frac{\partial^2 M_y}{\partial y^2} = -p(x,y) \tag{3.19}$$

The above equations are derived from equilibrium considerations alone. From Equation (2.64) and for the case of midplane symmetry ($B_{ij} = 0$) and no $(\ )_{16}$ and $(\ )_{26}$ terms, the constitutive relations are

$$M_x = D_{11}\kappa_x + D_{12}\kappa_y \tag{3.20}$$

$$M_y = D_{12}\kappa_x + D_{22}\kappa_y \tag{3.21}$$

$$M_{xy} = 2D_{66}\kappa_{xy} \tag{3.22}$$

where from Equation (2.49)

$$\kappa_x = \frac{\partial\overline{\alpha}}{\partial x},\ \kappa_y = \frac{\partial\overline{\beta}}{\partial y},\ \kappa_{xy} = \frac{1}{2}\left(\frac{\partial\overline{\alpha}}{\partial y} + \frac{\partial\overline{\beta}}{\partial x}\right)$$

It is well known that transverse shear deformation (that is $\varepsilon_{xz} \neq 0$, $\varepsilon_{yz} \neq 0$) effects are important for plates composed of composite materials and also sandwich plates in determining maximum deflections, vibration natural frequencies, and critical buckling loads. However, it is appropriate to use a simpler stress analysis involving classical theory that neglects transverse shear deformation for preliminary design to determine a "first cut" for stresses, the required overall stacking sequence, and required plate thickness.

If, in fact, transverse shear deformation is ignored, then from Equation (2.48)

$$\varepsilon_{xz} = 0 = \frac{1}{2}\left(\overline{\alpha} + \frac{\partial w}{\partial x}\right) \quad \text{and} \quad \varepsilon_{yz} = 0 = \frac{1}{2}\left(\overline{\beta} + \frac{\partial w}{\partial y}\right)$$

hence,

$$\overline{\alpha} = -\frac{\partial w}{\partial x} \quad \text{and} \quad \overline{\beta} = -\frac{\partial w}{\partial y} \tag{3.23}$$

and

$$\kappa_x = -\frac{\partial^2 w}{\partial x^2},\ \kappa_y = -\frac{\partial^2 w}{\partial y^2} \quad \text{and} \quad \kappa_{xy} = -\frac{\partial^2 w}{\partial x \partial y} \tag{3.24}$$

So, substituting Equation (3.23) into (3.20) through (3.22) results in

$$M_x = -D_{11}\frac{\partial^2 w}{\partial x^2} - D_{12}\frac{\partial^2 w}{\partial y^2} \tag{3.25}$$

$$M_y = -D_{12}\frac{\partial^2 w}{\partial x^2} - D_{22}\frac{\partial^2 w}{\partial y^2} \tag{3.26}$$

$$M_{xy} = -2D_{66}\frac{\partial^2 w}{\partial x \partial y} \tag{3.27}$$

Substituting these three equation in turn into Equation (3.19) results in

$$D_{11}\frac{\partial^4 w}{\partial x^4} + 2(D_{12} + 2D_{66})\frac{\partial^4 w}{\partial x^2 \partial y^2} + D_{22}\frac{\partial^4 w}{\partial y^4} = p(x,y) \tag{3.28}$$

The above coefficients can be simplified to:

$$D_{11} \equiv D_1,\ D_{22} \equiv D_2,\ (D_{12} + 2D_{66}) \equiv D_3 \tag{3.29}$$

With the simplified notation, Equation (3.28) becomes

$$D_1\frac{\partial^4 w}{\partial x^4} + 2D_3\frac{\partial^4 w}{\partial x^2 \partial y^2} + D_2\frac{\partial^4 w}{\partial y^4} = p(x,y) \tag{3.30}$$

This is the governing differential equation for the bending of a plate (as well as a sandwich plate) composed of a composite material, excluding transverse shear deformation, with no coupling terms (that is $B_{ij} = (\ )_{16} = (\ )_{26} = 0$), and no hygrothermal terms (that is, $\Delta T = \Delta m = 0$) subjected to a lateral distributed load $p(x,y)$. It can be shown that if the plate materials are isotropic then $D_1 = D_2 = D_3 = D$. As stated previously, neglecting transverse shear deformation and hygrothermal effects can lead to significant errors, as will be shown; but in many cases their neglect results in easier solutions that are useful in preliminary design to "size" the plate initially.

Solutions of Equation (3.30) can be obtained generally in two ways: direct solution of the governing differential Equation (3.30), or utilization of an energy principle solution. The latter offers more latitude in finding approximate solutions, often needed, and these will be discussed in detail in Chapter 5.

Direct solutions of the governing differential equations for plates of composite materials fall into three categories: Navier, Levy, and perturbation solutions. Each has its advantages and disadvantages. However, prior to that, boundary conditions need to be discussed.

3.4 CLASSICAL PLATE THEORY BOUNDARY CONDITIONS

In the "classical" (that is, ignoring transverse shear deformation) plate theory of Section 3.3, two and only two boundary conditions can be satisfied at each edge of the plate. The boundary conditions for a simply-supported edge and a clamped edge, shown below, are identical to those of classical beam theory, where n is the direction normal to the plate edge and t is the direction parallel or tangent to the edge:

3.4.1 Simply-Supported Edge

$$w = 0 \quad \text{and} \quad M_n = 0 \tag{3.31}$$

$M_n = 0$ implies $\partial^2 w/\partial n^2 = 0$, because in Equations (3.25) and (3.26), there is no curvature (that is $\partial^2 w/\partial t^2$) along the edge of the simply-supported plate because $w = 0$ along that edge.

3.4.2 Clamped Edge

$$w = 0 \quad \text{and} \quad \frac{\partial w}{\partial n} = 0 \tag{3.32}$$

3.4.3 Free Edge

For a free edge of a plate, that is, one on which there are no loads nor any displacement or slope requirements, it is clear that M_n, Q_n, and M_{nt} all should be zero. However, with classical plate theory only two boundary conditions can be satisfied. This was a major problem for the solid mechanicists of the early nineteenth century. Then Kirchhoff brilliantly formulated an approximate solution to the problem, which is developed in many texts [1,3]. He reasoned that for the free edge

$$M_n = 0 \tag{3.33}$$

where now, because there is curvature along the edge, the full expressions of Equations (3.25) and (3.26) must be utilized. The approximate expression for the second boundary condition is

$$V_n = Q_n + \frac{\partial M_{nt}}{\partial t} = 0 \tag{3.34}$$

where V_n is the "effective" shear resultant, and Q_n is given by Equation (3.16) or (3.17), M_n is given by Equation (3.25) or (3.26), and M_{nt} is given by Equation (3.27).

3.5 ANALYSIS OF COMPOSITE MATERIAL LAMINATED AND/OR SANDWICH PANELS INCLUDING TRANSVERSE SHEAR DEFORMATION EFFECTS

The derivations in Chapters 2 and 3 have involved "classical" plate theory, i.e., they have neglected transverse-shear deformation effects. Because in many composite material laminate plate and sandwich panel constructions, transverse-shear deformation effects are important, a more refined theory will now be developed. However, because of its simplicity, and the number of solutions available, classical theory is still useful for preliminary design and analysis to size the structure required in minimum time and effort.

In the simpler classical theory, the neglect of transverse shear deformation effects means that $\varepsilon_{xz} = \varepsilon_{yz} = 0$. To include transverse shear deformation effects, one uses

$$\varepsilon_{xz} = \frac{1}{2}\left(\frac{\partial u}{\partial z} + \frac{\partial w}{\partial x}\right) \tag{3.35}$$

$$\varepsilon_{yz} = \frac{1}{2}\left(\frac{\partial v}{\partial z} + \frac{\partial w}{\partial y}\right) \tag{3.36}$$

Now substituting the admissible forms of the displacement for a plate or panel, Equation (2.47) into Equations (3.35) and (3.36), results in

$$\varepsilon_{xz} = \frac{1}{2}\left(\overline{\alpha} + \frac{\partial w}{\partial x}\right) \tag{3.37}$$

$$\varepsilon_{yz} = \frac{1}{2}\left(\overline{\beta} + \frac{\partial w}{\partial y}\right) \tag{3.38}$$

No longer are the rotations $\overline{\alpha}$ and $\overline{\beta}$ explicit functions of the derivatives of the lateral deflection w, as shown by Equation (3.23) for classical plate theory. The result is that for this refined theory there are five geometric unknowns, u_0, v_0, w, $\overline{\alpha}$, and $\overline{\beta}$, instead of just the first three in classical theory.

Now one needs to look again at the equilibrium equations, the constitutive equations (stress-strain relations), the strain-displacement relations, and the compatibility equations. For the plate, the equilibrium equations are given by Equations (3.9) through (3.15), because they do not change from classical theory. The constitutive equations for a composite material laminated plate and sandwich panel are given by Equations (2.55) through (2.64). The new cogent strain-displacement (kinematic) relations are given in Equations (3.37) and (3.38). Because the resulting governing equations are in terms of displacements and rotations, any single valued, continuous solution will by definition satisfy the compatibility equations.

As an example, consider a plate that is midplane symmetric ($B_{ij} = 0$), and has no coupling terms [$(\ )_{16} = (\ )_{26} = (\ )_{45} = 0$]; the constitutive equations for this orthotropic plate are

$$N_x = A_{11}\varepsilon_x^0 + A_{12}\varepsilon_y^0 \tag{3.39}$$

$$N_y = A_{12}\varepsilon_x^0 + A_{22}\varepsilon_y^0 \tag{3.40}$$

$$N_{xy} = 2A_{66}\varepsilon_{xy}^0 \tag{3.41}$$

$$M_x = D_{11}\kappa_x + D_{12}\kappa_y \tag{3.42}$$

$$M_y = D_{12}\kappa_x + D_{22}\kappa_y \tag{3.43}$$

$$M_{xy} = 2D_{66}\kappa_{xy} \tag{3.44}$$

$$Q_x = 2A_{55}\varepsilon_{xz} = 2A_{55}\left(\overline{\alpha} + \frac{\partial w}{\partial x}\right) \tag{3.45}$$

$$Q_y = 2A_{44}\varepsilon_{yz} = 2A_{44}\left(\overline{\beta} + \frac{\partial w}{\partial y}\right) \tag{3.46}$$

Because the plate is midplane symmetric there is no bending-stretching coupling, hence the in-plane loads (N_x, N_y, N_{xy}) and deflections (u_0, v_0) are uncoupled (separate) from the lateral loads, deflections, and rotations. Hence, for the lateral distributed static loading, $p(x,y)$, Equations (3.16) through (3.18) and (3.42) through (3.46) are utilized: 8 equations and 8 unknowns.

Substituting Equations (3.42) through (3.46) into (3.16) through (3.18) and using Equation (2.46) results in the following set of governing differential equations for a laminated composite plate subjected to a lateral load, wherein ($B_{ij} = 0$), $(\ )_{16} = (\ )_{26} = (\ )_{45} = 0$, and there are no applied surface shear stresses (for simplicity)

$$D_{11}\frac{\partial^2\overline{\alpha}}{\partial x^2} + D_{66}\frac{\partial^2\overline{\alpha}}{\partial y^2} + (D_{12} + D_{66})\frac{\partial^2\overline{\beta}}{\partial x\partial y} - 2A_{55}\left(\overline{\alpha} + \frac{\partial w}{\partial x}\right) = 0 \tag{3.47}$$

$$(D_{12} + D_{66})\frac{\partial^2\overline{\alpha}}{\partial x\partial y} + D_{66}\frac{\partial^2\overline{\beta}}{\partial x^2} + D_{22}\frac{\partial^2\overline{\beta}}{\partial y^2} - 2A_{44}\left(\overline{\beta} + \frac{\partial w}{\partial y}\right) = 0 \tag{3.48}$$

$$2A_{55}\left(\frac{\partial\overline{\alpha}}{\partial x}+\frac{\partial^2 w}{\partial x^2}\right)+2A_{44}\left(\frac{\partial\overline{\beta}}{\partial y}+\frac{\partial^2 w}{\partial y^2}\right)+p(x,y)=0 \tag{3.49}$$

We see here that the inclusion of transverse shear deformation effects results in three coupled partial differential equations with three unknowns, $\overline{\alpha}$, $\overline{\beta}$, and w, contrasted to having one partial differential equation with one unknown, w, in classical plate (panel) theory; see Equation (3.30). Incidentally if one specified that $\overline{\alpha} = \partial w/\partial x$ and $\overline{\beta} = -\partial w/\partial y$, substituting that into Equations (3.47) through (3.49) reduces the three equations to Equation (3.30), the classical theory equation.

The classical plate theory governing partial differential equations is fourth order in both x and y, and therefore requires only two boundary conditions on each of the four edges, as discussed in Section 3.4. The refined theory, including transverse shear deformation, is really sixth order in both x and y, and therefore requires three boundary conditions on each edge as discussed in Section 3.6.

If the sandwich or laminated plate is orthotropic but not midplane symmetric, i.e., $B_{ij} \neq 0$, the governing equations are more complicated than Equations (3.47) through (3.49) and are given by Whitney in his text [3].

3.6 BOUNDARY CONDITIONS FOR A PLATE USING THE REFINED PLATE THEORY

3.6.1 Simply-Supported Edge

Again, Equation (3.31) holds, but a third boundary condition is required for plate bending. In addition, since the in-plane and lateral behavior are coupled, a fourth boundary condition enters the picture as well. This has resulted in the use of two different simply-supported boundary conditions, both of which are mathematically admissible as natural boundary conditions (to be discussed later) and are practical structural boundary conditions. By convention, the simply-supported boundary conditions are given as follows:

$$\begin{aligned}
&S1(x = \text{constant edge})\text{: } w = M_x = u_0 = N_{xy} = 0\\
&S1(y = \text{constant edge})\text{: } w = M_y = v_0 = N_{yx} = 0\\
&S2(x = \text{constant edge})\text{: } w = M_x = N_x = v_0 = 0\\
&S2(y = \text{constant edge})\text{: } w = M_y = N_y = u_0 = 0
\end{aligned} \tag{3.50}$$

where u_0 is the midsurface displacement in the x-direction and v_0 is the midsurface displacement in the y direction.

3.6.2 Clamped Edge

Similarly, for a clamped edge the lateral deflection w and the rotation $\overline{\alpha}$ or $\overline{\beta}$ (for an x = constant edge or a y = constant edge, respectively) are zero (note: the slope is not zero) and the other boundary conditions are analogous to Equation (3.50).

$$
\begin{aligned}
&C1(x = \text{constant edge}): w = \overline{\alpha} = u_0 = N_{xy} = 0 \\
&C1(y = \text{constant edge}): w = \overline{\beta} = v_0 = N_{yx} = 0 \\
&C2(x = \text{constant edge}): w = \overline{\alpha} = N_x = v_0 = 0 \\
&C2(y = \text{constant edge}): w = \overline{\beta} = N_y = u_0 = 0
\end{aligned}
\tag{3.51}
$$

3.6.3 Free Edge

The free edge requires three boundary conditions on each edge; therefore, it is no longer necessary to resort to the difficulties of the Kirchhoff boundary-conditions for the bending of the plate. The boundary conditions for the bending of the plate are simply:

$$M_n = Q_n = M_{nt} = 0 \tag{3.52}$$

where n and t are directions normal to and tangential with the edge. Again, the in-plane boundary conditions for the free edge are $N_c = N_{nt} = 0$.

3.6.4 Other Boundary Conditions

In addition to the above boundary conditions, widely used to approximate the actual structural boundary conditions, sometimes it is desirable to consider an edge whose lateral deflection is restrained, whose rotation is restrained, or both. The means by which to describe these boundary conditions is given in Reference [4, pp. 20–21].

3.7 LAMINATED OR SANDWICH PLATE ON AN ELASTIC FOUNDATION

Consider a composite material laminated or sandwich plate that is supported on an elastic foundation. In most cases an elastic foundation is modeled as an elastic medium with a constant foundation modulus, i.e., a spring constant per unit planform area, of k in units such as lb/in./in.^{2}. Therefore, this elastic foundation acts on the plate as a force in the negative direction proportional to the local lateral deflection $w(x,y)$. The force per unit area is $-kw$, because when w is positive the foundation modulus is acting in a negative direction, and vice versa. So to incorporate the effect of the elastic foundation modeled as above one simply adds another force to the $p(x,y)$ load term. The results are, that for classical theory, Equation (3.30) is modified to be Equation (3.53), and for the refined theory Equation (3.49) is modified to be Equation (3.54):

$$D_1 \frac{\partial^4 w}{\partial x^4} + 2D_3 \frac{\partial^4 w}{\partial x^2 \partial y^2} + D_2 \frac{\partial^4 w}{\partial y^4} + kw = p(x,y) \tag{3.53}$$

$$2A_{55}\left(\frac{\partial \overline{\alpha}}{\partial x} + \frac{\partial^2 w}{\partial x^2}\right) + 2A_{44}\left(\frac{\partial \overline{\beta}}{\partial y} + \frac{\partial^2 w}{\partial y^2}\right) - kw + p(x,y) = 0 \tag{3.54}$$

Of course in the above, the stiffness quantities are those used previously for a single-layer isotropic plate, a composite laminated plate, or a sandwich plate.

In addition, if one wishes to investigate the behavior of a face sheet of a sandwich only, then it can also be modeled as a plate on an elastic foundation, i.e., the core is the elastic foundation, and so the above holds, where the stiffness quantities are those of that face sheet only.

3.8 LAMINATED OR SANDWICH PLATES SUBJECTED TO DYNAMIC LOADS

To modify the governing differential equations above, derived for a static lateral load $p(x,y)$, in order to account for dynamic time-dependent loads $p(x,y,t)$, one does not need to completely rederive the equations or develop a new theory. Using D'Alembert's Principle, one can include inertia terms into the equations, remembering that the second derivative of a displacement with respect to time is the acceleration.

For the classical theory, Equation (3.30) therefore becomes

$$D_1 \frac{\partial^4 w}{\partial x^4} + 2D_3 \frac{\partial^4 w}{\partial x^2 y^2} + D_2 \frac{\partial^4 w}{\partial y^4} + \rho_m h \frac{\partial^2 w}{\partial t^2} = p(x,y,t) \tag{3.55}$$

where ρ_m is the mass density of the plate material and h is the plate thickness, so $\rho_m h$ is the mass per unit planform area. Note that the most common error in any dynamic load calculation is that the analyst mistakenly uses the weight density rather than the mass density.

In the case of the refined theory, including transverse shear deformation effects, Equations (3.47) through (3.49) become

$$D_{11}\frac{\partial^2\overline{\alpha}}{\partial x^2} + D_{66}\frac{\partial^2\overline{\alpha}}{\partial y^2} + (D_{12} + D_{66})\frac{\partial^2\overline{\beta}}{\partial x\partial y} - 2A_{55}\left(\overline{\alpha} + \frac{\partial w}{\partial x}\right) = I\frac{\partial^2\overline{\alpha}}{\partial t^2} \quad (3.56)$$

$$(D_{12} + D_{66})\frac{\partial^2\overline{\alpha}}{\partial x\partial y} + D_{66}\frac{\partial^2\overline{\beta}}{\partial x^2} + D_{22}\frac{\partial^2\overline{\beta}}{\partial y^2} - 2A_{44}\left(\overline{\beta} + \frac{\partial w}{\partial y}\right) = I\frac{\partial^2\overline{\beta}}{\partial t^2} \quad (3.57)$$

$$2A_{55}\left(\frac{\partial\overline{\alpha}}{\partial x} + \frac{\partial^2 w}{\partial x^2}\right) + 2A_{44}\left(\frac{\partial\overline{\beta}}{\partial y} + \frac{\partial^2 w}{\partial y^2}\right) - \rho_m h\frac{\partial^2 w}{\partial t^2} + p(x,y,t) = 0 \quad (3.58)$$

where

$$I = \rho_m h^3/12 \quad (3.59)$$

Equations (3.55) through (3.59) are written succinctly for a plate made up entirely of one material. For a sandwich plate or a laminated plate, the $\rho_m h$ and $\rho_m h^3/12$ terms need to be modified as follows, analogous to what was done previously in Equations (2.56) and (2.61).

$$\rho_m h = \sum_{k=1}^{N} \rho_{m_k}(h_k - h_{k-1}) \quad (3.60)$$

$$I = \frac{\rho_m h^3}{12} = \frac{1}{3}\sum_{k=1}^{N} \rho_{m_k}(h_k^3 - h_{k-1}^3) \quad (3.61)$$

The terms on the right-hand side of Equations (3.56) and (3.57) show the effects of rotatory inertia, which occurs because the rotation of the lineal element through the plate thickness is not "locked" into the vibratory motion of the lateral displacement $w(x,y)$, but moves independently. The result is that with this refined theory there are three natural frequencies associated with each mode shape: one is primarily lateral displacement motion with little $\overline{\alpha}$ and $\overline{\beta}$ motion; a second, much higher natural frequency associated with the first thickness shear mode in the x-direction ($\overline{\alpha}$), with very little $\overline{\beta}$ or w motion; and the third involving primarily $\overline{\beta}$ motion.

Quite often in dynamic problems, when one is primarily concerned with the dynamic effects on the lateral motion, one neglects the effects of rotatory inertia by setting the right-hand side of Equations (3.56) and (3.57) to zero, amounting to the neglect of rotatory inertia.

3.9 PROBLEMS

3.1 Consider a solid-core sandwich panel in which either one face has a different thickness or is made of a different material than the other face. The result is that the plate is not midplane symmetric, and the [B] matrix, although a constant, is not zero. Using classical theory, derive the three governing differential equations for this case using classical plate theory.

3.2 Redo Problem 3.1; use the refined theory, allowing for transverse-shear deformation.

3.10 REFERENCES

1. Vinson, J. R. and T-W. Chou, *Composite Materials and Their Use in Structures.* Applied Science Publishers, London, 1975.
2. Vinson, J. R. and R. L. Sierakowski, *The Behavior of Structures Composed of Composite Materials.* Martinus-Nijhoff Publishers (now Kluwer Academic Publishers) Dordrecht, The Netherlands, 1986.
3. Whitney, J. M., *Structural Analysis of Laminated Anisotropic Plates,* Technomic Publishing Co. Inc., Lancaster, PA, 1987.
4. Vinson, J. R., *The Behavior of Thin Walled Structures: Beams, Plates and Shells,* Kluwer Academic Publishers, Dordrecht, 1989.

CHAPTER 4

Beams, Columns, and Rods of Composite Materials

4.1 DEVELOPMENT OF CLASSICAL BEAM THEORY

A beam, column, or rod is a long thin structural component of width b, height h, and length L, where $b/L \ll 1$ and $h/L \ll 1$, as shown in Figure 4.1. Its loading and its response to that loading occur in the x-z plane only. It should be noted that if these are loads in the y-z plane, those are treated exactly analogous to the loadings and response in the x-z plane. Subsequently, superposition is used to analyze the response to both loadings if the loads occur simultaneously.

Furthermore, if the $(\ )_{16}$ and $(\ )_{26}$ coupling terms exist in the constitutive relations, then x-z, y-z plane coupling exist and the structure of Figure 4.1 does not behave as a beam, but behaves as a long, thin plate.

The term *beam* is utilized when the structure of Figure 4.1 is subjected to a lateral load in the x-z plane, in the z-direction, distributed and/or concentrated, applied to the upper and/or lower surface (that is, $z = \pm h/2$), such that bending (curvature, in the x-z plane) occurs. The term "rod" is used when the structure of Figure 4.1 is loaded in the axial direction (the x-direction) by tensile forces which try to "stretch" the structure. The term "column" is used when the structure shown is subjected to compressive forces in the x-direction which reduces the length of the structure, resulting in compressive stresses, and/or elastic instability (buckling) of the structure, which will be discussed later in this chapter. Combinations of these loads may occur, such as when lateral and in-plane compressive loads occur simultaneously, resulting in the structure being referred to as a "beam column."

As a simple first example, consider the beam to be loaded in the x-z plane only, whether the loading is lateral, in-plane tensile, in-plane compressive, or combinations thereof. Also, for simplicity, ignore the thermal and moisture

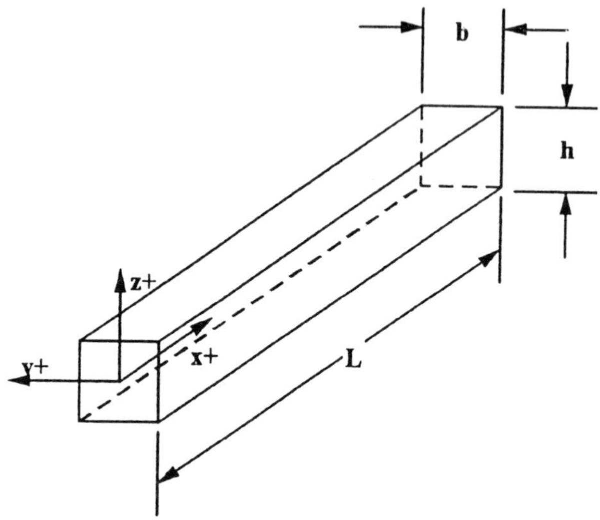

FIGURE 4.1. Nomenclature for a beam, column, or rod.

effects (that is, hygrothermal effects) which add on to the strain terms in the constitutive equations given in Chapter 2. Because a beam is so narrow ($b \ll L$), strains are ignored in the y-direction, implying that all Poisson's ratio effects can be ignored (that is a classical beam assumption). Lastly, as stated previously, there is no y-direction dependence of any quantity involved in the set of governing equations.

Looking then at Equation (2.64), the above assumptions dictate that the remaining constitutive equations for this beam are

$$\begin{Bmatrix} N_x \\ M_x \end{Bmatrix} = \begin{bmatrix} A_{11} & B_{11} \\ B_{11} & D_{11} \end{bmatrix} \begin{Bmatrix} \varepsilon_x^0 \\ \kappa_x \end{Bmatrix} \tag{4.1}$$

and from Equation (2.66), if transverse shear deformation effects are considered,

$$Q_x = 2A_{55}\varepsilon_{xz} \tag{4.2}$$

If one can utilize the simplest description of the midplane-symmetric sandwich beam stiffness quantities, then in Equations (4.1) and (4.2)

$$A_{11} = 2E_{11}t_f,\ D_{11} = \frac{1}{2}E_{11}t_f h_c^2,\ \text{and}\ A_{55} = G_{c_x}h_c \tag{4.3}$$

If the beam has midplane symmetry, as the majority of composite constructions do, then $B_{ij} = 0$, hence $B_{11} = 0$, and Equation (4.1) becomes uncoupled, i.e.,

$$N_x = A_{11}\varepsilon_x^0 \tag{4.4}$$

$$M_x = D_{11}\kappa_x \tag{4.5}$$

Note that without B_{11}, the in-plane load, N_x, and its resulting strain ε_x^0 are completely uncoupled from the stress couple, M_x, and its resulting curvature, κ_x.

In most composite constructions and in sandwich construction, transverse shear deformation cannot be ignored (that is, $\varepsilon_{xz} \neq 0$), because it significantly affects the lateral deformation, w; the natural frequencies of vibration; and the buckling loads. Yet, stress levels are only moderately affected by its inclusion and the simplified analytical methods, neglecting transverse shear deformation can be used for preliminary or approximate design to "size" the structure initially. So for this initial example, assume $\varepsilon_{xz} = 0$, and hence Equation (4.2) will be ignored in what follows.

Looking now to the plate equilibrium equations of the previous chapter, from the beam assumptions made, it is seen that $N_y = N_{xy} = M_y = M_{xy} = Q_y = \tau_{1y} = \tau_{2y} = 0$; hence the remaining equilibrium equations become

$$\frac{dN_x}{dx} + \tau_{1x} - \tau_{2x} = 0 \tag{4.6}$$

$$\frac{dQ_x}{dx} + p_1 - p_2 = 0 \tag{4.7}$$

$$\frac{dM_x}{dx} - Q_x + \frac{h}{2}[\tau_{1x} + \tau_{2x}] = 0 \tag{4.8}$$

Again, for simplicity of example, assume the surface shear stresses τ_{1x} and τ_{2x} are zero. Also, it is seen that the beam will only react to the difference between p_1 and p_2 the normal surface traction on the top and bottom surface, hence:

$$p(x) = p_1 - p_2 \tag{4.9}$$

The result is that Equations (4.6) through (4.9) become

$$\frac{dN_x}{dx} = 0 \tag{4.10}$$

$$\frac{dQ_x}{dx} + p(x) = 0 \tag{4.11}$$

$$\frac{dM_x}{dx} - Q_x = 0 \tag{4.12}$$

It should be remembered that in the above, the N_x, Q_x, and M_x quantities are each plate-type quantities which are per unit edge distance in the y-direction, and hence apply directly to a beam of "unit" width. However, since nothing varies in the y-direction for the beam in question, it is both traditional and easy to multiply all of the above equations by the beam width b. Hence, one can define the usual beam resultants, moments, and loadings to be:

$$P \equiv N_x b,\ V \equiv Q_x b,\ M_b \equiv M_x b,\ q(x) \equiv p(x)b \tag{4.13}$$

Therefore, the governing beam equations for a sandwich beam of composite materials (or even a simple isotropic single-layer beam) with midplane symmetry subjected to lateral and in-plane loads, ignoring hygrothermal effects and transverse-shear deformation are

$$P = bA_{11}\varepsilon_x^0 = bA_{11}\frac{du_0}{dx} \tag{4.14}$$

$$M_b = bD_{11}\kappa_x = -bD_{11}\frac{d^2w}{dx^2} \tag{4.15}$$

$$\frac{dP}{dx} = 0 \tag{4.16}$$

$$\frac{dV}{dx} + q(x) = 0 \tag{4.17}$$

$$\frac{dM_b}{dx} - V = 0 \tag{4.18}$$

From Equation (4.16) it is seen that P = constant (as can also be seen from a simple free-body diagram of a beam), and therefore integration of Equation (4.14) gives the solution for the midsurface in-plane displacement, u_0:

$$u_0(x) = \left(\frac{P}{bA_{11}}\right)x + C_0 \tag{4.19}$$

wherein C_0 is a constant of integration determined by where $u_0(x)$ is specified. If the rod is loaded by a tensile axial load P only, Equation (4.19) provides the displacement, from which all stresses in every ply can be determined by

$$(\sigma_x)_k = (\overline{Q}_{11})_k\,(\varepsilon_x^0) = (\overline{Q}_{11})_k\left(\frac{du_0}{dx}\right) \tag{4.20}$$

If the beam is a midplane-symmetric sandwich beam in which all in-plane loads are carried by the two identical faces only, then the face stress σ_{xf} is

$$\sigma_{xf} = \frac{P}{2bt_f} = E_{11}\varepsilon_x^0 \tag{4.21}$$

If the sandwich construction is more complicated, then Equation (4.20) can always be used. If P is a compressive load, the same equation applies. However, if the critical load P that would cause buckling is sought, a more refined theory is needed; that will be discussed later.

Analogously, substituting Equation (4.15) into (4.18) and the result into Equation (4.17) results in the following differential equation for the bending of the beam:

$$bD_{11}\frac{d^4w}{dx^4} = q(x) \tag{4.22}$$

This is the governing differential equation for a composite-material sandwich beam with $B_{11} = \Delta T = \Delta m = \varepsilon_{xz} = 0$. Four straightforward integrations of Equation (4.22) yields the entire solution including four constants of integration with which to satisfy the boundary conditions. It can easily be shown that under these conditions if the beam involved only a one layer, isotropic material, then $bA_{11} = EA = Ebh$ and $bD_{11} = EI = Ebh^3/12$, for a beam of rectangular cross-section.

From Equation (4.22), the governing differential beam equation for the midplane-symmetric sandwich beam can be written as

$$\frac{d^4w}{dx^4} = \frac{q(x)}{bD_{11}} \tag{4.23}$$

$$\frac{d^3w}{dx^3} = \frac{1}{bD_{11}}\int q(x)\,dx + C_1 \tag{4.24}$$

$$\frac{d^2w}{dx^2} = \frac{1}{bD_{11}} \iint q(x)dxdx + C_1x + C_2 \tag{4.25}$$

$$\frac{dw}{dx} = \frac{1}{bD_{11}} \iiint q(x)dxdxdx + \frac{C_1x^2}{2} + C_2x + C_3 \tag{4.26}$$

$$w(x) = \frac{1}{bD_{11}} \iiiint q(x)dxdxdxdx + \frac{C_1x^3}{6} + \frac{C_2x^2}{2} + C_3x + C_4 \tag{4.27}$$

Once the solution for $w(x)$ is found, the stresses in each of the laminae are found by

$$(\sigma_x)_k = z(\overline{Q}_{11})_k\,(\kappa_x) = -(\overline{Q}_{11})_k\, z\,\frac{d^2w}{dx^2} \tag{4.28}$$

If the beam is a sandwich beam, midplane symmetric, in which all bending is resisted only by the faces, then the bending face stress σ_{xf} is given by the following, using Equations (4.15) and (4.3),

$$\sigma_{x_f}\left(\pm\frac{h}{2}\right) = \pm\frac{M_b}{bt_fh_c} = \mp\frac{E_{11}h_c}{2}\frac{d^2w}{dx^2} \tag{4.29}$$

Obviously if both in-plane and lateral loads occur simultaneously, then the stresses in each face are found by the sum of Equations (4.20) and (4.28)

$$(\sigma_x)_k = (\overline{Q}_{11})_k\,\frac{du_0}{dx} - (\overline{Q}_{11})_k\, z\,\frac{d^2w}{dx^2} \tag{4.30}$$

which for the simplest case of a midplane-symmetric sandwich beam are merely the sum of Equations (4.21) and (4.29),

$$\sigma_{xf}\left(\pm\frac{h}{2}\right) = E_{11}\varepsilon_x^0 \mp \frac{E_{11}h_c}{2}\frac{d^2w}{dx^2} = \frac{P}{2bt_f} \pm \frac{M}{bt_fh_c} \tag{4.31}$$

Thus the theory for the classical-sandwich beam, i.e., no transverse-shear deformation has been developed; now for some solutions.

4.2 SOME SIMPLIFIED SANDWICH-BEAM SOLUTIONS

From the theory for a sandwich beam-rod-column developed in Section 4.1, solutions to all such problems can be found directly.

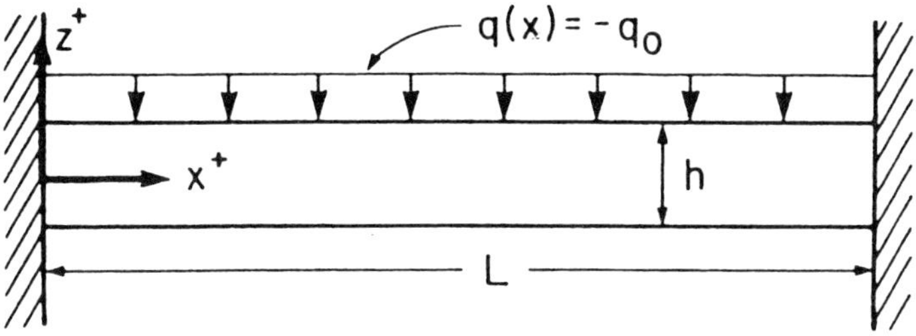

FIGURE 4.2. Clamped-clamped beam with a uniform lateral load.

For the bending of a beam, the Solutions (4.23) through (4.27) result in four constants of integration which are used to satisfy the boundary conditions at each end of the beam. Discussion of classical boundary conditions are found in any undergraduate mechanics textbook, and will not be developed here. There are three classical boundary conditions: simple-support, clamped, and free. For each, the explicit boundary conditions are

Simple Support:	*Clamped:*	*Free:*
$w = 0$	$w = 0$	$M_b = 0$
$M_b = 0$	$\dfrac{dw}{dx} = 0$	$V_b = 0$

(4.32)

As an example, consider a sandwich beam clamped at each end subjected to a uniform lateral load $q(x) = -q_0$, where q_0 is a constant as shown in Figure 4.2.

The clamped boundary conditions are used now to solve for the constants C_1 through C_4.

$$w(0) = 0 = C_4$$

$$\frac{dw}{dx}(0) = 0 = C_3$$

$$w(L) = 0 = -\frac{q_0 L^4}{24bD_{11}} + \frac{C_1 L^3}{6} + \frac{C_2 L^2}{2}$$

$$\frac{dw}{dx}(L) = 0 = -\frac{q_0 L^3}{6bD_{11}} + \frac{C_1 L^2}{2} + C_2 L$$

Therefore,

$$C_1 = \frac{q_0 L}{2bD_{11}} \quad \text{and} \quad C_2 = -\frac{q_0 L^2}{12bD_{11}}$$

Finally,

$$w(x) = -\frac{q_0}{24bD_{11}} [x^4 - 2Lx^3 + L^2x^2] \tag{4.33}$$

One objective of such analysis is to determine the maximum deflection, because some structures are deflection or stiffness critical. Therefore, it is seen (in this case through symmetry) that

$$w_{\max} = w(L/2) = -\frac{q_0 L^4}{384bD_{11}} \tag{4.34}$$

Similarly, it must be determined where the maximum stress is in order to calculate that value, because quite often structures are strength critical. From Equations (4.15) and (4.33),

$$M_b = -bD_{11}\frac{d^2w}{dx^2} = \frac{q_0}{12}[6x^2 - 6Lx + L^2] \tag{4.35}$$

Note that the bending moment M_b is not dependent on any elastic material property. It is seen that the maximum value occurs at $x = 0$ and L, and is

$$M_{b_{\max}} = M_b(0) = M_b(L) = \frac{q_0 L^2}{12} = -bD_{11}\frac{d^2w(0)}{dx^2} = -bD_{11}\frac{d^2w(L)}{dx^2} \tag{4.36}$$

To tie in the mechanics of single-layer isotropic classical beams with the mechanics of a symmetric laminated-composite beam, and the mechanics of a symmetric composite-sandwich beam in which the faces carry all of the bending loads, the following analyses are made.

Suppose the beam were of a one layer, isotropic material then the maximum stress would be at the top and bottom of the beam at each end, that is, $z = \pm h/2$, and the traditional equation would be used

$$\sigma_x = \frac{M_b z}{I}, \text{ where } I = bh^3/12$$

So,

$$\sigma_{x_{\max}} = \sigma_x(0, \pm h/2) = \sigma_x(L, \pm h/2) = \pm\frac{6M_{b_{\max}}}{bh^2} = \pm\frac{q_0 L^2}{2bh^2} \tag{4.37}$$

from which the maximum stress can be calculated, and compared to the allowable strength properties for the material.

For a beam of a given composite material, the calculation is not so simple. Having found $M_{b_{\max}}$ through Equation (4.36), Equation (4.15) must be utilized to find the maximum curvature $\kappa_{x_{\max}}$,

$$\kappa_{x_{\max}} = \kappa_x\begin{pmatrix}0\\L\end{pmatrix} = \frac{M_{b_{\max}}}{bD_{11}} = +\frac{q_0 L^2}{12bD_{11}} = \left(-\frac{d^2w}{dx^2}\right) \tag{4.38}$$

Then, and only then can the maximum stress be calculated for each lamina through Equations (4.28), (4.35), and (4.36)

$$[\sigma_x]_{k_{\max}} = [\overline{Q}_{11}]_k\,[\kappa_{x_{\max}}]z = -[\overline{Q}_{11}]_k\left(\frac{d^2w}{dx^2}\right)z_{\max} = \frac{[\overline{Q}_{11}]_k q_0 L^2 z}{12bD_{11}} \tag{4.39}$$

Then, this $[\sigma_x]_{k_{\max}}$ must be compared with the allowable strength values (that is, failure stress) for each lamina with its specific orientation and composite material system. Note that the maximum stress or failure may not occur in the outer laminae—it all depends upon the ply orientations (i.e., the stacking sequence).

For the simplest sandwich beam, utilizing Equations (4.29) and (4.36), at each end,

$$\sigma_{x_f}\left(\pm\frac{h}{2}\right) = \pm\frac{M_{b_{\max}}}{bt_f h_c} = \pm\frac{q_0 L^2}{12bt_f h_c} \tag{4.40}$$

Keep in mind that all of the above analysis could be much more complicated even for a composite or a sandwich beam, if the composite or sandwich beam were not midplane symmetric ($B_{11} \neq 0$) or if hygrothermal effects were present (that is, $\varepsilon_x^0 + z\kappa_x - \alpha_x\Delta T - \beta_x\Delta m$), all of which were discussed in Chapter 2.

As a second example of solutions to beam problems, consider a beam

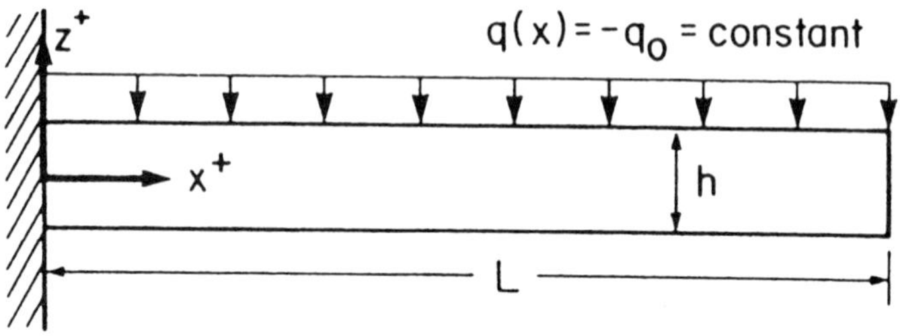

FIGURE 4.3. Clamped-free beam subjected to a uniform load.

cantilevered from the end $x = 0$, as shown in Figure 4.3, where the load $q(x) = -q_0$, a constant. Again, the solution is [see Equation 4.27],

$$w(x) = -\frac{q_0 x^4}{24bD_{11}} + \frac{C_1 x^3}{6} + \frac{C_2 x^2}{2} + C_3 x + C^4 \tag{4.41}$$

Again, as before, because $w(0) = 0$ and $dw/dx(0) = 0$,

$$C_3 = C_4 = 0$$

The other two boundary conditions are $M_b(L) = 0$ and $V(L) = 0$, which is analogous to saying

$$\frac{d^2w}{dx^2}(L) = 0 = -\frac{q_0 L^2}{2bD_{11}} + C_1 L + C_2$$

$$\frac{d^3w}{dx^3}(L) = 0 = -\frac{q_0 L}{bD_{11}} + C_1$$

Hence:

$$C_1 = \frac{q_0 L}{bD_{11}} \quad \text{and} \quad C_2 = -\frac{q_0 L^2}{2bD_{11}}$$

Thus,

$$w(x) = -\frac{q_0}{24bD_{11}}[x^4 - 4Lx^3 + 6L^2x^2] \tag{4.42}$$

It is easily seen that for maximum deflection and maximum stress couple (bending moment) that

$$w_{\max} = w(L) = -\frac{q_0 L^4}{8bD_{11}} \tag{4.43}$$

$$M_{b_{\max}} = M_b(0) = \frac{q_0 L^2}{2} \tag{4.44}$$

Again, one can use the above to investigate the maximum stresses in a single-layer isotropic beam, a midplane-symmetric composite laminate and a midplane-symmetric composite or isotropic sandwich beam.

If this beam were of a one ply, isotropic material

$$\sigma_{x_{\max}} = \pm\frac{6M_{b_{\max}}}{bh^2} = \pm\frac{3q_0 L^2}{bh^2} \tag{4.45}$$

If the beam is made of a laminated composite material then from Equations (4.15) and (4.43)

$$\kappa_{\max} = \kappa_x(0) = \frac{M_b}{bD_{11}} = \frac{q_0 L^2}{2bD_{11}}$$

and again from Equation (4.28)

$$\sigma_{x_{\max}} = [\overline{Q}_{11}]_k \, z\kappa_{x_{\max}} \tag{4.46}$$

For the simplest of sandwich beams, utilizing Equations (4.29) and (4.44),

$$\sigma_{x_{f_{\max}}}(\pm h/2) = \pm\frac{M_{b_{\max}}}{bt_f h_c} = \pm\frac{q_0 L^2}{2bt_f h_c} \tag{4.47}$$

Many more examples for the simplified sandwich beams could be given, but they would be repetitive; for these classical cases, many solutions are available in several texts, e.g., Gere and Timoshenko [1] which provides beam deflections and the location and magnitude of maximum deflections for over 20 different load-boundary condition combinations for beams. Then using the methods described above, one can determine maximum deflections and stresses in isotropic, composite, and sandwich beams. Some additional solutions of sandwich beam problems are also given by Swanson [2].

4.3 EIGENVALUE PROBLEMS OF SANDWICH BEAMS: NATURAL VIBRATIONS AND ELASTIC STABILITY

In Sections 4.1 and 4.2 the problems studied have concentrated on (1) finding the maximum deflection in sandwich beams to insure that they are not too large for a deflection-limited or stiffness-critical structure, and (2) determining the maximum stresses in the beam structure for those structures which are strength critical. However, there are two other ways in which a structure can become damaged or useless; one is through a dynamic response to time-dependent loads, resulting again in too large a deflection or too high stresses, and the other is through the occurrence of an elastic instability (buckling).

In the former, dynamic loading on a structure can vary from a recurring cyclic loading of the same repeated magnitude, such as a structure supporting an unbalanced motor that is turning at a specified number of revolutions per minute (for example), to the other extreme of a short time, intense, nonrecurring load, termed shock or impact loading, such as a bird striking an aircraft component during flight. A continuous infinity of dynamic loads exists between these extremes of harmonic oscillation and impact.

A whole volume could and should be written on the dynamic response of sandwich structures to time-dependent loads, but that is beyond the scope of this text. There are a number of texts dealing with dynamic response of isotropic structures. However, one common thread to all dynamic response will be presented—that of the natural frequencies of vibration.

Mathematically, any continuous structure has an infinity of natural frequencies and mode shapes. If a structure is oscillated at a frequency that corresponds to a natural frequency, it will respond by a rapidly growing amplitude with time, requiring very little input energy, until such time as the structure becomes overstressed and fails, or until the oscillations become so large that nonlinear effects may limit the amplitude to a large but usually unsatisfactory value, and then considerable fatigue damage can occur.

Thus, to insure the structural integrity of any structure being designed or analyzed, the natural frequencies should be determined in order to compare them with any time-dependent loadings to which the structure will be subjected. This is to make sure that the frequencies imposed and the natural frequencies differ considerably. Conversely, in designing a structure, over and above insuring that the structure will not overdeflect, or become overstressed, care should be taken to avoid resonances (that is, imposed loads having the same frequency as one or more natural frequencies).

The easiest example to illustrate this behavior is that of the bending of a beam previously studied, with midplane symmetry ($B_{ij} = 0$) so that there is no bending-stretching coupling and no transverse shear deformation ($\varepsilon_{xz} = 0$).

In this case the governing equation is given by Equation (4.22), as shown below:

$$bD_{11}\frac{d^4w}{dx^4} = q(x) \tag{4.48}$$

In Equation (4.48), it is seen that the imposed static load is written as a force per unit length. For dynamic loading, if D'Alembert's Principle were used then one can add a term to Equation (4.48) equal to the product of mass and acceleration per unit length. In that case Equation (4.48) becomes

$$bD_{11}\frac{\partial^4 w(x,t)}{\partial x^4} = q(x,t) - \frac{\rho_m A\partial^2 w(x,t)}{\partial t^2} \tag{4.49}$$

where w and q both become functions of time as well as space, and derivatives therefore become partial derivatives, ρ_m is the mass density of the beam material, and here A is the beam cross-sectional area. In the above, $q(x,t)$ is now the spatially varying time-dependent forcing function causing the dynamic response, and could be anything from a harmonic oscillation to an intense one-time impact.

For a sandwich beam $\rho_m^A = \rho_{mf}\, 2t_f + \rho_{mc}h_c$ for the case of two identical faces, otherwise using the terminology of laminated composites

$$\rho_m^A = \rho_m bh = \sum_{k=1}^{N} \rho_{mk}^b(h_k - h_{k-1}) \tag{4.50}$$

However, natural frequencies for the beam occur as functions of the material properties and the geometry and hence are not affected by the forcing functions; therefore, for their study let $q(x,t)$ be zero. Thus Equation (4.49) becomes

$$bD_{11}\frac{\partial^4 w}{\partial x^4} + \rho_m^A\frac{\partial^2 w}{\partial t^2} = 0 \tag{4.51}$$

We see that this is the homogeneous equation associated with Equation (4.49). For a simple case, (the easiest one) assume the sandwich beam to be simply supported at each end. One may assume a spatial component of the lateral deflection to satisfy the simply-supported boundary conditions, a harmonic temporal component, and an amplitude A_n such that

$$w(x,t) = \sum_{n=1}^{\infty} A_n \sin\frac{n\pi x}{L}\cos\omega_n t \tag{4.52}$$

Here ω_n is called the natural circular frequency in radians per unit time for the nth vibrational mode. Note that in this case there is one natural frequency for each natural mode shape, for $n = 1,2,3, \ldots$, etc.

For this to be true, then by substituting Equation (4.52) into (4.51) and solving for ω_n, it is seen that for each value of n,

$$\omega_n = \frac{n^2\pi^2}{L^2}\sqrt{\frac{bD_{11}}{\rho_m A}} \tag{4.53}$$

where for the simplest sandwich beam $D_{11} = (1/2)E_{11}t_f h_c^2$. Thus, for each n there is a different natural frequency, and a different mode shape, as shown in Figure 4.4.

For each value of n, the natural frequency is important because if the beam were forced to oscillate at that particular frequency, a resonance would occur, and the amplitude would grow until some form of structural failure would occur, as discussed previously.

The lowest natural frequency, corresponding to $n = 1$ in this simply-

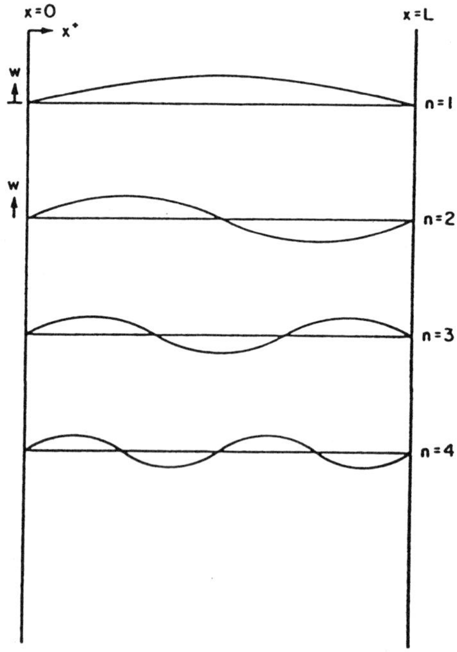

FIGURE 4.4. Simply-supported beam mode shape.

supported beam case, is termed the fundamental frequency. It should be noted that n could go from a value of one to infinity. However, the governing differential Equation (4.51) is only applicable over a portion of this range; i.e., it applies to those mode shapes in which the classical beam equations apply. For an isotropic single-layer beam, the equation breaks down when the half wavelength becomes close to the height, h, of the beam, because then transverse shear deformation effects ($\varepsilon_{xz} \neq 0$) become important. In composite material beams and sandwich beams, transverse-shear deformation effects can be important even for the fundamental natural frequency; but to include transverse shear deformation and rotatory inertia effects involves considerable analytical complications, that will be dealt with later in this chapter. Natural frequencies using classical beam theory, such as Equation (4.53) for the case of both ends simply-supported, are useful for preliminary design and forensic analyses.

It is handy to know the natural frequencies of beams for various practical boundary conditions in order to insure that no recurring forcing functions are close to any of the natural frequencies, because that would result almost certainly in a structural failure. In each case below, the natural circular frequency ω_n in radians/unit time are given as

$$\omega_n = \alpha^2(bD_{11}/\rho_m AL^4)^{1/2} \tag{4.54}$$

where all terms have been explained earlier except the coefficient α^2. Once ω_n is known then the natural frequency in cycles per second (Hertz) is given by $f_n = \omega_n/2\pi$. The values of α^2 have been catalogued by Warburton [3]. Young and Felgar [4], and Felgar [5]. Some values are given in Table 4.1 for ease of use.

Incidentally, the natural frequencies of a free-free beam equal those of a clamped-clamped beam. Also if transverse-shear deformation effects were included, each of the above natural frequencies would be lower. This will be discussed later.

TABLE 4.1. Tabulation of α^2 Values for Use in Determining Beam-Natural Frequencies.

n	Simple-Simple	Simple-Clamped	Clamped-Clamped	Clamped-Free
1	9.870	15.42	22.37	3.516
2	39.48	49.96	61.67	22.03
3	88.83	104.25	120.90	61.70

In the above introductory look at vibrations of the overall beam, it must be remembered that the sandwich construction is complex and that various components of the sandwich construction such as the faces and the core, also may be stimulated to vibrate in modes more complex than the overall beam vibration treated above.

When analyzing a structure, in addition to looking at maximum deflections, maximum stresses, and natural frequencies, one must investigate under what loading conditions an instability can occur, which is also generically referred to as buckling.

Looking at the governing equations for a sandwich beam with midplane symmetry ($B_{ij} = 0$) given by Equations (4.14) through (4.18), it is seen that there are only two equations dealing with in-plane loads P and in-plane deflections u_0, namely, Equations (4.14) and (4.16) with (4.19) being the resulting solution. The remaining three equations deal only with the lateral load $q(x)$ and the lateral deflection w that occurs with Equation (4.27) being the solution for the lateral loading $q(x)$. Therefore, it is seen that the in-plane action is entirely independent of the lateral action.

Yet it is well known and often observed that in-plane loads (through buckling) do cause lateral deflections, which are usually disastrous. The answer to the paradox is that up to this point in this chapter only linear elasticity theory has been used, and the physical event of buckling is a nonlinear theory problem. The development of the nonlinear theory will not be included herein because it is covered in so many other texts [6,7].

The result of including the terms to predict the advent or inception of buckling for the beam is

$$bD_{11}\frac{d^4w}{dx^4} = q(x) + P\frac{d^2w}{dx^2} \tag{4.55}$$

where clearly there is a coupling between the in-plane load P and the lateral deflection, w.

Some items should be noted. The buckling loads, like the natural frequencies in vibration, are independent of the lateral loads, which will be disregarded in what follows. However, in actual structural analysis, the effect of lateral loads, along with the in-plane loads could cause overstressing and failure before the in-plane buckling load is reached. Incidentally, common sense dictates that if one is designing a structure to withstand compressive loads, with the possibility of buckling being the failure mode, one had better design the structure to be midplane symmetric, so that $B_{ij} = 0$. Otherwise, the bending-stretching coupling would likely cause overstressing before the buckling load is reached, as well as reduce the buckling load itself; see Equation (4.60).

Looking at the buckling of a column only, Equation (4.55) can be written as

$$bD_{11}\frac{d^4w}{dx^4} + P\frac{d^2w}{dx^2} \tag{4.56}$$

One can assume that the buckling mode shapes $w(x)$ for the sandwich column, which is simply-supported at each end, are the same as those for the buckling modes of a monocoque column with the same boundary conditions discussed in many texts, namely,

$$w(x) = \sum_{n=1}^{\infty} A_n \sin\frac{n\pi x}{L} \tag{4.57}$$

Substituting Equation (4.57) into (4.56) yields

$$\sum_{n=1}^{\infty} A_n\left[bD_{11}\frac{n^4\pi^4}{L^4} + P\frac{n^2\pi^2}{L^2}\right]\sin\frac{n\pi x}{L} = 0$$

and for this to be true for all n, the critical values of P, called P_{cr} are

$$P_{cr} = -\frac{n^2\pi^2}{L^2}bD_{11} \tag{4.58}$$

Several things are clear: Equation (4.56) is a homogeneous equation; one cannot obtain the magnitude A_n of the buckling mode, and unlike in vibrations where many values of n are important, here only the one giving the lowest value of P_{cr} is important, because after buckling the column is usually permanently deformed or fails, and finally, P_{cr} is a negative value which means a compressive in-plane load.

Thus, it is seen that the eigenvalue problems differ from the considerations in previous sections because the event, natural vibration or buckling, occurs only at certain values. Hence, the natural frequencies and the buckling load are the eigenvalues; the vibrational mode and the buckling mode are the eigenfunctions.

In analyzing any structure, one should therefore determine four things: the maximum deflection, the maximum stresses, the natural frequencies (if there is any dynamic loading to the structure, or nearby to the structure), and the buckling loads (if there are any compressive loads).

Mathematically, one can see that the effects of transverse shear deformation occur because of the finite values of A_{55} in Equation (2.68). In classical theory, A_{55} is taken to be infinite since $\varepsilon_{xz} = 0$ and Q_x is finite. It is also seen that in the natural frequency calculation, the transverse shear deformation effects increase with the integer n.

All combinations of beam vibrational mode shapes applicable for use here have been developed by Warburton [3], and all derivatives and integrals of those functions catalogued conveniently by Young and Felgar [4] and Felgar [5] for easy use. Similar expressions for buckling modes are available in many texts, and can likewise be used to obtain buckling loads with other boundary conditions, analogous to Equation (4.58).

It must again be stated that the natural frequencies and buckling loads calculated in this section; do not include transverse-shear effects, and are therefore only approximate, but they are useful for preliminary design because of their relative simplicity. If transverse shear deformation were included (see the next section), both natural frequencies and buckling loads would be lower than those calculated in this section—so the buckling loads calculated, neglecting transverse shear deformation, are not conservative.

It has been shown that the simple expressions for approximate natural frequencies and buckling loads, which do not include transverse shear deformation (tsd) effects, can be modified to initiate such effects for the columns discussed previously. The result is an approximate but useful expression

$$\left.\begin{Bmatrix}\omega_n\\P_{cr}\end{Bmatrix}\right._{\text{with tsd}} = \frac{\left.\begin{matrix}\omega_n\\P_{cr}\end{matrix}\right\}\text{previously determined (classical)}}{\left(1+\dfrac{D_{11}}{A_{55}}\dfrac{n^2\pi^2}{L^2}\right)^{1/2}} \tag{4.59}$$

It has been shown that in the beam equation, if the laminate or sandwich is asymmetric with respect to the middle surface (that is, $B_{11} \neq 0$), then the D_{11} terms in those equations can be replaced by

$$\frac{A_{11}D_{11} - B_{11}^2}{A_{11}} \tag{4.60}$$

Equation (4.60) has been termed the "reduced" or "apparent" flexural stiffness.

4.4 OTHER CONSIDERATIONS

One important forced vibration problem of sandwich structures involves slamming loads occurring on the hulls of vessels during ocean travel. Foam core sandwich construction is excellent for the construction of small craft, such as lifeboats. Allen and Shenoi [9] have performed extensive research on sandwich beams, typical of that construction, involving 2×10^6 slam encounters, each lasting for a realistic one second duration. It has been pointed out that because of the particular loading due to the slamming of boat hulls, an important failure mode is cracking of the core due to shear stresses, generally followed by delamination. This is a rather unique loading condition and requires special testing and special core strength considerations.

Buene et al. [10] used four-point bend tests of sandwich beams, following the ASTM C393 Standard, to study the effects of slamming loads. They found that the core materials they tested were both viscoelastic and strain-rate dependent. Their results also agreed with Feichtinger [14]. They found that the beam stiffness increased only slightly under slamming loads, whereas the yield strength was markedly increased.

They state that it seems possible to create a single master *S/N* fatigue curve for the foam-core materials by normalizing the shear stresses to the static yield shear stress of the core material. They note that such a master curve cannot be obtained by normalizing the data to the core density. The *S/N* curves suggest the existence of a fatigue limit beyond 10^6 cycles, but more data is needed to confirm the speculation. Also, by way of their tests, they noted no degradation of sandwich beam properties until very shortly before failure. This can be good news or bad news.

Hyer, Anderson, and Scott [11] have studied the nonlinear large-amplitude vibration of three-layer beams with viscoelastic cores both analytically and experimentally.

Gordaninejad and Bert developed more inclusive methods of analysis for the bending of sandwich beams [12], while Frostig, Baruch, Vilnay, and Sheinman have investigated asymmetric sandwich beams with a flexible foam core [13].

4.5 PROBLEMS

4.1 Consider a sandwich beam composed of two identical faces of T300/5208 unidirectional graphite/epoxy and a Klegecell foam core.

The properties of the faces are

$E_1 = 14.5 \times 10^4$ MPa $= 21 \times 106$ psi

$E_2 = 1.17 \times 10^4$ MPa $= 1.76 \times 10^6$ psi

$G_{12} = 0.45 \times 10^4$ MPa $= 0.65 \times 10^6$ psi

$\nu_{12} = 0.21$, $\nu_{21} = 0.017$

$\rho_f = 0.06$ lbs/in^3, $g = 386$ in/sec^2

$\sigma_{\text{allowable}}$ (0° direction) $= 100{,}000$ psi

The core has properties of:

$G_c = 10{,}000$ psi

$\rho_c = 1$ lb/ft^3

$\nu_c = 0.4$

The beam faces are five laminae, each 0.005″ thick, and the core depth is 0.8 in. thick. The beam length is 15 in., its width is 1 in.

a. If the beam is simply supported at each end, and subjected to a uniform lateral load of 10 lb/inch of length, what is the maximum deflection using classical beam theory?

b. What are the maximum face stresses using classical beam theory?

c. What is the fundamental natural frequency in Hz (i.e., cycles per second)?
   (1) using classical beam theory?
   (2) including transverse shear deformation?
   (3) which is lower, by what percentage?

d. If the sandwich beam were subjected to a compressive end load, what would the overall critical buckling load P_{cr} be
   (1) using classical theory?
   (2) including transverse shear deformation effects?
   (3) which is lower, by what percentage?

4.2 Consider a sandwich beam composed of isotropic faces and core, clamped at $x = 0$, and simply supported at $x = L$. Using classical beam theory; if the sandwich beam is subjected to a constant uniform lateral load per unit length of $q(x) = -q_0$,

a. Where and what is the maximum stress?

b. Where and what is the maximum deflection?

4.3 Consider a sandwich beam symmetric with respect to its middle surface, of length L, face thickness t_f, core depth h_c, and width b. Consider the faces to be made of 2024-T3 aluminum: ($E = 10.6 \times 10^6$ psi, $\nu = 0.3$, $\rho = 0.1$ lb/in.^{3}, $\sigma_y = 50{,}000$ psi, $\sigma_u = 70{,}000$ psi) and the core to be made of a foam of Rigid Klegecell of density 15 lb/ft^3, and $G_c = 15{,}000$ psi. Consider the beam to be simply supported at each end. Consider a mandated factor of safety of 2 on yield strength and 3 on ultimate strength to determine the allowable stress. Now consider the face thickness to be $t_f = 0.125$ in., the core depth $h_c = 1.5$ in., the length of the beam to be $L = 36$ in., and the width of the beam $b = 3$ in.

a. What is the flexural stiffness EI of the beam?

b. What is the in-plane stiffness EA of the beam?

c. If the beam is subjected to a uniform lateral load q_0 (lb/in. of length), what q_0 can the beam withstand without being overstressed?

d. What uniform lateral load q_0 can the beam withstand without having the maximum deflection exceed the sandwich thickness?

e. If the beam is subjected to a uniform compressive load P (lb), evenly divided into $P/2$ acting on each face, using classical theory at what load P will the sandwich column undergo overall buckling?

f. Using classical theory, what is the fundamental flexural vibration frequency in Hz?

g. Now consider this same sandwich beam to be subjected to a concentrated lateral load W at the center ($x = L/2$) acting in the thickness direction. Using classical theory (i.e., no transverse shear deformation) what load W can be applied without the beam?

1. Being overstressed?
2. Being over deflected ($w_{max} = h_c + 2t_f$)?
3. Discuss what special problems might occur with the application of this concentrated load.
4. Discuss how the response of the sandwich beam would differ if instead of W being applied "statically," a weight W were dropped into the center of the beam from a considerable height that suddenly strikes the beam in its free fall.

h. For the beam under the lateral load situation of c and d above, name any one thing you would do to improve the sandwich design to take the same lateral load q_0, and reduce the weight.

i. For the beam subjected to an in-plane compressive load P of part e above, name any one thing you would do to improve the sandwich design to take the same load P and reduce the weight.

4.4 Consider a composite sandwich beam of constant flexural stiffness bD_{11}, simply supported at each end, and subjected to a lateral load per unit length given by

$$q(x) = \overline{q} \sin \frac{\pi x}{L}$$

where $\overline{q}$ is a constant. Solving the governing differential equation, and satisfying the boundary conditions,

a. What and where is the maximum lateral displacement w?

b. What and where is the maximum sandwich face stress?

c. For the same beam, what is the critical buckling load, P_{cr}, if it were subjected to an in-plane compressive load?

4.5 Consider a composite sandwich beam, with a content flexural stiffness bD_{11}, simply supported at each end, and subjected to a lateral load per unit length of

$$q(x) = \overline{q}\left[1 - \sin \frac{\pi x}{L}\right]$$

where $\overline{q}$ is a constant.

a. What and where is the maximum deflection w?
b. What and where is the maximum sandwich face stress?
c. For this beam, what is the critical buckling load, P_{cr}, if the beam were subjected to an in-plane compressive load?

4.6 Consider a sandwich beam of T300/934 graphite/epoxy faces, where the properties are given in Table 10.1, in which each face is composed of 4 unidirectional laminae, each 0.0055 in. thick, and a foam core 1/2 in. thick. The beam is 1 in. wide, and 20 in. long, simply supported at one end and clamped at the other. A lateral load of $q = 10$ lb/in. is imposed.
a. What and where is the maximum deflection?
b. What and where is the maximum stress?
c. Is the beam overstressed?
d. What is the critical buckling load, P_{cr}, if the beam were subjected to an in-plane compressive load?
e. If the densities of the face material and core material are $\rho_w =$ 0.05 lb/in.^{3} and 0.010 lb/in^3 respectively, what is the fundamental natural frequency in bending of this beam if it were simply supported at each end?

4.5 REFERENCES

1. Gere, J. and S. Timoshenko, *Mechanics of Materials,* Second Edition, PWS-Kent Publishing Company, Boston, MA, 1984, Appendix G.
2. Swanson, S. R., *Introduction to Design and Analysis with Advanced Composite Materials,* Prentice-Hall, 1997.
3. Warburton, G., "The Vibration of Rectangular Plates," *Proceedings of the Institute of Mechanical Engineers,* Institute of Mechanical Engineers, London, U.K., 1968, p. 371.
4. Young, D. and R. F. Felgar, Jr., "Tables of Characteristic Functions Representing Normal Modes of Vibration for a Beam," The University of Texas Engineering Research Series Report No. 44, July 1, 1949.
5. Felgar, R. P. Jr., "Formulas for Integrals Containing Characteristic Functions of a Vibrating Beam," The University of Texas Bureau of Engineering Research Circular No. 14, 1950.
6. Timoshenko, S. and J. Gere, *Elastic Stability,* McGraw-Hill Book Co., Inc., 2nd Edition, 1961.
7. Bleich, H. H., *Buckling of Metal Structures,* McGraw-Hill Book Co., Inc., 1952.
8. Simitses, G. J., *An Introduction to the Elastic Stability of Structures,* Prentice-Hall, Inc., 1976, R. E. Krieger Publishers, Melbourne, Florida, second printing, 1985.
9. Allen, H. G. and R. A. Shenoi, "Flexural Fatigue Tests on Sandwich Structures," in *Sandwich Constructions 2—Proceedings of the Second International Conference on Sandwich Construction,* Gainesville, Florida, Editors D. Weissman-Berman and K-A. Olsson, EMAS Publication, United Kingdom, 1992, pp. 499–517.
10. Buene, L., A. T. Echtenmeyer, B. Hayman, O. E. Sund and B. Engh, "Shear Properties of GRP Sandwich Beams Subjected to Slamming Loads," in *Sandwich Constructions 2—*

Proceedings of the Second International Conference on Sandwich Construction, Gainesville, Florida, Editors D. Weissman-Berman and K-A. Olsson, EMAS Publication, United Kingdom, 1992, pp. 737–756, 1992.

11. Hyer, M. W., W. J. Anderson and R. A. Scott, "Non-Linear Vibrations of Three-Layer Beams with Viscoelastic Cores Part 1—Theory," *Journal of Sound and Vibration,* Vol. 46 (1), pp. 121–136, 1976; "Part 2—Experiment," Vol. 61 (1), pp. 25–30, 1978.
12. Gordaninejad, F. and C. W. Bert, "A New Theory for Bending of Thick Sandwich Beams," *International Journal of Mechanical Science,* Vol. 31, pp. 925–934, 1989.
13. Frostig, Y., M. Baruch, O. Vilnay and I. Sheinman, "Sandwich Beams with Unsymmetrical Skins and a Flexible Core-Bending Behavior," *Journal of Engineering Mechanics Division ASCE,* Vol. 117, pp. 1931–1952, 1991.
14. Feichtinger, K. A., "Tests Methods and Performance of Structural Core Materials—IIA Strain Rate Dependence of Shear Properties," *5th Autumn INERN Conference, Composite Materials and Sandwich Structures,* Lorient, October 1990.

CHAPTER 5

Energy Methods for Sandwich Structures

5.1 INTRODUCTION

Before discussing solutions of sandwich-plate structures (Chapter 6), it is both convenient and advantageous to introduce solutions employing energy methods. Because sandwich structures are usually comparatively weak in shear, it is necessary to include transverse-shear deformation effects into the methods to analyze or design such structures. This complicates the set of governing equations, raises their order, and adds new boundary conditions. Due to the complications, it may be more efficient to use energy methods in such analyses. Moreover, if the loads are complicated, or there are environmental loads such as thermal or moisture considerations—which cause nonhomogeneous boundary conditions—then sandwich structures are certainly more easily analyzed by energy methods, rather than solving partial differential equations and their associated boundary conditions. Finally, with energy methods, one can always obtain a good approximate solution, no matter what the complications are in the structure, the loads, or the boundary conditions.

In structural mechanics three energy principles are used: Minimum Potential Energy, Minimum Complementary Energy, and Reissner's Variational Theorem. The first two are discussed at length in Sokolnikoff, [1] and many other references. The Reissner Variational Theorem, likewise, is widely referenced. In solid mechanics, Minimum Complementary Energy is rarely used, because it requires assuming functions that insure stresses satisfy boundary conditions and equilibrium. It is usually far easier to make assumptions about the functions to represent displacements.

Minimum Potential Energy is widely used in solutions to problems involving beams and plates. In fact, the more complicated the loading, the more complicated the material (e.g., laminated, anisotropic composite material), the more complicated the geometry (e.g., sandwich and stiffened panel constructions), and the more complicated the boundary conditions (e.g., discontinuous or concentrated boundary conditions), the more desirable it is to use Minimum Potential Energy to obtain approximate solutions, compared to attempting to solve the governing differential equations and satisfy boundary conditions exactly.

In addition, in some cases energy principles can be useful for eigenvalue problems such as in the buckling and vibration problems which are the subjects of Chapters 7 and 9.

The Theorem of Minimum Potential Energy is discussed in Sections 5.2 and 5.3, and Reissner's Variational Theorem is discussed in Sections 5.4 through 5.7, in each case illustrated by beam theory, as was done in References [3,4] in Chapter 2. There are numerous books dealing with energy theorems and variational methods. One of the newest is Mura and Koya [2].

5.2 THEOREM OF MINIMUM POTENTIAL ENERGY

For any generalized elastic body, the potential energy of that body can be written as follows:

$$V = \int_R W dR - \int_{S_T} T_i u_i dS - \int_R F_i u_i dR \tag{5.1}$$

where

W = strain energy density function, defined in Equation (5.4) below
R = volume of the elastic body
T_i = ith component of the surface traction
u_i = ith component of the deformation
F_i = ith component of a body force
S_T = portion of the body surface over which tractions are prescribed

We see that the first term on the right-hand side of Equation (5.1) is the strain energy of the elastic body. The second and third terms are the work done by the surface tractions; and the body forces, respectively. The Theorem of Minimum Potential Energy can be stated as [1]: "Of all the displacements satisfying compatibility and the prescribed boundary conditions, those that satisfy the equilibrium equations make the potential energy a minimum."

Mathematically, the operation is simply

$$\delta V = 0 \tag{5.2}$$

The lowercase delta is a mathematical operation known as a variation. Operationally, it is analogous to partial differentiation. To employ variational operations in structural mechanics, only the following three operations are usually needed:

$$\frac{d(\delta y)}{dx} = \delta\left(\frac{dy}{dx}\right),\ \delta(y^2) = 2y\delta y,\ \int \delta y dx = \delta \int y dx \tag{5.3}$$

In Equation (5.1) the strain energy density function, W, is defined as follows in a Cartesian coordinate frame:

$$W = \frac{1}{2}\sigma_{ij}\,\varepsilon_{ij} = \frac{1}{2}\sigma_x\,\varepsilon_x + \frac{1}{2}\sigma_y\,\varepsilon_y + \frac{1}{2}\sigma_z\,\varepsilon_z$$

$$+ \sigma_{xy}\varepsilon_{xy} + \sigma_{xz}\varepsilon_{xz} + \sigma_{yz}\varepsilon_{yz} \tag{5.4}$$

To utilize the Theorem of Minimum Potential Energy, the stress-strain relations for the elastic body are employed to change the stresses in Equation (5.4) to strains, and the strain-displacement relations are employed to change all strains to displacements. Thus, it is necessary for the analyst to select the proper stress-strain relations and strain-displacement relations for the problem being solved.

Although this text is dedicated to sandwich structures of all types, it is best to use isotropic monocoque beams, a much simpler structural component, to first illustrate the energy principles.

5.3 ANALYSIS OF A BEAM USING THE THEOREM OF MINIMUM POTENTIAL ENERGY

As the simplest example of the use of Minimum Potential Energy, consider a beam in bending, shown in Figure 5.1.

From Figure 5.1 it is seen that the beam is of length L, in the x-direction, width b and height h. It is subjected to a lateral distributed load, $q(x)$ in the positive z-direction, in units of force per unit length. The modulus of elasticity of the isotropic beam material is E, and the stress strain relation is

$$\sigma_x = E\varepsilon_x \tag{5.5}$$

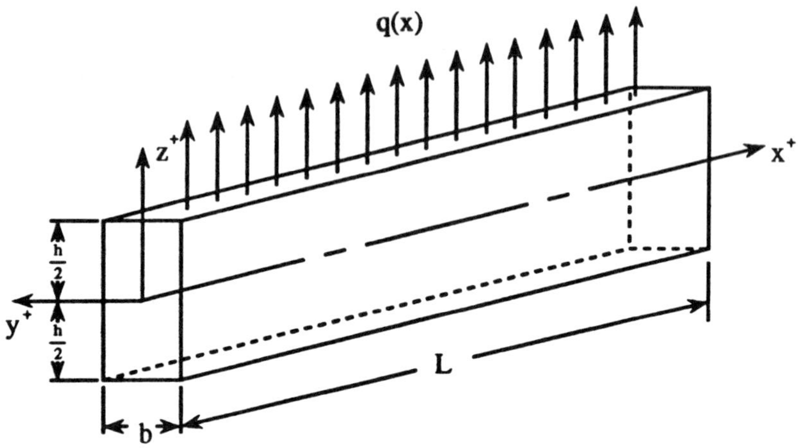

FIGURE 5.1. Beam in bending.

and the strain displacement relation is

$$\epsilon_x = \frac{du_x}{dx} = -\frac{d^2w}{dz^2}z \tag{5.6}$$

since in the bending of beams, $u = -z(dw/dx)$ only, as discussed in Chapter 4.

Looking at Equations (5.4) - (5.6) and remembering that in elementary beam theory

$$\sigma_y = \sigma_z = \sigma_{xy} = \varepsilon_{xz} = \varepsilon_{yz} = \sigma_{xy} = 0$$

then

$$W = \frac{1}{2}\sigma_x\varepsilon_x = \frac{1}{2}E\,\varepsilon_x^2 = \frac{1}{2}E\left(\frac{d^2w}{dx^2}\right)^2 z^2 \tag{5.7}$$

Therefore, the strain energy, U, which is the volume integral of the strain energy density function, W, is

$$U = \int_0^L \int_{-b/2}^{+b/2} \int_{-h/2}^{+h/2} \frac{1}{2}E\left(\frac{d^2w}{dx^2}\right)^2 z^2\,dzdydx$$

$$= \frac{EI}{2}\int_0^L \left(\frac{d^2w}{dx^2}\right)^2 dx \tag{5.8}$$

where here, $I = bh^3/12$, the flexural stiffness for a beam of rectangular cross-section.

Similarly, from the surface traction work term in Equation (5.1) it is seen that

$$\int_{S_T} T_i\, u_i\, ds = \int_0^L q(x)\, w(x) dx$$

Equation (5.1) then becomes

$$V = \frac{EI}{2} \int_0^L \left(\frac{d^2w}{dx^2}\right)^2 dx - \int_0^L q(x)\, w(x)\, dx \tag{5.9}$$

Following Equation (5.2) and remembering Equation (5.3) then

$$\delta V = 0 = \frac{EI}{2} \int_0^L \delta \left(\frac{d^2w}{dx^2}\right)^2 dz - \int_0^L q(x)\, \delta w(x)\, dx \tag{5.10}$$

The variation δ can be included under the integral, because the order of variation and integration can be interchanged. Also, there is no variation of E, I or $q(x)$ because they are all specified quantities.

Integrating by parts the first term on the right-hand side of Equation (5.10),

$$\frac{EI}{2} \int_0^L \delta \left(\frac{d^2w}{dx^2}\right)^2 dx = EI \int_0^L \left(\frac{d^2w}{dx^2}\right) \delta \left(\frac{d^2w}{dx^2}\right) dx$$

$$= EI \int_0^L \frac{d^2w}{dx^2} \frac{d^2(\delta w)}{dx^2} dx$$

$$= \left[EI \frac{d^2w}{dx^2} \delta \left(\frac{dw}{dx}\right) \right]_0^L - EI \int_0^L \frac{d^3w}{dx^3} \frac{d(\delta w)}{dx}$$

$$= \left[EI \frac{d^2w}{dx^2} \delta \left(\frac{dw}{dx}\right) \right]_0^L - \left[EI \frac{d^3w}{dx^3} \delta w \right]_0^L$$

$$+ EI \int_0^L \frac{d^4w}{dx^4} \delta w dx \tag{5.11}$$

Substituting Equation (5.11) into (5.10) and rearranging, it is seen that:

$$\delta V = 0 = \left[EI\frac{d^2w}{dx^2}\,\delta\left(\frac{dw}{dx}\right)\right]_0^L - \left[EI\frac{d^3w}{dx^3}\,\delta w\right]_0^L + \int_0^L \left[EI\frac{d^4w}{dx^4} - q(x)\right]\delta w dx \qquad (5.12)$$

For this to be true, the following equation must be satisfied for the integral above to be zero:

$$EI\frac{d^4w}{dx^4} = q(x) \qquad (5.13)$$

This is obviously the governing equation for the bending of a beam under a lateral load. So it is seen that if one considers a beam-type structure, uses beam assumptions, and uses proper stress-strain relations and strain-displacement relations, the result is the beam bending equation. However, it can be emphasized that if a nonclassical-shaped elastic structure were being analyzed, by using physical intuition, experience, or some other reasoning to formulate stress-strain relations, and strain-displacement relations for the body, then through the Theorem of Minimum Potential Energy one can formulate the governing differential equations for the structure and load analogous to Equation (5.13). Incidentally, the resulting governing differential equations derived from the Theorem of Minimum Potential Energy are called the Euler-Lagrange equations.

Note also for Equation (5.12) to be true, each of the first two terms must be zero. Hence, at $x = 0$ and $x = L$ (at each end) either $EI(d^2w/dx^2) = -M_x = 0$ or (dw/dx) must be specified (that is, its variation must be zero), also either $EI(d^3w/dx^3) = -V_x = 0$ or w must be specified. These are the natural boundary conditions. All of the classical boundary conditions,; including simply-supported, clamped, and free edges are contained in the above "natural boundary conditions." This is a nice byproduct from using the variational approach for deriving governing equations for analyzing any elastic structure.

5.4 REISSNER'S VARIATIONAL THEOREM AND ITS APPLICATIONS

Because of its broad application in the analysis of isotropic or anisotropic thin-walled structures accounting for transverse shear deformation and transverse-normal stress, Reissner's Variational Theorem is treated herein. A general discussion of the Variational Principle is presented, followed by a treatment of the theory of moderately thick beams, which represents a simple example of the power of this technique. The first application is the development of the governing equations for the static deformations of moderately thick rectangular beams, including the effects of transverse-shear deformation and transverse normal stress. The second application involves the use of the Theorem, together with Hamilton's Principle, to develop a theory of beam vibrations including rotatory inertia, in addition to the other effects listed above. Although dynamic effects have not previously been discussed herein, the mathematics involved is straightforward and it is convenient to discuss vibrations at this time.

The principle of Minimum Potential Energy was discussed in Sections 5.2 and 5.3. It was noted that, in carrying out the variations to minimize V, the class of admissible variations are displacements satisfying the boundary conditions. The resulting Euler-Lagrange equations of the variational problem are then equilibrium equations. When the Principle is used to formulate approximate theories (i.e., beam, plate, or shell theory), it can therefore only yield approximate equilibrium equations and the stress-strain or stress-displacement relations must be obtained independently.

In 1950, Reissner [8] developed a new variational theorem of elasticity that yields as its Euler-Lagrange equations *both* the equilibrium equations *and* the stress-displacement relations. The result is the Reissner Variational Theorem, that may be stated as follows: "Of all the stress and displacement states satisfying the boundary conditions, those which also satisfy the equilibrium equations and the stress displacement relations correspond to a minimum of the functional ψ defined as,

$$\psi = \int_R H dR - \int_R F_i u_i dR - \int_{S_T} T_i u_i dS'' \tag{5.14}$$

Here,

$$H = \sigma_{ij}\varepsilon_{ij} - W(\sigma_{ij}) \tag{5.15}$$

S_t is the portion of S on which stresses are prescribed, and $W(\sigma_{ij})$ is the strain-energy density function in terms of stresses only. In a rectangular coordinate system $W(\sigma_{ij})$ is written as follows for an isotropic material:

$$W(\sigma_{ij}) = \frac{1}{2E}[\sigma_x^2 + \sigma_y^2 + \sigma_z^2 - 2\nu(\sigma_x\sigma_y + \sigma_y\sigma_z + \sigma_z\sigma_x)$$

$$+ 2(1 + \nu)(\sigma_{xy}^2 + \sigma_{zx}^2 + \sigma_{yz}^2)] \tag{5.16}$$

The proof of the Theorem is as follows. Taking the variation of ψ in Equation (5.14) and equating it to zero, one obtains:

$$\delta\psi = \int_R \left[\sigma_{ij}\delta\varepsilon_{ij} + \varepsilon_{ij}\,\delta\sigma_{ij} - \frac{\partial W}{\partial\sigma_{ij}}\,\delta\sigma_{ij}\right] dR$$

$$- \int_R F_i\delta u_i dR - \int_{S_T} T_i\delta u_i\, ds = 0 \tag{5.17}$$

where

$$\varepsilon_{ij} = \frac{1}{2}\left(\frac{\partial u_i}{\partial x_j} + \frac{\partial u_j}{\partial x_i}\right)$$

It should be noted that all stress and strain components have been varied, while the body force and the surface traction components F_i and T_i, are not varied because they are prescribed functions. Rearranging the above expression, one obtains

$$\delta\psi = \int_R \left\{\left[\varepsilon_{ij} - \frac{\partial W}{\partial\sigma_{ij}}\right]\delta\sigma_{ij} + \frac{1}{2}\,\sigma_{ij}\left[\frac{\partial}{\partial x_j}(\delta u_i) + \frac{\partial}{\partial x_i}(\delta u_j)\right]\right\} dR$$

$$- \int_R F_i\sigma u_i dR - \int_{S_T} T_i\delta u_i dS = 0 \tag{5.18}$$

The terms $\sigma_{ij}\dfrac{\partial}{\partial x_j}(\delta u_i)$ and $\sigma_{ij}\dfrac{\partial}{\partial x_i}(\delta u_i)$ are symmetric with respect to i and j, and one may therefore interchange these indices and obtain

$$\delta\psi = \int_R \left\{\left[\varepsilon_{ij} - \frac{\partial W}{\partial\sigma_{ij}}\right]\delta\sigma_{ij} + \sigma_{ij}\frac{\partial}{\partial x_j}(\delta u_i)\right\} dR$$

$$- \int_R F_i\delta u_i dR - \int_{S_T} T_i\delta u_i dS = 0 \tag{5.19}$$

Also note that $\dfrac{\partial}{\partial x_j}(\sigma_{ij}\delta u_i) = \sigma_{ij}\dfrac{\partial}{\partial x_j}(\delta u_i) + \dfrac{\partial\sigma_{ij}}{\partial x_j}\delta u_i$, so that Equation (5.19) may be written

$$\delta\psi = \int_R \left\{\left[\varepsilon_{ij} - \frac{\partial W}{\partial\sigma_{ij}}\right]\delta\sigma_{ij} + \frac{\partial}{\partial x_j}(\sigma_{ij}\delta u_i) - \frac{\partial\sigma_{ij}}{\partial x_j}\delta u_i\right\} dR$$

$$- \int_R F_i\delta u_i\, dR - \int_{S_T} T_i\delta u_i\, dS = 0$$

Applying the Green-Gauss divergence theorem and remembering that $\delta u_i = 0$ on all surfaces where displacements are prescribed, one obtains:

$$\int_R \frac{\partial}{\partial X_j}(\sigma_{ij}\delta u_i)\, dR = \int_{S_T} \sigma_{ij}\nu_j\, \delta u_i\, dS = \int_{S_T} T_i \delta u_i\, dS \tag{5.20}$$

where here ν_j is the direction cosine. Finally, substituting Equation (5.20) into (5.19),

$$\delta\psi = \int_R \left\{ \left[\varepsilon_{ij} - \frac{\partial W}{\partial \sigma_{ij}} \right] \delta\sigma_{ij} - \left[\frac{\partial \sigma_{ij}}{\partial x_j} + F_i \right] \delta u_i \right\} dR = 0 \tag{5.21}$$

Since $\delta\sigma_{ij}$ and δu_i are arbitrary variations, Equation (5.21) is satisfied only if the stresses σ_{ij} and strains ε_{ij} satisfy the equations,

$$\frac{\partial \sigma_{ij}}{\partial x_j} + F_i = 0 \tag{5.22}$$

$$\varepsilon_{ij} = \frac{\partial W(\sigma_{ij})}{\partial \sigma_{ij}} \tag{5.23}$$

We see that Equations (5.22) and (5.23) are the equilibrium and stress-strain relations of elasticity. Thus, the Reissner Variational Theorem is found to be equivalent to the three-dimensional equations of elasticity and is, therefore, established. Now consider typical applications of the Theorem to the static and dynamic deformations of beams, as they permit the simplest example.

5.5 STATIC DEFORMATION OF MODERATELY THICK BEAMS

As a first illustration, consider the development of the theory for the static deformations of moderately thick beams in which the effects of transverse shear deformation and transverse normal stress are taken into account following Reissner [8]. Consider a beam of rectangular cross-section of width b, height h and length L, as shown in Figure 5.1, subjected to a distributed load $q(x)$ acting on the surface $z = +h/2$.

In order to apply the variational theorem, one must first assume functions for the stresses in the beams. In this case the following are assumed for a beam of rectangular cross-section,

$$\sigma_x = \frac{Mz}{I}, \text{ where } I = \frac{bh^3}{12} \tag{5.24}$$

$$\sigma_{xz} = \frac{3Q}{2A}\left[1 - \left(\frac{z}{h/2}\right)^2\right], \text{ where } A = bh \tag{5.25}$$

$$\sigma_z = \frac{3q}{4b}\left[\frac{z}{h/2} + \frac{2}{3} - \frac{1}{3}\left(\frac{z}{h/2}\right)^3\right] \tag{5.26}$$

$$\sigma_y = \sigma_{yx} = \sigma_{yz} = 0 \tag{5.27}$$

In the above, M is the beam bending moment, and Q is the beam transverse-shear resultant. It should be noted that the form of the stress components σ_x and σ_{xz} is identical to that of classical theory. The form of the transverse normal stress σ_z may easily be derived from the stress equation of equilibrium in the thickness direction, as a consequence of the assumptions made above for σ_x and σ_{xz}. The expression shown in Equation (5.26) is derived for

$$\sigma_z\,(+h/2) = q \text{ and } \sigma_z\,(-h/2) = 0$$

An analogous expression can be derived easily if there were a normal stress on the lower surface.

The stress couple M and shear resultant Q are defined in the usual manner by the equations

$$M(x) = \int_{-h/2}^{h/2} b\sigma_x z dz \tag{5.28}$$

$$Q(x) = \int_{-h/2}^{h/2} b\sigma_{xz} dz \tag{5.29}$$

It should be noted that Equations (5.24) through (5.29) satisfy all of the stress boundary conditions.

As in classical beam bending theory, one assumes that beam cross-sections undergo at most a translation and a rotation. Such displacements are of the following form for bending only (no stretching), which is the simplest case to study:

$$u = z\overline{\alpha}\,(x) \tag{5.30}$$

$$w = w(x) \tag{5.31}$$

It should be noted that the cross-sections are not assumed to remain normal to the deformed middle surface. If they did, it would result in classical beam theory, which neglects transverse-shear deformation.

The appropriate strain displacement relations may be written

$$\varepsilon_x = \frac{\partial u}{\partial x} = z\,\overline{\alpha}'(x) \tag{5.32}$$

$$\varepsilon_{xz} = \frac{1}{2}\left(\frac{\partial u}{\partial z} + \frac{\partial w}{\partial x}\right) = \frac{1}{2}(\overline{\alpha} + w') \tag{5.33}$$

$$\varepsilon_z = \frac{\partial w}{\partial z} = 0 \tag{5.34}$$

where the primes denote differentiation with respect to x.

For the present case, the Reissner functional ψ, Equation (5.14), takes the form

$$\psi = \int_0^L \int_{-h/2}^{h/2} b\,\{\sigma_x z\overline{\alpha}' + \sigma_{xz}(\overline{\alpha} + w')$$

$$- \frac{1}{2E}[\sigma_x^2 + \sigma_z^2 - 2\nu\sigma_x\sigma_z + 2(1+\nu)\,\sigma_{xz}^2]\}\,dzdx - \int_0^L wqdx \tag{5.35}$$

Substituting Equations (5.24) through (5.34) into Equation (5.35) and carrying out the integrations with respect to z, one obtains,

$$\psi = \int_0^L \left\{M\overline{\alpha}' + Q(\overline{\alpha} + w') - \frac{M^2}{2EI} + \frac{6\nu qM}{5EA} - \frac{3Q^2}{5GA} - qw\right\}dx$$

$$+ \int_0^L \int_{-h/2}^{h/2} \frac{\sigma_z^2}{2E}\,bdzdx \tag{5.36}$$

It should be noted that the integration of the term in σ_z^2 has not been carried out because this term depends only on q and not on the basic unknown stresses and displacements, $\overline{\alpha}$, W, M, and Q. Thus, when variations to minimize ψ are taken, the term in σ_z^2 will not contribute to the result. One may now obtain the governing equations by minimizing the functional ψ of Equation (5.36). Taking the variation of this equation and setting it equal to zero results in

$$\delta\psi = \int_0^L \left\{ M\delta(\overline{\alpha}') + \overline{\alpha}'\delta M + Q(\delta\overline{\alpha} + \delta(w')) + (\overline{\alpha} + w')\,\delta Q - \frac{M}{EI}\,\delta M \right.$$

$$\left. + \frac{6\nu q}{5EA}\,\delta M - \frac{6Q}{5GA}\,\delta Q - q\delta w \right\} dx = 0 \tag{5.37}$$

Integrating by parts and rearranging, Equation (5.37) may be written in the form

$$\delta\psi = [M\delta\overline{\alpha} + Q\delta w]_0^L + \int_0^L \left\{ [Q - M']\,\delta\overline{\alpha} - [Q' + q]\,\delta w \right.$$

$$\left. + \left[\overline{\alpha}' - \frac{M}{EI} + \frac{6\nu q}{5EA}\right]\delta M + \left[\overline{\alpha} + w' - \frac{6Q}{5GA}\right]\delta Q \right\} dx = 0 \tag{5.38}$$

Setting the first term equal to zero yields the natural boundary conditions; for the beam. It is seen that, either $M = 0$ or $\overline{\alpha}$ must be precribed at $x = 0$ and L, and either $Q = 0$ or w must be prescribed at $x = 0$ and L.

Finally, since the variations $\delta\overline{\alpha}$, δw, δM, and δQ are all independent arbitrary functions of x, the only way in which the definite integral of Equation (5.38) can be made to vanish is by requiring the unknowns M, Q, $\overline{\alpha}$, and w to satisfy the equations.

$$-\frac{dM}{dx} + Q = 0 \tag{5.39}$$

$$\frac{dQ}{dx} + q = 0 \tag{5.40}$$

$$\frac{d\overline{\alpha}}{dx} - \frac{M}{EI} + \frac{6\nu q(x)}{5EA} = 0 \tag{5.41}$$

$$\overline{\alpha} + \frac{dw}{dx} - \frac{6Q}{5GA} = 0 \tag{5.42}$$

Note that Equations (5.39) and (5.40) are identical to the equilibrium equations of classical-beam theory. This is as expected since no new stress resultants or stress couples were introduced. Considering Equation (5.42), it is seen that the quantity $\bar{\alpha} + w'$ is precisely the change in the angle between the beam cross-section and the middle surface occurring during the deformation; Equation (5.42) shows that this angular change, which is a measure of the shear deformation, is proportional to Q/A which is the average shear stress. In addition, it should be noted that as $G \to \infty$, the transverse-shear deformation tends to vanish as assumed in classical beam theory. Finally, observe that the third term of Equation (5.41) depends on the lateral load q and the Poisson's ratio ν. This term would vanish if one assumed $\nu = 0$ as in classical beam theory; it is identified as the effect of the normal stress σ_z which is proportional to q, according to the initial assumptions, as seen in Equation (5.26).

Solutions of Equations (5.39) through (5.42) may easily be obtained for typical loading and boundary conditions. These solutions reveal that for beams; of isotropic materials, the effects of the transverse shear deformation and transverse normal stress are negligible for sufficiently large values of L/h and become important as L/h decreases to less than ten.

As stated previously, for simplicity's sake the example treated in detail here is for a beam subjected to a lateral load. Analogous procedures could be followed to develop the equations and solutions for plates, shells, and structures of other shapes, whether in a Cartesian coordinate system or any other coordinate system, such as cylindrical or spherical coordinates.

One interesting use of beam theory including transverse shear deformation is in the development of ASTM Standard C393-62, in which the sandwich bending stiffness and shear modulus are measured using 3-point and 4-point bending tests. For each test the midspan deflection is measured, and is given by the following equations, respectively:

$$w_{\max} = \frac{PL^3}{48D} + \frac{PL}{4G_c(t_f + h_c)^2/h_c b} \tag{5.43}$$

$$w_{\max} = \frac{11PL^3}{768D} + \frac{PL}{8G_c(t_f + h_c)^2/h_c b} \tag{5.44}$$

From these simultaneous equations the sandwich flexural stiffness D, and the shear modulus of the core G_c are found.

5.6 FLEXURAL VIBRATIONS OF MODERATELY THICK BEAMS

As a second example, a theory, developed by Reissner [8], of free flexural vibrations for moderately-thick beams of rectangular cross-section is treated; this will include the effects of transverse-shear deformation and rotatory inertia. Although beam vibrations were discussed previously in Section 4.3, here the mathematical operations involved in the Variational Principle are presented.

In order to derive the equations of motion, one now applies Hamilton's principle in conjunction with the Reissner Variational Theorem. It will be remembered that Hamilton's principle is nothing but a variational statement of Newton's Laws of Motion. Thus, one may state that the motion of the beam of Figure 5.1 will be such as to minimize the integral

$$\Phi = \int_{t_1}^{t_2} (T - \psi)\, dt \tag{5.45}$$

where T is the kinetic energy of the system, ψ is the Reissner functional, and t is time. The quantity $(T - \psi) = L$ is often called the Lagrangian. The equations of motion will now be obtained from the condition,

$$\delta\Phi = 0 \tag{5.46}$$

for a conservative system, and it must be remembered that all stresses, strains, and displacements are now functions of time, as well as the spatial coordinates x and z.

The kinetic energy, T, for the beam of Figure 5.1 may be written in the form

$$T = \int_0^L \int_{-h/2}^{h/2} \frac{\rho_m b}{2} \left[\left(\frac{\partial u}{\partial t} \right)^2 + \left(\frac{\partial w}{\partial t} \right)^2 \right] dz dx \tag{5.47}$$

where ρ_m is the mass density of the beam material. Substituting Equations (5.30) and (5.31) in Equation (5.47) and integrating with respect to z gives

$$T = \int_0^L \frac{\rho_m}{2} \left[I \left(\frac{\partial \alpha}{\partial t} \right)^2 + A \left(\frac{\partial w}{\delta t} \right)^2 \right] dx \tag{5.48}$$

The substitution of Equations (5.36) and (5.48) into Equation (5.45) then yields

$$\Phi = \int_{t_1}^{t_2} \int_0^L \left\{ \frac{\rho_m}{2} \left\{ I \left(\frac{\partial \overline{\alpha}}{\partial t} \right)^2 + A \left(\frac{\partial w}{\partial t} \right)^2 \right\} \right.$$

$$\left. - M \frac{\partial \overline{\alpha}}{\partial x} - Q \left(\overline{\alpha} + \frac{\partial w}{\partial x} \right) + \frac{M^2}{2EI} - \frac{6\nu q M}{5EA} + \frac{3Q^2}{5GA} + qw \right\} dxdt \quad (5.49)$$

where again the term in σ_z^2 has been dropped since it will not contribute to the variations (as explained previously).

The governing equations are then obtained by taking the variation of Equation (5.49) and setting the result equal to zero. It is found that the natural-boundary conditions are the same as for the static case, while the initial deflection and velocity will also be specified. The equations of motion are obtained in the form

$$Q - \frac{\partial M}{\partial x} + \rho_m I c \frac{\partial^2 \overline{\alpha}}{\partial t^2} = 0 \quad (5.50)$$

$$\frac{\partial Q}{\partial x} - \rho_m A \frac{\partial^2 w}{\partial t^2} + q(x,t) = 0 \quad (5.51)$$

$$\frac{\partial \overline{\alpha}}{\partial x} - \frac{M}{EI} + \frac{6\nu q(x,t)}{5EA} = 0 \quad (5.52)$$

$$\overline{\alpha} + \frac{\partial w}{\partial x} - \frac{6Qk}{5GA} = 0 \quad (5.53)$$

In the above equations, two tracing constants c and k are introduced for the purpose of identifying terms. Note that Equation (5.50) is identical to the corresponding moment-equilibrium condition of classical beam theory, except for the term $\rho_m Ic\ (\partial^2 \overline{\alpha}/\partial t^2)$ that represents the contribution of rotatory inertia. Thus, when $c = 1$ in the resulting solutions, rotatory inertia effects are included; when $c = 0$, the theory neglects the effect of rotatory inertia. Equation (5.51) is identical to the classical beam theory equation for transverse force equilibrium with the transverse inertia term added. Equation (5.52) exhibits the term $6\nu q/5EA$, which is the contribution of transverse normal stress; since this is the only term in which ν appears explicitly, setting $\nu = 0$ is equivalent to neglecting the transverse normal stress. Equation (5.53) is nearly identical to the corresponding equation of classical theory with the term $6Qk/5GA$

representing the effect of transverse shear deformation that is included when $k = 1$, and neglected when $k = 0$. Now consider a simple application of this theory.

5.7 FLEXURAL NATURAL FREQUENCIES OF A SIMPLY-SUPPORTED BEAM INCLUDING TRANSVERSE SHEAR DEFORMATION AND ROTATORY INERTIA EFFECTS

For free vibrations, Equations (5.50) through (5.53) reduce to

$$\left.\begin{aligned} &Q - \frac{\partial M}{\partial x} + \rho_m I c \frac{\partial^2 \alpha}{\partial t^2} = 0 \\ &\frac{\partial Q}{\partial x} - \rho_m A \frac{\partial^2 w}{\partial t^2} = 0 \\ &\frac{\partial \overline{\alpha}}{\partial x} - \frac{M}{EI} = 0 \\ &\overline{\alpha} + \frac{\partial w}{\partial x} - \frac{6Qk}{5GA} = 0 \end{aligned}\right\} \tag{5.54}$$

It is convenient to reduce these equations to a system of two equations in the unknown displacements w and $\overline{\alpha}$.

From the first and the third of Equations (5.54), one obtains

$$M = EI \frac{\partial \overline{\alpha}}{\partial x}, \qquad Q = EI \frac{\partial^2 \overline{\alpha}}{\partial x^2} - \rho_m I c \frac{\partial^2 \overline{\alpha}}{\partial t^2}$$

and the substitution of these expressions in the second and fourth of Equations (5.54) yields

$$\left.\begin{aligned} &\frac{\partial^3 \overline{\alpha}}{\partial x^3} - \frac{\rho_m c}{E} \frac{\partial^3 \overline{\alpha}}{\partial x \partial t^2} - \frac{\rho_m A}{EI} \frac{\partial^2 w}{\partial t^2} = 0 \\ &\overline{\alpha} + \frac{\partial w}{\partial x} - \frac{kh^2}{10} \left[\frac{E}{G} \frac{\partial^2 \overline{\alpha}}{\partial x^2} - \frac{\rho_m c}{G} \frac{\partial^2 \overline{\alpha}}{\partial t^2} \right] = 0 \end{aligned}\right\} \tag{5.55}$$

For simply-supported beam of length L, the boundary conditions are $w =$

$M = 0$ at $x = 0$, and L. When the beam is oscillating in a normal mode, the motion is harmonic so that the solutions for $\overline{\alpha}$ and w may be taken in the form:

$$\left.\begin{aligned} w &= W_n \sin \frac{n\pi x}{L} \cos \omega_n t \\ \overline{\alpha} &= \Upsilon_n \cos \frac{n\pi x}{L} \cos \omega_n t \end{aligned}\right\} \tag{5.56}$$

Here, W_n and Υ_n are the amplitudes of the displacement and rotation respectively, and ω_n is the natural frequency of the nth mode of vibration. It is easily verified that these expressions satisfy the boundary conditions. The substitution of Equations (5.56) in Equations (5.55) yields two simultaneous homogeneous algebraic equations for the amplitude W_n and Υ_n. These are

$$\left.\begin{aligned} &\left[\left(\frac{n\pi^3}{L} - \frac{\rho_m c}{E}\left(\frac{n\pi}{L}\right)\omega_n^2\right)\right]\Upsilon_n + \frac{\rho_m A}{EI}\omega_n^2 W_n = 0 \\ &\left[1 + \frac{kh^2}{10}\left(\frac{E}{G}\frac{n^2\pi^2}{L^2} - \frac{\rho_m c}{G}\omega_n^2\right)\right]\Upsilon_n + \left(\frac{n\pi}{L}\right)W_n = 0 \end{aligned}\right\} \tag{5.57}$$

Since Equations (5.57) form a homogeneous system, the condition for a nontrivial solution is that the determinant of the system vanish; this yields the frequency equation, while the amplitude ratios are obtained from satisfying either of the two equations. Thus, the amplitude ratio is given by

$$\frac{\Upsilon_n}{W_n} = -\frac{(\rho_m A/EI)\,\omega_n^2}{(n\pi/L)^3 - (\rho_m c/E)\left(\frac{n\pi}{L}\right)\omega_n^2} \tag{5.58}$$

while the frequency equation may be written in the form,

$$\omega_n^4 - \left[\frac{10G}{\rho_m h^2 kc} + \left(\frac{n\pi}{L}\right)^2\left(\frac{E}{\rho_m c} + \frac{5}{6}\frac{G}{\rho_m k}\right)\right]\omega_n^2 + \frac{10}{h^2 kc}\left(\frac{G}{\rho_m}\right)\left(\frac{EI}{\rho_m A}\right)\left(\frac{n\pi}{L}\right)^4 = 0 \tag{5.59}$$

The natural frequencies for the case where both transverse-shear deformation and rotatory inertia are included may be obtained by solving Equation (5.59) with $k = c = 1$. The frequency equation is then of the form,

$$\omega_n^4 - \left[\frac{10G}{\rho_m h^2} + \left(\frac{n\pi}{L}\right)^2 \left(\frac{E}{\rho_m} + \frac{5}{6}\frac{G}{\rho_m}\right)\right]\omega_n^2 + \frac{10}{h^2}\left(\frac{G}{\rho_m}\right)\left(\frac{EI}{\rho_m A}\right)\left(\frac{n\pi}{L}\right)^4 = 0 \quad (5.60)$$

It is seen that there are two solutions of Equation (5.60) for every wave number n. The lower value is a frequency that is primarily a lateral displacement, corresponding to a low ratio of Y_n/W_n in Equation (5.58). The higher value is primarily a thickness shear mode in which the ratio Y_n/W_n is high.

To obtain a simplified theory neglecting the effect of rotatory inertia, but retaining transverse shear deformation, set $k = 1$ and $c = 0$ after multiplying the frequency Equation (5.59) by c. The resulting simplified frequency equation may be written

$$\left[\frac{10G}{\rho_m h^2} + \left(\frac{n\pi}{L}\right)^2 \frac{E}{\rho_m}\right]\omega_n^2 - \frac{10}{h^2}\left(\frac{G}{\rho_m}\right)\left(\frac{EI}{\rho_m A}\right)\left(\frac{n\pi}{L}\right)^4 = 0 \quad (5.61)$$

and the natural frequencies; are given by

$$\omega_n^2 = \frac{\left(\dfrac{EI}{\rho_m A}\right)\left(\dfrac{n\pi}{L}\right)^4}{1 + \dfrac{n^2\pi^2}{10}\left(\dfrac{E}{G}\right)\left(\dfrac{h}{L}\right)^2} \quad (5.62)$$

Finally, to obtain a frequency equation in which both transverse-shear deformation and rotatory inertia are neglected, multiply Equation (5.59) by kc and set $k = c = 0$; the natural frequencies are then given by

$$\omega_n^2 = \left(\frac{EI}{\rho_m A}\right)\left(\frac{n\pi}{L}\right)^4 \quad (5.63)$$

Equation (5.63) is easily recognized as the well-known solution of the flexural natural circular frequencies of classical-beam theory for a simply-supported beam. A few calculations using Equations (5.60) and (5.62) will show that most of the error (approximately 90 percent) in the classical theory frequency

[Equation (5.63)] is due to the neglect of transverse shear deformation. Therefore, accurate results may be obtained by using Equation (5.62) which still neglects rotatory inertia, but has the advantage of greater simplicity. Comparison of Equations (5.62) and (5.63) reveals that the effect of shear deformation is to reduce the square of the frequencies by a factor equal to $1 + \frac{n^2\pi^2}{10}\left(\frac{E}{G}\right)\left(\frac{h}{L}\right)^2$. It is seen that this factor increases with increasing h/L, so that the errors from using classical theory tend to become larger as the beam becomes "stubbier." This factor also increases as n increases, indicating that classical theory is only adequate for lower modes of vibration, and it becomes increasingly inaccurate for higher modes.

5.8 MINIMUM POTENTIAL ENERGY FOR RECTANGULAR PLATES

Returning now to the more complicated problem of rectangular plates, and employing the Theory of Minimum Potential Energy, the strain-energy-density function, W, for a three-dimensional solid in rectangular coordinates is given by Equation (5.4). The assumptions associated with the classical plate theory of Chapter 2 are now employed in Equation (5.4) for a rectangular plate. If transverse shear deformation is neglected, then $\varepsilon_{xz} = \varepsilon_{yz} = 0$. If there is no plate thickening, then $\varepsilon_z = 0$. Stresses are written in terms of strains, such that for the classical plate

$$\sigma_x = \frac{E}{(1-\nu^2)}[\varepsilon_x + \nu\varepsilon_y];\ \sigma_y = \frac{E}{(1-\nu^2)}[\varepsilon_y + \nu\varepsilon_x];$$

$$\sigma_{xy} = \frac{E}{(1+\nu)}\varepsilon_{xy} \tag{5.64}$$

Therefore, Equation (5.4) becomes

$$W = \frac{E\varepsilon_x}{2(1-\nu^2)}(\varepsilon_x + \nu\varepsilon_y) + \frac{E\varepsilon_y}{2(1-\nu^2)}(\varepsilon_y + \nu\varepsilon_x) + \frac{E}{(1+\nu)}\varepsilon_{xy}^2 \tag{5.65}$$

If the plate is subjected to bending and stretching, the deflection functions are given by Equations (2.24) through (2.28). Substituting these into Equation (5.65) results in the following:

$$W = \frac{1}{2}\frac{E}{(1-\nu^2)}\left\{\left(\frac{\partial u_0}{\partial x}\right)^{22} + \left(\frac{\partial v_0}{\partial y}\right)^2 + 2\nu\left(\frac{\partial u_0}{\partial x}\right)\left(\frac{\partial v_0}{\partial y}\right) + \frac{1+\nu}{2}\left[\frac{\partial u_0}{\partial y} + \frac{\partial v_0}{\partial x}\right]^2\right\}$$

$$+ \frac{Ez^2}{2(1-\nu^2)}\left[\left(\frac{\partial^2 w}{\partial x^2}\right)^2 + \left(\frac{\partial^2 w}{\partial y^2}\right)^2 + 2\nu\left(\frac{\partial^2 w}{\partial x^2}\right)\left(\frac{\partial^2 w}{\partial y^2}\right)\right] + \frac{Ez^2}{(1+\nu)}\left(\frac{\partial^2 w}{\partial x \partial y}\right)^2 \tag{5.66}$$

From this the strain energy $U(=\int_R WdR)$ is found for the isotropic classical rectangular plate.

$$U = \frac{K}{2}\int_0^a\int_0^b \left\{\left(\frac{\partial u_0}{\partial x} + \frac{\partial v_0}{\partial y}\right)^2 - 2(1-\nu)\frac{\partial u_0}{\partial x}\frac{\partial v_0}{\partial y} + \frac{1-\nu}{2}\left(\frac{\partial u_0}{\partial y} + \frac{\partial v_0}{\partial x}\right)^2\right\}dxdy$$

$$+ \frac{D}{2}\int_0^a\int_0^b \left\{\left(\frac{\partial^2 w}{\partial x^2} + \frac{\partial^2 w}{\partial y^2}\right)^2 - 2(1-\nu)\left[\left(\frac{\partial^2 w}{\partial x^2}\right)\left(\frac{\partial^2 w}{\partial y^2}\right) - \left(\frac{\partial^2 w}{\partial x \partial y}\right)^2\right]\right\}dxdy \tag{5.67}$$

It is seen that the first term is the extensional or in-plane strain energy of the plate, and the second is the bending strain energy of the plate. In the latter, it is seen that the first term is proportional to the square of the average plate curvature, while the second term is known as the Gaussian curvature. If Equation (5.67) is used for the usual sandwich plate one may use the following for the extensional stiffness $K = 2E_f t_f/(1-\nu^2)$ and the flexural stiffness $D = E_f t_f h_c^2/2(1-\nu^2)$, as discussed previously.

For the plate the total work term due to surface tractions is seen to be

$$\int_{S_T} T_i u_i dS = \int_0^a\int_0^b p(x,y)\, w(x,y)\, dx\, dy - \int_0^a\int_0^b \left\{N_x\left[\frac{\partial u_0}{\partial x} + \frac{1}{2}\left(\frac{\partial w}{\partial x}\right)^2\right]\right.$$

$$\left. + N_y\left[\frac{\partial v_0}{\partial y} + \frac{1}{2}\left(\frac{\partial w}{\partial y}\right)^2\right] + N_{xy}\left[\left(\frac{\partial u_0}{\partial y} + \frac{\partial v_0}{\partial x}\right) + \left(\frac{\partial w}{\partial x}\right)\left(\frac{\partial w}{\partial y}\right)\right]\right\}dxdy \tag{5.68}$$

Hence, in Equations (5.67) and (5.68) if one considers a plate subjected only to a lateral load $p(x,y)$, one assumes $u_0 = v_0 = N_x = N_y = N_{xy} = 0$. Considering in-plane loads only (except for buckling) assume $w(x,y) = p(x,y) = 0$. If one is looking for buckling loads, assume $p(x,y) = u_0 = v_0 = 0$. The rationale for each of these will be discussed in subsequent sections.

5.9 A RECTANGULAR COMPOSITE MATERIAL PLATE SUBJECTED TO LATERAL AND HYGROTHERMAL LOADS

A detailed study by Sloan [9] shows clearly what is involved in analyzing composite panels to accurately account for the effects of anisotropy, transverse shear deformation, thermal, and hygrothermal effects. The results are also shown in the text by Vinson and Sierakowski [10]

The stress-strain equations to be considered are given by Equation (2.40), wherein the $\overline{Q}_{ij}$ are given by Equation (2.41). The strain-displacement equations are given by Equation (2.45), the form of the panel displacements by Equation (2.47), from which Equation (2.48) results. By neglecting ε_z and σ_z the constitutive relations for the laminate reduce to Equation (2.64). Sloan [9] has shown that even for this problem, the σ_z and ε_z quantities can be neglected.

Employing the Theorem of Minimum Potential Energy, Equation (5.1), for the plate under discussion it is seen that summing the strain-energy-density functions for each lamina across the N laminae that comprise the plate gives the total potential energy as

$$
\begin{aligned}
V = \frac{1}{2}\sum_{k=1}^{N} \iint_A \int_{h_{k-1}}^{h_k} & \{\sigma_x\,[\varepsilon_x - \alpha_x \Delta T - \beta_x \Delta_{\mathrm{m}}] \\
& + \sigma_y[\varepsilon_y - \alpha_y \Delta T - \beta_y \Delta m] + \sigma_{xz}[2\varepsilon_{xz}] \\
& + \sigma_{yz}[2\varepsilon_{yz}] + \sigma_{xy}[2\varepsilon_{xy} - \alpha_{xy}\Delta T - \beta_{xy}\Delta m]\}dzdA \\
& - \iint_A p(x,y)w(x,y)dA
\end{aligned}
\tag{5.69}
$$

Here A refers to the planform area of the plate whose dimensions are $0 \leq x \leq a$, $0 \leq y \leq b$ and $h/2 \leq z \leq h/2$. It is noted that the strains used in the strain-energy relations are the isothermal strains, hence one sees the differences between total strain and the thermal and hygrothermal strains in Equation (5.69).

Now, substituting the constitutive Equations (2.64) and the strain-displacement Relations (2.48) into Equation (5.69) results in

$$V = \iint_A \left\{ \frac{A_{11}}{2}\left(\frac{\partial u_0}{\partial x}\right)^2 + B_{11}\frac{\partial u_0}{\partial x}\frac{\partial \overline{\alpha}}{\partial x} + \frac{D_{11}}{2}\left(\frac{\partial \overline{\alpha}}{\partial x}\right)^2 + A_{12}\frac{\partial u_0}{\partial x}\frac{\partial v_0}{\partial y} \right.$$

$$+ B_{12}\left[\frac{\partial u_0}{\partial x}\frac{\partial \overline{\beta}}{\partial y} + \frac{\partial v_0}{\partial y}\frac{\partial \overline{\alpha}}{\partial x}\right] + D_{12}\frac{\partial \overline{\beta}}{\partial y}\frac{\partial \overline{\alpha}}{\partial x} + A_{16}\left[\frac{\partial u_0}{\partial x}\frac{\partial u_0}{\partial y} + \frac{\partial u_0}{\partial x}\frac{\partial v_0}{\partial x}\right]$$

$$+ D_{16}\left[\frac{\partial \overline{\alpha}}{\partial x}\frac{\partial \overline{\alpha}}{\partial y} + \frac{\partial \overline{\alpha}}{\partial x}\frac{\partial \overline{\beta}}{\partial x}\right] + B_{16}\left[\frac{\partial u_0}{\partial x}\frac{\partial \overline{\alpha}}{\partial y} + \frac{\partial u_0}{\partial x}\frac{\partial \overline{\beta}}{\partial x} + \frac{\partial u_0}{\partial y}\frac{\partial \overline{\alpha}}{\partial x} + \frac{\partial v_0}{\partial x}\frac{\partial \overline{\alpha}}{\partial x}\right]$$

$$+ \frac{A_{22}}{2}\left(\frac{\partial v_0}{\partial y}\right)^2 + B_{22}\frac{\partial v_0}{\partial y}\frac{\partial \overline{\beta}}{\partial y} + \frac{D_{22}}{2}\left(\frac{\partial \overline{\beta}}{\partial y}\right)^2 + A_{26}\left[\frac{\partial v_0}{\partial y}\frac{\partial u_0}{\partial y} + \frac{\partial v_0}{\partial y}\frac{\partial v_0}{\partial x}\right]$$

$$+ B_{26}\left[\frac{\partial v_0}{\partial x}\frac{\partial \overline{\alpha}}{\partial y} + \frac{\partial v_0}{\partial y}\frac{\partial \overline{\beta}}{\partial x} + \frac{\partial u_0}{\partial y}\frac{\partial v_0}{\partial y} + \frac{\partial v_0}{\partial x}\frac{\partial \overline{\beta}}{\partial y}\right] + D_{26}\left[\frac{\partial \overline{\alpha}}{\partial y}\frac{\partial \overline{\beta}}{\partial y} + \frac{\partial \overline{\beta}}{\partial x}\frac{\partial \overline{\beta}}{\partial y}\right]$$

$$+ A_{45}\left[\overline{\alpha}\overline{\beta} + \overline{\alpha}\frac{\partial w}{\partial y} + \overline{\beta}\frac{\partial w}{\partial x} + \frac{\partial w}{\partial x}\frac{\partial w}{\partial y}\right] + A_{55}\left[\frac{\overline{\alpha}^2}{2} + \overline{\alpha}\frac{\partial w}{\partial x} + \frac{1}{2}\left(\frac{\partial w}{\partial x}\right)^2\right]$$

$$+ A_{44}\left[\frac{\overline{\beta}^2}{2} + \overline{\beta}\frac{\partial w}{\partial y} + \frac{1}{2}\left(\frac{\partial w^2}{\partial y}\right)\right] + A_{66}\left[\frac{1}{2}\left(\frac{\partial u_0}{\partial y}\right)^2 + \frac{\partial u_0}{\partial y}\frac{\partial v_0}{\partial x} + \frac{1}{2}\left(\frac{\partial v_0}{\partial x}\right)^2\right]$$

$$+ B_{66}\left[\frac{\partial u_0}{\partial y}\frac{\partial \overline{\alpha}}{\partial y} + \frac{\partial u_0}{\partial y}\frac{\partial \overline{\beta}}{\partial x} + \frac{\partial v_0}{\partial x}\frac{\partial \overline{\alpha}}{\partial y} + \frac{\partial v_0}{\partial x}\frac{\partial \overline{\beta}}{\partial x}\right]$$

$$+ D_{66}\left[\frac{1}{2}\left(\frac{\partial \overline{\alpha}}{\partial y}\right)^2 + \frac{\partial \overline{\alpha}}{\partial y}\frac{\partial \overline{\beta}}{\partial x} + \frac{1}{2}\left(\frac{\partial \overline{\beta}}{\partial x}\right)^2\right] - \frac{\partial u_0}{\partial x}\left(N_{1j}^T + N_{1j}^m\right)$$

$$- \frac{\partial v_0}{\partial y}(N_{2j}^T + N_{2j}^m) - \left(\frac{\partial u_0}{\partial y} + \frac{\partial v_0}{\partial x}\right)(N_{6j}^T + N_{6j}^m) - \frac{\partial \overline{\alpha}}{\partial x}(M_{1j}^T + M_{1j}^m)$$

$$- \frac{\partial \overline{\beta}}{\partial y}(M_{2j}^T + M_{2j}^m) - \left(\frac{\partial \overline{\alpha}}{\partial y} + \frac{\partial \overline{\beta}}{\partial x}\right)(M_{6j}^T + M_{6j}^m)$$

$$\left. + T^* + M^* + \overline{MT}^* - p(x,y)w(x,y) \right\} dA \tag{5.70}$$

In the above, all quantities (except displacements) are defined by Equations (2.56) through (2.59), (2.61) through (2.63), and the following (note that the α_i and β_i below are the coefficients of thermal and hygrothermal expansion):

$$T^* = \frac{1}{2}\sum_{k=1}^{N}\int_{h_{k-1}}^{h_k} [\overline{Q}_{ij}]_k\,[\alpha_i]_k\,[\alpha_j]_k\,(\Delta T)^2(z,t)dz$$

$$M^* = \frac{1}{2}\sum_{k=1}^{N}\int_{h_{k-1}}^{h_k} [\overline{Q}_{ij}]_k\,[\beta_i]_k\,[\beta_j]_k\,(\Delta m)^2\,(z,t)dz$$

$$\overline{MT^*} = \sum_{k=1}^{N}\int_{h_{k-1}}^{h_k} [\overline{Q}_{ij}]_k\,[\alpha_i]\,[\beta_j]\Delta T(z,t)\Delta m(z,t)dz$$

As written Equation (5.70) provides the expression to use in the analysis of monocoque or sandwich rectangular plates of constant thickness, wherein one uses the appropriate values of the $[A]$, $[B]$, $[D]$ stiffness matrix quantities discussed in Chapter 2.

Equation (5.70) is the most general formulation and it is seen that without the surface load term there are 30 terms to represent the strain energy in the composite material panel. Referring to Equation (2.62) and the ensuing discussion it is seen that if the laminate has no stretching-shearing coupling ($A_{16} = A_{26} = 0$) then two terms would be dropped; if no twisting-stretching coupling ($B_{16} = B_{26} = 0$) two more would be dropped; likewise two more are dropped if there were no bending-twisting coupling ($D_{16} = D_{26} = 0$). If the laminate were symmetric about the midplane, a very common construction, then five B_{ij} terms would be canceled out, because there would be no bending-stretching coupling.

For tapered sandwich plates with laminated composite faces and an orthotropic core, Hong and Leon [3] used Minimum Potential Energy and the Rayleigh-Ritz method to obtain solutions for various boundary conditions. Also with the increased use of full-depth low-density honeycomb sandwich in the empennage and control surfaces of aircraft, Libove and Lu [4,5] have studied the static and vibrational behavior of variable-thickness sandwich plates.

Foam and nonmetallic honeycomb cores are being used more often, and are replacing the metallic honeycomb cores, due to greater ease in manufacturing, lower weight, and price considerations [6]. Although metallic honeycombs are very flexible in the longitudinal direction, they are relatively stiff in the transverse (z) direction, so that the core depth, h_c, can be considered as a constant.

However, these newer cores are softer in the transverse direction than the metallic cores, and under localized loads, h_c may vary due to compression in the core.

In this case one may use the Theorem of Minimum Potential Energy, in which one writes out the strain energy for the faces and the core, but in the core permits stresses and strains in the transverse direction. Then one can use separate approximate expressions for the displacements for each face as well as the core, in which one constructs each to insure compatibility of displacements at each core-face interface as well as satisfying boundary conditions at the edges of the beam, plate, ring or shell. By applying the loads on the appropriate face, and taking the variation with respect to each undetermined amplitude, one can solve the problem. As a result, one determines the effects of the loads on the transversely-flexible core.

5.10 IN-PLANE SHEAR STRENGTH DETERMINATION OF COMPOSITE MATERIALS IN LAMINATED AND SANDWICH PANELS

As an illustrative problem using the Theorem of Minimum Potential Energy, consider a simple test procedure to determine the in-plane shear strength of laminated composite materials, as well as other orthotropic and isotropic advanced material systems; see the recent publication by Vinson [7]. The test apparatus shown in Figure 5.2 is simple, inexpensive, and the flat rectangular plate test specimen is not restricted in size or aspect ratio. In addition to its use for laminated composite materials, the test can also be used for foam-core sandwich panels. In sandwich panels the tests can be used to determine

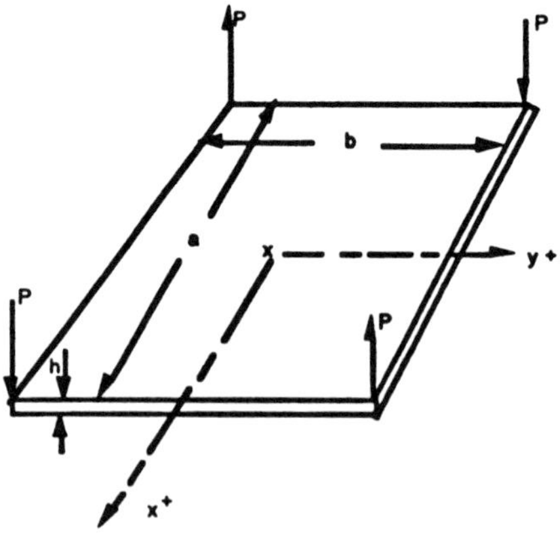

FIGURE 5.2. Test method and geometry for the determination of in-plane shear strength. (Reprinted from Reference [11].)

the in-plane shear strength of the faces, the core, and/or the adhesive bond between face and core. The shear stresses developed vary linearly in the thickness direction and are constant over the entire planform area.

Consider a panel of the material to be tested to be rectangular in planform, with dimensions a in the x-direction and b in the y-direction. The panel is of constant thickness h and if anisotropic, the material principal axes 1-2, should be aligned with the structural axes x-y. The rectangular panel is placed in a test machine such that the loads P are applied at each of the four corners as shown in Figure 5.2. The loads P are recorded as applied until the test specimen fails. As a result, the in-plane shear strength of the material is easily calculated from the equations developed below.

The Theorem of Minimum Potential Energy is utilized to obtain solutions. It is assumed that the material system is specially orthotropic, i.e., the material axes (1-2) are aligned with the structural axes (x-y). It is further assumed that the test specimen is midplane symmetric (i.e., no bending-stretching coupling, $B_{ij} = 0$), no stretching-shearing coupling ($A_{16} = A_{26} = 0$), and no bending-twisting coupling ($D_{16} = D_{26} = 0$). The following methods of analysis can be altered to include laminates or sandwich panels asymmetric to the midsurface (i.e., $B_{ij} \neq 0$).

The potential energy V for the structure and loading of Figure 5.2 is given by Equation (5.70), which for this case is

$$
\begin{aligned}
V = \int_{-a/2}^{a/2} \int_{-b/2}^{b/2} \Bigg\{ & \frac{A_{11}}{2}\left(\frac{\partial u_0}{\partial x}\right)^2 + \frac{D_{11}}{2}\left(\frac{\partial \overline{\alpha}}{\partial x}\right)^2 \\
& + A_{12}\frac{\partial u_0}{\partial x}\frac{\partial v_0}{\partial y} + D_{12}\frac{\partial \overline{\beta}}{\partial y}\frac{\partial \overline{\alpha}}{\partial x} \\
& + \frac{A_{22}}{2}\left(\frac{\partial v_0}{\partial y}\right)^2 + \frac{D_{22}}{2}\left(\frac{\partial \overline{\beta}}{\partial y}\right)^2 \\
& + A_{66}\left[\frac{1}{2}\left(\frac{\partial u_0}{\partial y}\right)^2 + \frac{\partial u_0}{\partial y}\frac{\partial v_0}{\partial x} + \frac{1}{2}\left(\frac{\partial v_0}{\partial x}\right)^2\right] \\
& + D_{66}\left[\frac{1}{2}\left(\frac{\partial \overline{\alpha}}{\partial y}\right)^2 + \frac{\partial \overline{\alpha}}{\partial y}\frac{\partial \overline{\beta}}{\partial y} + \frac{1}{2}\left(\frac{\partial \overline{\beta}}{\partial x}\right)^2\right]\Bigg\} dydx \\
& - Pw({}^a/_2, {}^b/_2) \\
& - Pw(-{}^a/_2, -{}^b/_2) + Pw({}^a/_2, -{}^b/_2) + Pw(-{}^a/_2, {}^b/_2)
\end{aligned}
\tag{5.71}
$$

where u_0 and v_0 are the midplane in-plane displacements in the x- and y-direction respectively, and w is the lateral displacement. All displacements are positive in the positive coordinate direction. The means to portray the concentrated loads at the corners is clearly seen in the last two lines of Equation (5.71).

Neglecting transverse-shear deformation effects, then

$$\overline{\alpha} = -\frac{\partial w}{\partial x} \text{ and } \overline{\beta} = -\frac{\partial w}{\partial y} \tag{5.72}$$

Again, the in-plane and flexural stiffness matrix quantities are given by

$$A_{ij} = \sum_{k=1}^{N} (\overline{Q}_{ij})_k \, (h_k - h_{k-1}) \tag{5.73}$$

$$D_{ij} = \frac{1}{3} \sum_{k=1}^{N} (\overline{Q}_{ij})_k \, (h_k^3 - h_{k-1}^3) \tag{5.74}$$

where $i, j = 1, 2, 6$ in the usual notation. Note that these quantities are valid for both laminate and sandwich construction.

To insure complete generality, the following forms for the displacements are assumed as trial functions, where the numbered coefficients are constants to be determined by boundary conditions and the variational operation:

$$\begin{aligned}
w(x, y) &= C_1 + C_2x + C_3y + C_4x^2 + C_5xy + C_6y^2 \\
&\quad + C_7x^3 + C_8x^2y + C_9xy^2 + C_{10}y^3 + C_{11}x^4 \\
&\quad + C_{12}x^3y + C_{13}x^2y^2 + C_{14}xy^3 + C_{15}y^4 \\
u_0(x, y) &= A_1 + A_2x + A_3y + A_4x^2 + A_5xy + A_6y^2 \\
v_0(x, y) &= B_1 + B_2x + B_3y + B_4x^2 + B_5xy + B_6y^2
\end{aligned} \tag{5.75}$$

When, as an analyst, one has trouble deciding on a suitable trial function for the deflection, include as many terms in the polynomial such that the highest power in the polynomial is equal to the highest order in the differential equations.

The following physical conditions are used to simplify the above assumed functions:

$$w(0, 0) = 0; \ u_0(0, 0) = 0;$$

$$v_0(0, 0) = 0; \ \frac{\partial w}{\partial x}(0, 0) = 0; \ \frac{\partial w}{\partial y}(0, 0) = 0;$$

$$N_x = N_{xy} = 0 \text{ on } x = \pm(^a/_2) \text{ edges}; \ M_x = M_{xy} = 0$$

$$\text{on } x = \pm(^a/_2) \text{ edges}$$

$$N_y = N_{xy} = 0 \text{ on } y = \pm(^b/_2)\text{edges}; \ M_{xy} = M_y = 0$$

$$\text{on } y = \pm(^b/_2) \text{ edges}$$

$$w(a / 2, b / 2) = w(-a / 2, -b / 2)$$

$$= -w(a / 2, -b / 2) = -w(-a / 2, b / 2)$$

The result of satisfying all of the above is that the trial functions [Equation (5.75)] are reduced to the following:

$$w(x, y) = C_5 xy$$

$$u_0(x, y) = A_2 x + A_3 y \qquad v_0(x, y) = B_2 x + B_3 y \tag{5.76}$$

Substituting Equation (5.76) into (5.71), and setting the variation equal to zero ($\delta V = 0$) results in the following relationships:

$$C_5 = \frac{P}{4D_{66}}, \qquad A_3 = -B_2, \qquad A_2 = B_3 = 0$$

Therefore:

$$w(x, y) = \frac{P}{4D_{66}} xy, \qquad u_0(x, y) = A_3 y \qquad v_0(x, y) = -Ax \tag{5.77}$$

From Equation (5.77) it is seen that no curvature exists in the loaded panel, and that if the panel is a monocoque construction or a laminate in which each lamina is oriented the same as all other laminae, then $D_{66} = \frac{G_{12}h^3}{12}$, where G_{12} is the in-plane shear modulus of the material, and h is the panel thickness.

For a laminated plate the in-plane shear stress for the kth lamina is given by

$$(\sigma_{xy})_k = [2\overline{Q}_{66}]_k\,[\varepsilon^0_{xy} + \kappa_{xy}z] = -\frac{[2\overline{Q}_{66}]_{kq}Pz}{4D_{66}} \tag{5.78}$$

since for this test, from Equation (5.77), $\varepsilon^0_{xy} = 0$ and $\kappa_{xy} = -C_5 = -P/4D_{66}$.

In Equation (5.78), for a specially orthotropic material $\overline{Q}_{66} = G_{12}$, the in-plane shear modulus of the material, and D_{66} is given by

$$D_{66} = \frac{1}{3}\sum_{k=1}^{N} (\overline{Q}_{66})_k\,(h_k^3 - h_{k-1}^3)$$

Equation (5.78) provides an easy way to calculate the shear strength of the failed material simply by measuring the load P at failure, and the location z (i.e., the distance from the midsurface of the panel) of the initial failure site. Likewise, if one is only interested in overall panel in-plane shear strength, then knowing the load P at panel failure, and using $z = h/2$, provides the "panel" in-plane shear strength.

For instance if the plate is an isotropic single-layer material then

$$\sigma_{xy}(\pm h/2) = \mp\frac{3P}{h^2}, \text{ or } |\sigma_{xy}| = \frac{3P}{h^2} \tag{5.79}$$

Equation (5.79) is also applicable if the plate is composed of an orthotropic single-layer material. In either case the shear modulus is not needed to calculate the in-plane shear strength.

If the in-plane shear strength of the materials in a midplane symmetric, specially orthotropic or isotropic sandwich panel is desired Equation (5.78) may again be used, where $\overline{Q}_{66_k} = G_{12_k}$ for the material being investigated and for the sandwich

$$D_{66} = \frac{G_{12_f}h_c^2t_f}{2}\left[1 + (t_f/h_c) + \frac{4}{3}(t_f/h_c)^3\right] + \frac{G_{12_c}h_c^3}{12} \tag{5.80}$$

where subscripts f and c refer to face and core respectively, h_c is the depth of the core and t_f is the thickness of each face. For the usual sandwich construction in which $t_f \ll h_c$

$$D_{66} = \frac{G_{12_f}h_c^2t_f}{2} + \frac{G_{12_c}h_c^3}{12} \tag{5.81}$$

For many sandwich panels in which $G_{12_c} \ll G_{12_f}$,

$$D_{66} = G_{12_f} h_c^2 t_f / 2 \tag{5.82}$$

For sandwich constructions in which Equation (5.82) applies, the following shear strengths can be determined.

For the face material:

$$\sigma_{xy_f} = \pm P/2t_f h_c \tag{5.83}$$

For the core material:

$$\sigma_{xy_x} = \pm \left(\frac{P}{2h_c t_f}\right)\left(\frac{G_{12_c}}{G_{12_f}}\right) \tag{5.84}$$

For the adhesive material:

If the adhesive being used is isotropic such that one can assume the shear strength in the x-y plane is the same as the shear strengths in the x-z and y-z planes (i.e., the transverse planes), then Equation (5.85) can be used to determine the shear strength of the adhesive material

$$\sigma_{xy_a} = \pm \frac{P}{2t_f h_c}\left(\frac{G_{12_a}}{G_{12_f}}\right) \tag{5.85}$$

where the subscript a refers to the adhesive material.

In the above, after a test is conducted, one can determine where the failure first occurred (face, core, or adhesive), then Equation (5.83), (5.84), or (5.85) can be used to determine that material strength.

Again, to determine the in-plane shear strength of the face material in a conventional sandwich panel, where Equation (5.82) applies, the in-plane shear moduli of the face, core, and adhesive material are not required. However, to determine the in-plane shear strength of the core or adhesive material, the moduli are required, as in Equations (5.84) and (5.85).

5.11 PROBLEMS

5.1 Consider a sandwich beam of flexural stiffness

$$bD_{11} = \frac{b}{2} E t_f h_c^2$$

clamped on each end and subjected to a uniform lateral load $q(x) = q_0$ lb/in. Using the Theorem of Minimum Potential Energy (MPE) and an assumed deflection function of

$$w(x) = A\left[1 - \cos\frac{2\pi x}{L}\right]$$

where A is the amplitude, and L the beam length, find, using classical beam theory:

a. the magnitude and location of the maximum deflection
b. the magnitude and location of the maximum face stresses
c. Does this form of the deflection function satisfy all of the boundary conditions necessary to use MPE for this problem?

5.2 Consider an orthotropic composite sandwich panel shown in Figures 2.11 and 2.14, which has the following boundary conditions:

$x = 0$ clamped
$x = a$ simply supported
$y = 0$ clamped
$y = b$ free

Select a suitable function for the lateral deflection $w(x, y)$ with which to utilize the Theorem of Minimum Potential Energy for this sandwich plate to determine the deflection of the panel when it is subjected to a laterally-distributed loading $p(x, y)$.

5.3 Consider the sandwich beam of Problem 4.4 once again, subjected to the load

$$q(x) = \overline{q} \quad \sin\frac{\pi x}{L}$$

where $\overline{q}$ is a constant. Using the Theorem of Minimum Potential Energy, and letting the lateral deflection be

$$w(x) = W \sin\frac{\pi x}{L}$$

a. Determine W.
b. What and where is the maximum sandwich face stress?
c. Compare the above results with those obtained in Problem 4.4, i.e., what are the percentage differences and which method produces the greater deflection and stress?

5.4 Consider the sandwich beam of Problem 4.5 once again, subjected to the load

$$q(x) = \overline{q}\left[1 - \sin\frac{\pi x}{L}\right]$$

where $\overline{q}$ is a constant. Using the Theorem of Minimum Potential Energy, and letting the lateral deflection be

$$w(x) = W\sin\frac{\pi x}{L}$$

a. Determine W.
b. What and where is the maximum sandwich-face stress?
c. Compare the above results with those obtained in Problem 4.5, i.e., what are the percentage differences and which method produces the greater deflection and stress?

5.5 The Theorem of Minimum Potential Energy is to be used to analyze an orthotropic composite rectangular sandwich plate. The plate is midplane symmetric, has no moisture or thermal loading, does include transverse shear deformation effects, has no in-plane displacements, and is subjected to a lateral distributed load $p(x, y)$ only. What is the explicit expression for the potential energy V to use in solving this problem?

5.6 The exact solution for a simply-supported beam subjected to a uniform lateral load per unit length of q_0 is:

$$w(x) = \frac{q_0 x}{24EI}(x^3 - 2Lx^2 + L^3)$$

Using the Theorum of Minimum Potential Energy and a deflection function of

$$w(x) = \sum_{n=1}^{m} A_n \sin\left(\frac{n\pi x}{L}\right)$$

a. Determine A_n.
b. Where and what is the maximum deflection?
c. Where and what is the maximum face stress?
d. Compare these results with the exact values. What are the percentage differences?

5.12 REFERENCES

1. Sokolnikoff, I. S., *The Mathematical Theory of Elasticity*, Second Edition, McGraw-Hill Book Company, Inc., New York, 1956.
2. Mura, T. and T. Koya, *Variational Methods in Mechanics*, Oxford University Press, March 1992.
3. Hong, C. S. and J. S. Leon, "An Approximate Solution for Tapered Sandwich Plate with Composite Laminated Faces," *Sandwich Constructions 2, Volume 1*, EMAS Publication, United Kingdom, pp. 35–47, 1992.
4. Libove, C. and C. H. Lu, "Beam Type Bending of Variable Thickness Sandwich Plates," *AIAA Journal*, Volume 27, No. 4, pp. 500–507, April 1989.
5. Lu, C. H. and C. Libove, "Beam Like Harmonic Vibration of Variable-Thickness Sandwich Plates," *AIAA Journal*, Vol. 29, No. 2, February 1991, p. 299.
6. Shenhai, Y., Y. Frostig and E. Altus, "Stresses and Failure Patterns in the Bending of Sandwich Beams with Transversely Flexible Cores and Laminated Composite Skins," *Composite Structures*, Vol. 35, 1996, pp. 143–152.
7. Vinson, J. R. "In-Plane Shear Strength Determination of Composite Materials in Laminated and Sandwich Panels," Applied Mechanics Reviews (forthcoming).
8. Reissner, E., "On a Variational Theorem in Elasticity," *J. Math. Phys.*, Vol. 29, 1950, p. 90.
9. Sloan, J. G., "The Behavior of Rectangular Composite Material Plates Under Lateral and Hygrothermal Loads," MMAE Thesis, University of Delaware, 1979 (also, AFOSR-TR-78-1477, July 1978).
10. Vinson, J. R. and R. L. Sierakowski, *The Behavior of Structures Composed of Composite Materials*, Martinus-Nijhoff Publishers (now Kluwer Academic Publishers) Dordrecht, The Netherlands, 1986.
11. Vinson, J. R., "On Determining the In-Plane Shear Strength of Laminated Composite Materials," *Applied Mechanics in the Americas, Vol. 4*, eds. L. A. Godoy, M. Rysz and L. E. Suarez, 1997, pp. 179–182.

CHAPTER 6

Solutions for Rectangular Sandwich Plates

6.1 INTRODUCTION

The basic plate theory was discussed and the governing differential equations and boundary conditions were formulated in Chapter 3. In Chapter 5, the energy formulations were developed in order to obtain solutions for rectangular sandwich plates. If solutions are obtained from both Chapter 3 and Chapter 5, using the same plate assumptions in both, the solutions will be identical. However, the energy formulation can be used to obtain solutions to problems for which analytical solutions are not possible or are inappropriately difficult. For instance, energy methods are definitely suggested for problems with discontinuous boundary conditions; numerous discontinuous or concentrated loads; general anisotropy with $(\ )_{16}$ and $(\ )_{26}$ terms; or thermoelastic problems wherein the thermoelastic considerations require nonhomogeneous boundary conditions.

In the following sections solutions are presented for a variety of increasingly complex problems which on one hand provide the techniques of solutions available for rectangular sandwich plates, and on the other hand provide useful information for the design and analysis of rectangular sandwich plates.

6.2 NAVIER SOLUTIONS FOR RECTANGULAR SANDWICH PLATES

The Navier approach to these solutions involves separable solutions, as shown below, of the governing differential-plate equation shown in Equation (3.30). Thus

$$w(x,y) = \sum_{m=1}^{\infty}\sum_{n=1}^{\infty} A_{mn}\,\overline{X}_m(x)\overline{Y}_n(y)$$

$$p(x,y) = \sum_{m=1}^{\infty}\sum_{n=1}^{\infty} B_{mn}\,\overline{X}_m(x)\overline{Y}_n(y) \tag{6.1}$$

In the above, the functions $\overline{X}_m(x)$ and $\overline{Y}_n(y)$ are a uniformly convergent, complete, orthogonal set of functions that satisfy the boundary conditions. Thus, when the complete summation is taken, the exact solution is obtained. From a practical point of view, because of uniform convergence, a finite number of terms is sufficient to provide any desired accuracy.

If one assumes a plate that is simply supported on all four edges, then Equation (6.1) can be written as

$$w(x,y) = \sum_{m=1}^{\infty}\sum_{n=1}^{\infty} A_{mn} \sin\left(\frac{m\pi x}{a}\right) \sin\left(\frac{n\pi y}{b}\right) \tag{6.2}$$

$$p(x,y) = \sum_{m=1}^{\infty}\sum_{n=1}^{\infty} B_{mn} \sin\left(\frac{m\pi x}{a}\right) \sin\left(\frac{n\pi y}{b}\right) \tag{6.3}$$

These half-range sine series (the original Navier problem) satisfy the simply-supported boundary conditions on all four edges of the plate shown in Figure 2.11. If the plate has other boundary conditions, then other functions must be used for $\overline{X}_m(x)$ and $\overline{Y}_n(y)$, in which case the approach is labeled the Generalized Navier approach.

To proceed it is first necessary to determine B_{mn} in Equation (6.3) in order to describe the load $p(x,y)$, whether it is continuous or discontinuous. One simply multiplies both the left- and right-hand side of Equation (6.3) by sin $\frac{r\pi x}{a}$ sin $\frac{s\pi y}{b}$ $dxdy$ and integrates both sides from 0 to a in the x-direction and 0 to b in the y-direction (i.e., over the planform area of the plate). It must be remembered that

$$\int_0^a \sin\left(\frac{m\pi x}{a}\right) \sin\left(\frac{r\pi x}{a}\right) dx = \begin{Bmatrix} a/2 & \text{if} & m = r \\ 0 & \text{if} & m \neq r \end{Bmatrix} \tag{6.4}$$

Therefore,

$$B_{mn} = \frac{4}{ab} \int_0^a \int_0^b p(x,y) \sin\left(\frac{m\pi x}{a}\right) \sin\left(\frac{n\pi y}{b}\right) dydx \tag{6.5}$$

For example if $p(x,y) = p_0 =$ a constant, a commonly occurring load,

$$B_{mn} = \frac{4p_0}{mn\pi^2} [1 - (-1)^m] [1 - (-1)^n] \tag{6.6}$$

Now the above can be used for an isotropic or orthotropic plate, classical theory or advanced theory (including transverse-shear deformation) and a sandwich, laminated, or single-layer plate. Considering an orthotropic sandwich panel and classical plate theory, Equation (3.30) is repeated here for convenience:

$$D_1 \frac{\partial^4 w}{\partial x^4} + 2D_3 \frac{\partial^4 w}{\partial x^2 \partial y^2} + D_2 \frac{\partial^4 w}{\partial y^4} = p(x,y) \tag{6.7}$$

Simply substituting Equations (6.2) and (6.3) into (6.7), the equality requires that each term be equated, thus

$$A_{mn} = \frac{B_{mn}}{D_1 \left(\frac{m\pi}{a}\right)^4 + 2D_3 \left(\frac{m\pi}{a}\right)^2 \left(\frac{n\pi}{b}\right) + D_2 \left(\frac{n\pi}{b}\right)^4} \tag{6.8}$$

where the flexural stiffness quantities D_1, D_2 and D_3 are given in Chapter 3 for both a laminated panel and a sandwich panel. For the isotropic case they are all equal.

After obtaining the solution for $w(x,y)$ using Equations (6.2), (6.6), and (6.8), one may calculate the magnitude and location of the maximum deflection. By taking derivatives of $w(x,y)$, and using Equations (3.25) through (3.27), the stress couples M_x, M_y and M_{xy} are determined to find the maximum values and their location. Finally, depending upon whether the panel is a sandwich, a laminate, or a single layer, the maximum stresses are determined using the appropriate expressions. The number of terms necessary to attain a desired accuracy depends upon the particular load $p(x,y)$, the aspect ratio of the plate (b/a), and the material system of which the plate is fabricated.

6.3 LEVY SOLUTION FOR PLATES OF COMPOSITE MATERIALS

The second direct method of solution for the bending of rectangular plates due to lateral loads, is due to Maurice Levy [1] who, in 1899, introduced a single infinite-series method of solution for isotropic plate problems. The method can also be used to solve Equation (3.30) for an orthotropic composite material plate and a sandwich plate.

Consider the plate, shown in Figure 2.11, with edges $y = 0$ and $y = b$ simply supported. The boundary conditions for those edges are

$$\begin{aligned} w(x,0) &= w(x,b) = 0 \\ M_y(x,0) &= M_y(x,b) = 0 \end{aligned} \tag{6.9}$$

The latter implies that

$$\frac{\partial^2 w(x,0)}{\partial y^2} = \frac{\partial^2 w(x,b)}{\partial y^2} = 0 \tag{6.10}$$

Levy assumed the following solution form: a single infinite half-range sine series that satisfies the simply-supported boundary conditions on both y edges:

$$w(x,y) = \sum_{n=1}^{\infty} \phi_n(x) \sin\left(\frac{n\pi y}{b}\right) \tag{6.11}$$

where $\phi_n(x)$ is at this point an unknown function of x. A laterally distributed load $p(x,y)$ can be expressed as follows:

$$p(x,y) = g(x)h(y) \tag{6.12}$$

where $g(x)$ and $h(y)$ are specified. The form of Equation (6.11) requires that the $h(y)$ portion of the load also be expanded in terms of a half range sine series, such as

$$h(y) = \sum_{n=1}^{\infty} A_n \sin\left(\frac{n\pi y}{b}\right) \tag{6.13}$$

where

$$A_n = \frac{2}{b}\int_0^b h(y) \sin\left(\frac{n\pi y}{b}\right) dy \tag{6.14}$$

Substituting Equations (6.11) through (6.13) into (6.7) and observing that the equation exists only if it is true term by term, it is seen that, after dividing by D_1:

$$\frac{d^4\phi_n(x)}{dx^4} - \frac{2D_3}{D_1}\lambda_n^2\frac{d^2\phi_n(x)}{dx^2} + \frac{D_2}{D_1}\lambda_n^4\phi_n(x) = \frac{A_n g_n(x)}{D_1} \tag{6.15}$$

where $\lambda_n = n\pi/b$. Note that Equation (6.15) was derived without specifying any boundary conditions on the x-edges. In fact, the homogeneous solution of Equation (6.15) yields four constants of integration which are determined through satisfying the boundary conditions on those edges.

To obtain the homogeneous solution of Equation (6.15), the right-hand side is set equal to zero:

$$\frac{d^4\phi_n(x)}{dx^4}(x) - \frac{2D_3}{D_1}\lambda_n^2\frac{d^2\phi_n(x)}{dx^2} + \frac{D_2}{D_1}\lambda_n^4\phi_n(x) = 0 \tag{6.16}$$

After letting $\phi_n(x) = e^{sx}$, and dividing the result by e^{sx}, Equation (6.16) becomes:

$$s^4 - \frac{2D_3}{D_1}\lambda_n^2 s^2 + \frac{D_2}{D_1}\lambda_n^4 = 0 \tag{6.17}$$

Unlike the case of an isotropic plate where $D_1 = D_2 = D_3$, such that the roots are easily seen to be $\pm\lambda_n$ and $\pm\lambda_n$ (repeated roots), for this case there are three sets of roots depending upon whether $(D_2/D_1)^{1/2}$ is greater than, equal to or less than D_3/D_1. Hence, for the orthotropic composite plate, sandwich or laminate, using the Levy-type solution requires three different forms for the homogeneous solution of $\phi_n(x)$ to be put in Equation (6.11) depending on the relative stiffness of the plate in various directions.

For the case, $(D_2/D_1)^{1/2} < (D_3/D_1)$

$$\phi_{n_h}(x) = C_1\cosh(\lambda_n s_1 x) + C_2\sinh(\lambda_n s_1 x)$$

$$+ C_3\cosh(\lambda_n s_2 x) + C_4\sinh(\lambda_n s_2 x) \tag{6.18}$$

where the roots are

$$s_1 = \sqrt{\left(\frac{D_3}{D_1}\right) + \sqrt{\left(\frac{D_3}{D_1}\right)^2 - \frac{D_2}{D_1}}},\quad s_2 = \sqrt{\frac{D_3}{D_1} - \sqrt{\left(\frac{D_3}{D_1}\right)^2 - \frac{D_2}{D_1}}}$$

For the case, $(D_2/D_1)^{1/2} = (D_3/D_1)$

$$\phi_{n_h}(x) = (C_5 + C_6x)\cosh(\lambda_n s_3 x) + (C_7 + C_8 x)\sinh(\lambda_n s_3 x) \quad (6.19)$$

where the roots are

$$s_3 = \pm\sqrt{\frac{D_3}{D_1}}$$

For the case $(D_2/D_1)^{1/2} > (D_3/D_1)$

$$\phi_{nh}(x) = (C_9 \cos\lambda_n s_5 x + C_{10}\sin\lambda_n s_5 x)\cosh(\lambda_n s_4 x)$$
$$+ (C_{11}\cos\lambda_n s_5 x + C_{12}\sin\lambda_n s_5 x)\sinh(\lambda_n s_4 x) \quad (6.20)$$

where the roots are

$$s_4 = \sqrt{\frac{1}{2}\left[\left(\frac{D_2}{D_1}\right)^{1/2} + \frac{D_3}{D_1}\right]},\ s_5 = \sqrt{\frac{1}{2}\left[\left(\frac{D_2}{D_1}\right)^{1/2} - \frac{D_3}{D_1}\right]}$$

Obviously, for a given plate whose materials and orientation have already been specified (the analysis problem) only one of the three cases needs to be solved. However, if one is trying to determine the best material and orientation (the design problem), then more than one case may need to be solved, with the necessity of determining not just four constants, but eight or all twelve to satisfy the edge boundary conditions to determine which construction is best for the design.

Concerning the particular solution, it is noted that if the lateral load, $p(x,y)$, is at most linear in x, hence from Equation (6.12), $g_n(x)$ is at most linear in x, then from Equation (6.15) the particular solution is

$$\phi_{n_p}(x) = \frac{A_n g_n(x)}{\lambda_n^4 D_2}$$

Otherwise, one must seek another particular solution. In any case, one must then add the relevant homogeneous ϕ_{nh} to the particular ϕ_{np} to satisfy any set of boundary conditions on the x-edges of the plate. For example, suppose the $x = 0$ edge is simply supported, then from Equation (3.30) the boundary conditions are

$$w(0,y) = 0, \quad \text{and} \quad M_x(0,y) = 0 \rightarrow \frac{\partial^2 w}{\partial x^2}(0,y) = 0 \tag{6.21}$$

However, when $w(x, y)$ has the form of Equation (6.11) this then implies that:

$$\phi_n(0) = \phi_n''(0) = 0 \tag{6.22}$$

where primes denote differentiation with respect to x.

Similarly, appropriate expressions can be found if the x-edges are clamped or free. Then whatever the relevant form of the boundary conditions on $x = 0$ and $x = a$, the total $\phi_n(x) = \phi_{nh} + \phi_{np}$ and hence $w(x,y)$ is known from Equation (6.11). Then, for a composite-material laminated or sandwich plate, one must calculate the curvatures

$$\kappa_x = -\frac{\partial^2 w}{\partial x^2}, \; \kappa_y = -\frac{\partial^2 w}{\partial y^2}, \quad \text{and} \quad \kappa_{xy} = -\frac{\partial^2 w}{\partial x \partial y}$$

Knowing these, one can calculate the stresses in each of the k laminae through the following:

$$\begin{bmatrix} \sigma_x \\ \sigma_y \\ \sigma_{xy} \end{bmatrix}_k = \begin{bmatrix} \overline{Q}_{11} & \overline{Q}_{12} & 0 \\ \overline{Q}_{12} & \overline{Q}_{22} & 0 \\ 0 & 0 & \overline{Q}_{66} \end{bmatrix}_k \begin{bmatrix} \kappa_x \\ \kappa_y \\ 2\kappa_{xy} \end{bmatrix} z \tag{6.23}$$

In the case of a sandwich plate the face stresses are determined for $k = 1$ and 3, while the core is lamina $k = 2$. If the faces are themselves laminates, then this too must be taken into account.

The stresses thus derived for each lamina must then be compared with the allowable stresses, determined through some failure criterion to see if structural integrity is retained under a given load (the analysis problem), or if this set of materials and orientation is sufficient for an objective load (the design problem).

Finally, the Levy-type solution is fine for a composite plate with no bending-stretching coupling, with midplane symmetry, and with two opposite edges simply supported. If two opposite edges are not simply supported then the complexity of the functions necessary to satisfy the boundary conditions on the y = constant edges cause problems.

6.4 PERTURBATION SOLUTIONS FOR THE BENDING OF A COMPOSITE MATERIAL SANDWICH PLATE, WITH MIDPLANE SYMMETRY AND NO BENDING-TWISTING COUPLING

As shown in the previous two sections, the Navier approach is excellent for sandwich plates with all four edges simply supported, and the Levy approach is fine for sandwich plates with two opposite edges simply supported, regardless of the boundary conditions on the other two edges. But for a composite sandwich plate with two opposite edges simply supported, even the Levy approach yields three distinct solutions depending on the relative magnitudes of D_1, D_2 and $D_3 = D_{12} + 2D_{66}$. In addition, there are numerous books and papers available for the solution of isotropic plate problems, a list of which is readily available [2,3].

Aware of all of the above, and based upon the fact that the solution of the second case of the Levy solution of (6.19) has the same form as that of the isotropic case, Vinson showed that the cases of (6.18) and (6.20) can be dealt with as perturbations about the solution of the same plates composed of isotropic materials [7,8].

Consider the governing equation for the bending of a composite material plate exhibiting mid-plane symmetry ($B_{ij} = 0$), no bending-twisting coupling ($D_{16} = D_{26} = 0$), and no transverse shear deformation (classical theory). Then Equation (6.7) becomes, after dividing both sides by D_1:

$$\frac{\partial^4 w}{\partial x^4} + \frac{2D_3}{D_1}\frac{\partial^4 w}{\partial x^2 \partial y^2} + \frac{D_2}{D_1}\frac{\partial^4 w}{\partial y^4} = \frac{p(x, y)}{D_1} \tag{6.24}$$

If one uses coordinate stretching such that the following is defined, let

$$\bar{y} = \left(\frac{D_2}{D_1}\right)^{-1/4} y, \quad \bar{b} = \left(\frac{D_2}{D_1}\right)^{-1/4} b \tag{6.25}$$

Substituting Equation (6.25) into (6.24) yields

$$\frac{\partial^4 w}{\partial x^4} + 2\left(\frac{D_2}{D_1}\right)^{-1/2}\left(\frac{D_3}{D_1}\right)\frac{\partial^4 w}{\partial x^2 \partial \bar{y}^2} + \frac{\partial^4 w}{\partial \bar{y}^4} = \frac{p(x,\bar{y})}{D_1} \tag{6.26}$$

Next defining a quantity α to be

$$\alpha = 2\left[1 - \left(\frac{D_2}{D_1}\right)^{-1/2}\left(\frac{D_3}{D_1}\right)\right] \tag{6.27}$$

it is seen that substituting Equation (6.27) in (6.26) yields

$$\frac{\partial^4 w}{\partial x^4} + \frac{2\partial^4 w}{\partial x^2 \partial \bar{y}^2} + \frac{\partial^4 w}{\partial \bar{y}^4} - \alpha \frac{\partial^4 w}{\partial x^2 \partial \bar{y}^2} = \frac{p(x,\bar{y})}{D_1} \tag{6.28}$$

If one defines the biharmonic operator, used in all isotropic plate problems, to be (in the stretched coordinate system)

$$\nabla^4 w = \frac{\partial^4 w}{\partial x^4} + \frac{2\partial^4 w}{\partial x^2 \partial \bar{y}^2} + \frac{\partial^4 w}{\partial \bar{y}^4} \tag{6.29}$$

then Equation (6.28) becomes

$$\nabla^4 w - \alpha \frac{\partial^4 w}{\partial x^2 \partial \bar{y}^2} = \frac{p(x,y)}{D_1} \tag{6.30}$$

Finally, assume the form of the solution for $w(x,\bar{y})$ to be

$$w(x,\bar{y}) = \sum_{n=0}^{\infty} w_n\,(x,\bar{y})\alpha^n \tag{6.31}$$

which is a perturbation solution employing the "small" parameter α defined in Equation (6.27). Substituting Equation (6.31) into (6.30) and equating all coefficients of α^n to zero, it is easily found that:

$$\nabla^4 w_0 = \frac{p(x,\bar{y})}{D_1} \tag{6.32}$$

$$\nabla^4 w_n = -\frac{\partial^4 w_{n-1}(x,\bar{y})}{\partial x^2 \partial \bar{y}^2} \qquad n \geq 1 \tag{6.33}$$

We see that Equation (6.32) is the governing differential equation for an isotropic sandwich plate of stiffness D_1, subjected to the actual lateral load $p(x,\bar{y})$ with the stretched coordinate $\bar{y}$ defined in Equation (6.25). It is probable that, regardless of boundary conditions on any edge, the solution of Equation (6.32) is available, either exactly or approximately. Subsequently w_1, w_2, w_3, and so on are available from solving Equation (6.33), a plate whose lateral load is $-D_1(\partial^4 w_{n-1}/\partial x^2 \partial \bar{y}^2)$, where w_{n-1} is known, whose flexural stiffness is D_1 and whose boundary conditions are the same as those of the actual plate if the boundary conditions are homogeneous; if not the nonhomogeneity should be accounted for in the w_0 solution.

This technique is very useful because the "small" perturbation parameter need not be so small; it has been proven that when $|\alpha| < 1$, Equation (6.31) is another form of the exact solution, and $|\alpha| < 1$ covers much of the practical composite material construction regime. Also from a computational point of view, it is seldom necessary to include terms past $n = 1$, namely from Equation (6.31):

$$x(x,\overline{y}) = w_0 + w_1\alpha \tag{6.34}$$

6.5 ISOTROPIC SANDWICH PANELS SUBJECTED TO A UNIFORM LATERAL LOAD

For isotropic flat monocoque plates several textbooks such as Timoshenko and Woinowsky-Krieger [2] and Vinson [9] provide expressions for the maximum deflection, w_{max}, and the maximum stress couple, M, a plate attains when subjected to a constant laterally distributed load p_0, such as,

$$w_{\text{max}} = \frac{C_0 p_0 a^4}{Eh^3} \tag{6.35}$$

$$M_{\text{max}} = C_1 p_0 a^2 \tag{6.36}$$

where a is the smaller plate dimension; b is the longer plate dimension, $b \geq a$; E is the modulus of elasticity of the plate; h is the plate thickness; and $\nu = 0.3$.

The dimensionless constants C_0 and C_1 are given in tabular form for various boundary conditions, and these are repeated herein for completeness in Tables 6.1 through 6.4. Table 6.5 also provides the case wherein the plate is subjected to a hydraulic head. These tables and procedures are well known and well used.

TABLE 6.1. Coefficients for Determining Maximum Deflections and Maximum Stresses for a Rectangular Plate, with b > a, *Simply Supported at the Edges, under Uniform Pressure Loading* p_0 *with Sufficient Corner Forces to Hold It Down on the Foundation (ν = 0.3).*

b/a	1	1.2	1.4	1.6	1.8	2	3	4	5	∞
C_0	0.044	0.062	0.077	0.091	0.102	0.111	0.134	0.140	0.142	0.142
C_1	0.048	0.063	0.075	0.086	0.095	0.102	0.119	0.124	0.125	0.125

TABLE 6.2. Coefficients for Determining Maximum Deflections and Maximum Stresses for a Rectangular Plate, under Uniform Load p_0*, with the* a *Edges Clamped and* b *Edges Simply Supported* ($\upsilon = 0.3$).

b/a	∞	3	2	1.6	1.3	1	0.75	0.50	0.25
C_0	0.142	0.128	0.099	0.066	0.042	0.021	0.0081	0.00177	0.00011
C_1	0.125	0.125	0.119	0.109	0.094	0.070	0.045	0.021	0.0052

TABLE 6.3. Coefficients for Determining Maximum Deflections and Maximum Stresses for a Rectangular Plate, under Uniform Loading p_0*, Clamped along the* a *Edge, and Simply Supported Along the Three Remaining Edges*($\upsilon = 0.3$).

b/a	∞	2	1.5	1.2	1	0.75	0.50	0.25
C_0	0.142	0.101	0.070	0.047	0.030	0.0133	0.0033	0.0002
C_1	0.125	0.122	0.112	0.098	0.084	0.058	0.031	0.0077

TABLE 6.4. Coefficients for Determining Maximum Deflections and Maximum Stresses for a Rectangular Plate, under Uniform Loading p_0*, Clamped on All Four Edges* ($\upsilon = 0.3$).

b/a	1	1.2	1.4	1.6	1.8	2.0	2.2
C_0	0.0138	0.0188	0.0226	0.0251	0.0267	0.0277	0.0285
C_1	0.0513	0.0639	0.0726	0.0780	0.0812	0.0829	0.0833

TABLE 6.5. Coefficients for Determining Maximum Deflections and Maximum Stresses for a Rectangular Plate, Simply Supported on All Four Sides, Subjected to a Linearly Increasing Hydraulic Pressure along the a *Edges, One* b *Side Having Zero Pressure, the Opposite* b *Side Having* p_0*. [The Maximum Deflection Occurs Just Off the Middle of the Plate toward the* p_0 *side (at about 0.55a), the Maximum Stress Somewhat Farther Off to the Side* ($\upsilon = 0.3$).*]*

b/a	∞	4	3	2	1.5	1	0.75	0.50	0.25
C_0	0.071	0.070	0.067	0.055	0.042	0.022	0.012	0.0037	0.0004
C_1	0.064	0.063	0.061	0.053	0.043	0.026	0.021	0.0139	0.0051

Of course, for the isotropic plate, the flexural stiffness is given by

$$D = \frac{Eh^3}{12(1 - \nu^2)} \tag{6.37}$$

and the maximum bending stress, which occurs on the top and bottom surfaces of the plate, is

$$\sigma_{max}\ (\pm h/2) = \pm\frac{6M_{max}}{h^2} \tag{6.38}$$

Also for Tables 6.1 through 6.5, the numerical coefficients correspond to a Poisson's ratio of $\nu = 0.3$ wherein $1 - \nu^2 = 0.91$. Therefore, for materials with other Poisson ratios, ν, Equation (6.35) must be changed to be as follows:

$$w_{max} = \frac{C_0 p_0 a^4}{Eh^3}\left(\frac{1 - \nu^2}{0.91}\right) \tag{6.39}$$

To consider a sandwich plate shown in Figure 2.14, for this midplane symmetric isotropic case, the faces are of thickness t_f, the core depth is h_c, the modulus of elasticity of the faces is E, and Poisson's ratio of the faces is ν. For honeycomb core and foam core it is quite customary to assume (as stated previously) that the faces resist all of the bending, in-plane tensile and compressive loads, while the core's functions are to keep the faces in place, and resist the transverse shear loads, analogous to the web of I-beams.

In that case the flexural stiffness D of the isotropic sandwich is given by Equations (1.3) and (2.74), but is repeated here

$$D = \frac{1}{2}\frac{Eh_c^2 t_f}{(1 - \nu_f^2)} \tag{6.40}$$

Also for the usual sandwich construction, $t_f \ll h_c$, and the maximum bending stress $\sigma_{f_{max}}$ is determined by Equation (1.7), repeated here

$$\sigma_{f_{max}} \pm \frac{h}{2} = \pm\frac{M_{max}}{t_f h_c} \tag{6.41}$$

where one face is in tension, the other in compression.

For the case of sandwich panels subjected to a uniform lateral load, p_0, with various boundary conditions, Tables 6.1 through 6.4, can be used, wherein

M_{max} can be used directly and w_{max} can be determined from Equation (6.39) by first using Equation (6.37) such that

$$w_{max} = \frac{C_0 p_0 a^4}{Eh^3}\left(\frac{(1-\nu^2)}{0.91}\right)\frac{Eh^3/12\,(1-\nu^2)}{D} = \frac{C_0 p_0 a^4}{12D\,(0.91)} \quad (6.42)$$

Now the Tables can be used for sandwich panels by simply substituting Equation (6.40) into (6.42) resulting in

$$w_{max} = \frac{C_0 p_0 a^4}{6E_f h_c^2 t_f}\left(\frac{1-\nu_f^2}{0.91}\right) \quad (6.43)$$

Similarly for the sandwich panel, using Equations (6.36) and (6.41)

$$\sigma_{max} = \frac{C_1 p_0 a^2}{t_f h_c} \quad (6.44)$$

With these equations, the well-used Tables 6.1 through 6.4 for monocoque plates can be used to analyze sandwich plates as well. It must be remembered that these are classical theory solutions, i.e., no transverse-shear deformation effects are included. Table 6.5 can be used to analyze and design sandwich plates subjected to a hydraulic head.

It is seen that for a *monocoque* plate design,

(1) The plate cannot be overstressed, i.e., the maximum stress determined by Equation (6.38) cannot exceed some allowable stress, σ_{all}, defined by the material's ultimate stress or yield stress divided by a factor of safety on ultimate stress or yield stress, whichever is smaller. This requires a certain value of plate thickness, h.

(2) The monocoque plate cannot be overdeflected determined by Equation (6.39). This is sometimes specified, but in other cases the plate deflection cannot exceed the plate thickness or some fraction thereof. If the maximum plate deflection reaches a value of the plate thickness, h, the equations discussed herein become inapplicable because the plate behavior becomes increasingly nonlinear which requires that other equations be used. Again, to prevent overdeflection, a plate thickness, h, is required as determined by (6.39).

Therefore, in monocoque plate design, the plate thickness, h, is determined either from a strength or a stiffness requirement, whichever requires the larger thickness.

With a *sandwich* plate, however, the stiffness requirement is determined by

Equation (6.43) and the strength requirement is determined by Equation (6.44), where σ_{max} is the allowable face stress discussed above. However, for these two requirements there are also two design parameters to manipulate, i.e., the core depth h_c and the face thickness t_f. This suggests an optimization to determine a minimum-weight sandwich panel.

6.6 MINIMUM-WEIGHT OPTIMIZATION FOR A SANDWICH PANEL SUBJECTED TO A DISTRIBUTED LATERAL LOAD

For brevity, only a uniform lateral load, p_0, will be discussed herein, but the philosophy is the same for the sandwich panel subjected to any lateral load, distributed or concentrated, uniform or variable. In those cases, solutions must be obtained to determine new coefficients C_0 and C_1 that are analogous to those given in Tables 6.1 through 6.4, to use in Equations (6.43) and (6.44) for the particular load system and boundary conditions of the problem being studied.

To attain minimum weight for a sandwich panel of a given material system, h_c and t_f can be manipulated so that the panel's stiffness restricts the maximum deflection of Equation (6.43) while simultaneously restricting the maximum stress of Equation (6.44) to the allowable stress of the face material. Otherwise, material (which obviously has weight) can be removed or shifted from face to core or the reverse until the above optimum situation occurs. Thus to attain minimum weight Equations (6.43) and (6.44) are written as

$$w_{max} = \frac{C_0 p_0 a^4}{6\ E_f h_c^2 t_f}\left(\frac{1 - \nu_f^2}{0.91}\right) \tag{6.45}$$

$$\sigma_{all} = \frac{C_1 p_0 a^2}{t_f h_c} \tag{6.56}$$

In these two equations there are two and only two unknowns. The equations can be solved simultaneously with the result that:

$$h_c = \left(\frac{C_0}{C_1}\right)\left(\frac{1 - \nu_f^2}{0.91}\right)\frac{a^2\ \sigma_{all}}{6E_f w_{max}} \tag{6.47}$$

$$t_f = \left(\frac{C_1^2}{C_0}\right)\left(\frac{0.91}{1 - \nu_f^2}\right)\frac{6E_f w_{max}}{\sigma_{all}^2} \tag{6.48}$$

So if there are no other restrictions on the choice of t_f and h_c, the values determined by Equations (6.47) and (6.48) will result in a minimum-weight isotropic sandwich panel, given here explicitly for a constant lateral load p_0. It is probable that h_c can be fabricated exactly. For the face thickness, the commercially available thickness or an integer number of plies in a laminate may result in Equation (6.48) providing the minimum thicknesses, but the actual thickness being the next larger plausible thickness available.

The sandwich panel weight per unit planform area, W, can be written as

$$W - W_{ad} = 2\rho_f t_f + \rho_c h_c \tag{6.49}$$

where ρ_f and ρ_c are the weight densities of the face and core materials respectively. W_{ad} is the weight per unit planform area of the adhesive or brazing material joining the face to the core; it is on the left-hand side because it is a nonanalytical quantity, often depending on the experience, diligence, and care of the workman.

Substituting Equations (6.47) and (6.48) into Equation (6.49) results in

$$W - W_{ad} = \frac{2\rho_f C_1^2 p_0 6E_f w_{max}}{C_0 \sigma_{all}^2}\left(\frac{0.91}{1 - \nu_f^2}\right) + \frac{\rho_c C_0 a^2 \sigma_{all}}{6E_f C_1 w_{max}}\left(\frac{1 - \nu_f^2}{0.91}\right) \tag{6.50}$$

The first term is the specific weight of the face, the second term for the core. Since the face weight is the predominant weight for the sandwich panel, it follows that for face material selection the face material with the highest value of

$$\text{Factor of Merit} = \frac{(1 - \nu_f^2)(\sigma_{all}^2)}{\rho_f E_f} \tag{6.51}$$

will result in the minimum weight panel. This factor of merit then provides a means to compare weight savings with cost, manufacturability, etc., in making a rational design decision.

6.7 ANALYSIS OF AN ISOTROPIC-SANDWICH PLATE ON AN ELASTIC FOUNDATION SUBJECTED TO A UNIFORM LATERAL LOAD

The governing differential equation for an isotropic plate on an elastic foundation using classical theory subjected to a uniform lateral load is shown below. It is a simpler form of Equations (3.53) and (3.54).

$$D\left[\frac{\partial^4 w}{\partial x^4} + 2\frac{\partial^4 w}{\partial x^2 \partial y^2} + \frac{\partial^4 w}{\partial y^4}\right] + kw = p_0(x,y) \tag{6.52}$$

where D is the flexural stiffness. In the case a sandwich plate, $D = \frac{1}{2}\frac{E_f t_f h_c^2}{(1 - \nu_f^2)}$, and for a monocoque plate, $D = \frac{E_f h^3}{12(1 - \nu^2)}$ where w is the lateral deflection, k is the foundation modulus in terms of (lb./in.)/in.^{2}, for example, and $p_0(x,y)$ is the lateral uniform load in psi, for example.

For a Navier solution for a plate simply supported on all four edges, let

$$w(x,y) = \sum_{m=1}^{\infty}\sum_{n=1}^{\infty} A_{mn} \sin\left(\frac{m\pi x}{a}\right) \sin\left(\frac{n\pi y}{b}\right) \tag{6.53}$$

The lateral uniform load expanded in the same trigonometric functions is

$$p_0 = \sum_{m=1}^{\infty}\sum_{n=1}^{\infty} B_{mn} \sin\left(\frac{m\pi x}{a}\right) \sin\left(\frac{n\pi y}{b}\right) \tag{6.54}$$

where it can be shown that $B_{mn} = \frac{16p_0}{mn\pi^2}$ (for m and n odd only) for a uniform lateral load p_0. Substituting Equations (6.53) and (6.54) into (6.52) determines that for m and n odd only

$$A_{mn} = \frac{16p_0}{mn\pi^2 D} \frac{1}{\left[\left(\frac{m\pi}{a}\right)^2 + \left(\frac{n\pi}{b}\right)^2\right]^2 + \frac{k}{D}} \tag{6.55}$$

Because of the elastic foundation it is seen that the amplitude of the lateral deflection is greatly reduced by the presence of the constant quantity (k/D). Also, far more terms in both m and n are needed to have the maximum value converge to a constant value (i.e., terminate the series after an acceptable accuracy is attained), compared to the case of no elastic foundation (i.e., $k = 0$). If $b > a$, then the curvature in the shorter dimension direction is

$$\frac{\partial^2 w}{\partial x^2} = -\frac{16p_0}{Da^2}\sum_{m=1}^{\infty}\sum_{n=1}^{\infty} \frac{m}{n} \frac{\sin(m\pi x/a)\sin(n\pi y/b)}{\left[\left(\frac{m\pi}{a}\right)^2 + \left(\frac{n\pi}{b}\right)^2\right]^2 + \frac{k}{D}} \tag{6.56}$$

Again, the same characteristics as discussed above hold, particularly that if (k/D) is large, the curvature produced in the plate can be quite small, and therefore the bending stresses will also be small.

It is seen that the above analysis for an isotropic plate simply supported on all four edges and subjected to a uniform lateral load p_0 can also be applied to the following:

(1) A monocoque plate on an elastic foundation

(2) A sandwich plate on an elastic foundation

(3) The behavior of one sandwich face on a core (that acts as an elastic foundation for the face)

6.8 STATIC ANALYSIS OF SANDWICH PLATES OF COMPOSITE MATERIALS INCLUDING TRANSVERSE-SHEAR DEFORMATION EFFECTS

In the preceding sections of Chapter 6, all of the analyses involve classical plate theory and therefore have neglected transverse shear deformation effects. Now these effects will be taken into account for a composite material plate subjected to static loads. The derivation of the governing equations comprises Section 3.5. The final equations, (3.47) through (3.49), are repeated here as follows. In the following κ with no subscript is a transverse shear factor, often given as $\pi^2/12$ or 5/6.

$$D_{11}\frac{\partial^2\overline{\alpha}}{\partial x^2} + D_{66}\frac{\partial^2\overline{\alpha}}{\partial y^2} + (D_{12} + D_{66})\frac{\partial^2\overline{\beta}}{\partial x\partial y} - \kappa A_{55}\left(\overline{\alpha} + \frac{\partial w}{\partial x}\right) = 0 \quad (6.57)$$

$$(D_{12} + D_{66})\frac{\partial^2\overline{\alpha}}{\partial x\partial y} + D_{66}\frac{\partial^2\overline{\beta}}{\partial x^2} + D_{22}\frac{\partial^2\overline{\beta}}{\partial y^2} - \kappa A_{44}\left(\overline{\beta} + \frac{\partial w}{\partial y}\right) = 0 \quad (6.58)$$

$$\kappa A_{55}\left(\frac{\partial\overline{\alpha}}{\partial x} + \frac{\partial^2 w}{\partial x^2}\right) + \kappa A_{44}\left(\frac{\partial\overline{\beta}}{\partial y} + \frac{\partial^2 w}{\partial y^2}\right) + p(x,y) = 0 \quad (6.59)$$

Dobyns [3] has employed the Navier approach to solving these equations for a composite plate simply supported on all four edges subjected to a lateral load, using the following functions:

$$w(x,y) = \sum_{m=1}^{\infty}\sum_{n=1}^{\infty} C_{mn}\sin\left(\frac{m\pi x}{a}\right)\sin\left(\frac{n\pi y}{b}\right) \quad (6.60)$$

$$\overline{\alpha}(x,y) = \sum_{m=1}^{\infty}\sum_{n=1}^{\infty} A_{mn}\cos\left(\frac{m\pi x}{a}\right)\sin\left(\frac{n\pi y}{b}\right) \quad (6.61)$$

$$\overline{\beta}(x,y) = \sum_{m=1}^{\infty} \sum_{n=1}^{\infty} B_{mn} \sin\left(\frac{m\pi x}{a}\right) \cos\left(\frac{n\pi y}{b}\right) \tag{6.62}$$

$$p(x,y) = \sum_{m=1}^{\infty} \sum_{n=1}^{\infty} q_{mn} \sin\left(\frac{m\pi x}{a}\right) \sin\left(\frac{n\pi y}{b}\right) \tag{6.63}$$

It is seen that Equations (6.60) through (6.63) satisfy the boundary conditions on all edges such that

$$w = 0, \frac{\partial \overline{\alpha}}{\partial x} = 0 \text{ on } x = 0, a \quad \text{and} \quad \frac{\partial \overline{\beta}}{\partial y} = 0 \text{ on } y = 0,b$$

Substituting these functions into the governing differential equations above results in the following:

$$\begin{bmatrix} L_{11} & L_{12} & L_{13} \\ L_{12} & L_{22} & L_{23} \\ L_{13} & L_{23} & L_{33} \end{bmatrix} \begin{Bmatrix} A_{mn} \\ B_{mn} \\ C_{mn} \end{Bmatrix} = \begin{Bmatrix} 0 \\ 0 \\ q_{mn} \end{Bmatrix} \tag{6.64}$$

if $\lambda_m \equiv m\pi/a$, $\lambda_n \equiv n\pi/b$, and q_{mn} is the lateral load coefficient, then the operators L_{ij} $(i,j = 1,2,3)$ are given by the following:

$$L_{11} = D_{11}\lambda_m^2 + D_{66}\lambda_n^2 + \kappa A_{55} \qquad L_{12} = (D_{12} + D_{66})\lambda_m\lambda_n$$

$$L_{13} = \kappa A_{55}\lambda_m \qquad L_{22} = D_{66}\lambda_m^2 + D_{22}\lambda_n^2 + \kappa A_{44}$$

$$L_{23} = \kappa A_{44}\lambda_n \qquad L_{33} = \kappa A_{55}\lambda_m^2 + \kappa A_{44}\lambda_n^2.$$

Solving Equation (6.64), one obtains

$$A_{mn} = \frac{(L_{12}L_{23} - L_{22}L_{13})q_{mn}}{\det} \tag{6.65}$$

$$B_{mn} = \frac{(L_{12}L_{13} - L_{11}L_{23})q_{mn}}{\det} \tag{6.66}$$

$$C_{mn} = \frac{(L_{11}L_{22} - L_{12}^2)q_{mn}}{\det} \tag{6.67}$$

where det is the determinant of the $[L]$ matrix in Equation (6.64).

Having solved the problem to obtain $\overline{\alpha}$, $\overline{\beta}$, and w, the curvatures $\kappa_x = (\partial\overline{\alpha}/\partial x)$, $\kappa_y = (\partial\overline{\beta}/\partial y)$, and $\kappa_{xy} = 1/2[(\partial\overline{\alpha}/\partial y) + (\partial\overline{\beta}/\partial y)]$ may be obtained. These then can be substituted back into Equations (3.20) through (3.22) to obtain the stress couples M_x, M_y, and M_{xy}. To then find the face stresses in a sandwich panel, $\sigma_{ij}(i,j = x,y)$, one employs Equation (6.41);

$$\sigma_{ij} = \pm\frac{M_{ij}}{t_f h_c} \tag{6.68}$$

where if $i = j$ one usually uses the single subscript.

For a laminated composite plate the bending stresses in each lamina may be obtained by

$$\begin{bmatrix} \sigma_x \\ \sigma_y \\ \sigma_{xy} \end{bmatrix}_k = \begin{bmatrix} \overline{Q}_{11} & \overline{Q}_{12} & 0 \\ \overline{Q}_{12} & \overline{Q}_{22} & 0 \\ 0 & 0 & \overline{Q}_{66} \end{bmatrix}_k \begin{bmatrix} \kappa_x \\ \kappa_y \\ \kappa_{xy} \end{bmatrix} z \tag{6.69}$$

Looking at the load $p(x,y)$ in Equations (6.59) and (6.63), if the lateral load $p(x,y)$ is distributed over the entire lateral surface, then the Euler coefficient, q_{mn}, is found to be

$$q_{mn} = \frac{4}{ab}\int_0^a\int_0^b p(x,y)\sin\left(\frac{m\pi x}{a}\right)\sin\left(\frac{n\pi y}{b}\right)dxdy \tag{6.70}$$

If that load is uniform then,

$$q_{mn} = \frac{4p_0}{mn\pi^2}(1 - \cos m\pi)(1 - \cos n\pi). \tag{6.71}$$

For a concentrated load located at $x = \xi$ and $y = \eta$,

$$q_{mn} = \frac{4P}{ab}\sin\left(\frac{m\pi\xi}{a}\right)\sin\left(\frac{n\pi\eta}{b}\right) \tag{6.72}$$

where P is the total load.

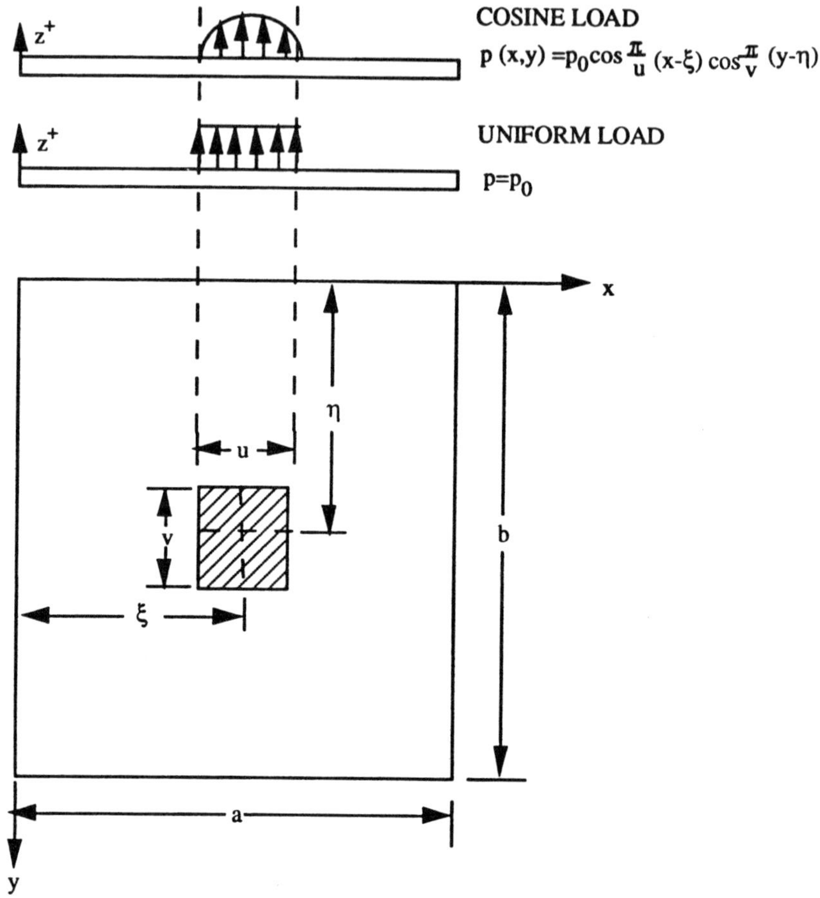

FIGURE 6.1. Loads over a rectangular area. (Reprinted from Reference [8].)

For loads over a rectangular area of side lengths u and v whose center is at ξ and η, as shown in Figure 6.1, q_{mn} is given as follows:

For $\dfrac{n}{b} \neq \dfrac{1}{v}$ and $\dfrac{m}{a} \neq \dfrac{1}{u}$, then

$$q_{mn} = \frac{4P \sin\left(\dfrac{m\pi\eta}{b}\right) \sin\left(\dfrac{m\pi\xi}{a}\right) \cos\left(\dfrac{n\pi v}{2b}\right) \cos\left(\dfrac{m\pi u}{2a}\right)}{abu^2v^2 \left(\dfrac{n}{b} - \dfrac{1}{v}\right)\left(\dfrac{n}{b} + \dfrac{1}{v}\right)\left(\dfrac{m}{a} - \dfrac{1}{u}\right)\left(\dfrac{m}{a} + \dfrac{1}{u}\right)} \tag{6.73}$$

where P is the total load. Note that when $n/b = 1/v$, $m/a = 1/u$, then $q_{mn} = 0$. Of course, any other lateral load can be characterized by the use of Equation (6.70).

6.9 EXACT SOLUTION

Pagano [4] provides an exact solution to the problem of a rectangular orthotropic sandwich plate subjected to a laterally-distributed load $p(x,y)$. His governing field equations are written in terms of displacement components and an assumed exponential-trigonometric solution is substituted into the equations. This provides a set of equations with six unknowns that are tractable. The solution unknowns are determined by the boundary conditions and the interface-continuity conditions.

6.10 OTHER CONSIDERATIONS

Of course, in all practical engineering problems there are special problems that require detailed treatment, consideration, and analysis. One of these is the stress singularities that occur in the sandwich core due to the inclusion of inserts. Such problems have been studied in detail at Aalborg in Denmark. Kristensen [5], Thomsen et al. [6], and others have written extensively on this subject, which is beyond the scope of this text.

6.11 PROBLEMS

6.1 A rectangular sandwich hull plate comprises a part of the flat bottom of a ship. Consider this as a rectangular plate under a uniform lateral distributed loading, p_0, from the water pressure, and clamped along all edges. The steel sandwich faces are 0.125″ thick, and the foam-core depth $h_c = 2.5''$. The panel is 4 ft. wide and 6 ft. long, and the ship draws 14 ft. of water, where seawater weighs 64 lb/ft^3. The steel face properties are: $E = 30 \times 10^6$ psi, $\nu = 0.3$, $\sigma_{\text{all}} = \pm 20{,}000$ psi. What is maximum stress in the sandwich panel? What is the maximum deflection? Use Table 6.4.

6.2 A rectangular wing panel component, 16″ x 12″ is made of a foam-core sandwich with identical aluminum faces ($E = 10 \times 10^6$ psi, $\nu = 0.3$, $\sigma_{\text{all}} = \pm 20{,}000$ psi). The panel is considered to be simply supported on all four edges. With a face thickness of 0.064 in., and a core depth of 1.25 in., if a maximum design pressure of 20 psi is used.

a. What and where is the maximum stress?

b. What and where is the maximum deflection?
Use Table 6.1.

6.3 A flat portion of a wind tunnel measuring 30″ × 54″ is subjected to a uniform pressure of 20 psi. If the sandwich faces are steel [$E = 30 \times 10^6$ psi, $\nu = 0.3$, $\sigma_{\text{all}} = 30{,}000$ psi] and the foam-core depth is 1.0″, what face thicknesses are needed if all four sides of the panel are
a. Simply supported?
b. Clamped?
The panel cannot be overstressed and the maximum deflection cannot exceed 0.125″. Use Tables 6.1 and 6.4.

6.4 A portion of the cover of a hovercraft is approximated by a rectangular sandwich plate measuring 8 ft. by 4 ft. in planform, and is simply supported on all four edges. It is subjected to a uniform lateral pressure of 20 psi. Assume that in the design $h_c = 20t_f$. The choices for face material are aluminum ($E = 10 \times 10^6$ psi, $v = 0.3$, $\sigma_{\text{all}} = 30{,}000$ psi, $\rho = 0.1$ lb/in.^{3}) or steel ($E = 30 \times 10^6$ psi, $v = 0.3$, $\sigma_{\text{all}} = 60{,}000$ psi, $\rho = 0.283$ lb/in^3). Using Table 6.1 for each material:
a. How thick must the faces be in order to not be overstressed?
b. Which panel has the larger deflection?
c. Assuming $\rho_c = 15$ lb/ft.^{3} for both, which panel will weigh less?

6.5 A designer must design a rectangular plate cover over an opening measuring 9 ft. by 3 ft. The maximum design-load uniform pressure is 10 psi. If steel-faced sandwich is used ($E = 30 \times 10^6$ psi, $\nu = 0.3$, $\sigma_{\text{all}} = 35{,}000$ psi and $\rho_f = 0.283$ lb/in.^{3}, $\rho_c = 15$ lb/ft^3) and assume $h_c = 20t_f$, using Table 6.1 through 6.4, what will the panel weigh if all four edges are
a. simply-supported?
b. clamped?
c. two opposite edges clamped, two edges simply-supported?
d. reverse the boundary conditions of c above.
e. one edge clamped, three edges simply-supported?
f. one edge clamped (an edge perpendicular to the one in e above), the other three edges simply-supported?
g. Rank the results of the above from lightest to heaviest panel.

6.6 In Section 6.6, Equation (6.51), it is seen that the factor of merit for materials selection to attain least weight is given by

$$FM = \frac{\sigma_{\max}{}^2}{\rho_f E_f}$$

ignoring Poisson's ratio effects in 6.51. Consider using each of the materials in Table 2.1 to make a symmetric, isotropic sandwich to be subjected to a uniform lateral load p_0. Rank them in descending order of Factor of Merit.

6.7 Consider a sandwich panel of dimensions $0 \leq x \leq a$, $0 \leq y \leq b$, if overall thickness h, with E-glass/epoxy faces and a foam core, simply supported on all four edges, subjected to a lateral load

$$p(x,y) = Cy, \text{ where } C \text{ is a constant.}$$

Determine explicitly the Euler coefficient q_{mn} given in Equation (6.70).

6.12 REFERENCES

1. Levy, M., "Sur L'equilibrie Elastique d'une Plaque Rectangulaire," *Compt. Rend* 129, pp. 535–539, 1899.
2. Timoshenko, S. and A. Woinowski-Krieger, *Theory of Plates and Shells*, McGraw-Hill Book Co., Inc., New York, 2nd Edition, 1959.
3. Dobyns, A. L., "The Analysis of Simply-Supported Orthotropic Plates Subjected to Static and Dynamic Loads," *AIAA Journal*, pp. 642–650, May 1981.
4. Pagano, N. J., "Exact Solutions for Rectangular Bidirectional Composites and Sandwich Plates," *Journal of Composite Materials*, Vol. 4, pp. 20–34, 1970.
5. Kristensen, K. B., "Stress Singularities in Sandwich Structures," *Sandwich Construction 3—Proceedings of the Third International Conference on Sandwich Construction*, Southampton, Great Britain, Editor, H. G. Allen, EMAS Publications, United Kingdom, 1996, pp. 185–194.
6. Thomsen, O. T., S. C. Liu and K. M. Hsaio, "Analysis of Local Bending Effects in Sandwich Plates with Orthopedic Face Layers Subjected to Localized Loads," *Composite Structures*, Vol. 25, No. 1, 1993, p. 571.
7. Vinson, J. R. and T-W. Chou, *Composite Materials and Their Use in Structures*. Applied Science Publishers, London, 1975.
8. Vinson, J. R. and R. L. Sierakowski, *The Behavior of Structures Composed of Composite Materials*, Martinus-Nijhoff Publishers (now Kluwer Academic Publishers) Dordrecht, The Netherlands, 1986.
9. Vinson, J. R., *The Behavior of Thin Walled Structures: Beams, Plates and Shells*, Kluwer Academic Publishers, Dordrecht, 1989.

CHAPTER 7

Dynamic Effects on Sandwich Panels

7.1 INTRODUCTION

Seldom in real life is a structure subjected only to static loads. More often products and structures are subjected to vehicular, impact, crash, earthquake, handling, or fabricating dynamic loads. In the linear-elastic range, dynamic effects can be divided into two categories: natural vibrations and forced vibrations, and the latter can be further subdivided into one-time events (an impact) or receiving loads (such as cyclic loading). These will be discussed in turn.

Physically every elastic continuous body has an infinity of natural frequencies, only a few of which are of practical significance. When a structure is excited cyclically at a natural frequency, it takes little input energy for the amplitude to grow until one of four things happens:

(1) The amplitude of vibration grows until the ultimate strength of a brittle material is exceeded and the structure fails.

(2) Portions of the structure exceed the yield strength, plastically deform and the behavior changes drastically.

(3) The amplitude grows until nonlinear effects become significant, and there is no natural frequency.

(4) Due to damping or other mechanism the amplitude is limited, but as the natural vibration continues, fatigue failures may occur.

Physically, when a structure is undergoing a natural vibration the sum of the potential energy and kinetic energy remains constant if no damping is present. This can be termed a conservative system. However, the energy is compartmentalized, i.e., if a structure is truly vibrating in one mode of vibration it will not change and commence vibrating in some other mode at some other natural frequency. In a complex structure if two components have vibrational natural frequencies that are identical, then when one component is excited, the other component will also be excited. It is for this very reason that duplicative natural frequencies are to be avoided.

Mathematically, natural vibration problems are called eigenvalue problems. They are represented by homogeneous equations, for which nontrivial solutions only occur at certain characteristic (*eigen*, in German) values of a parameter, from which the natural frequencies are determined. In a natural vibration the displacement field comprises a normal mode. At any two different natural frequencies the corresponding normal modes are orthogonal to each other (hence the compartmentalization of the energy). The normal modes comprise the solutions to the homogeneous governing differential equations, which are now zero only at the eigenvalues for those equations and boundary conditions.

If there is a forcing function, then the particular solution for the specific forcing function (which can be cyclical or a one time dynamic impact load) is added onto the homogeneous solution, which involves the natural frequencies and mode shapes. Physically, any dynamic load excites each and every one of the normal modes and corresponding natural frequencies. Usually, only a relatively few are large enough to be of concern. The largest amplitude of response will be in those mode shapes whose natural frequency is closest to the oscillatory component of the forcing functions.

When there are no natural frequencies close to the oscillatory portion of the dynamic load, then the structure will respond in deflection and stresses corresponding only to the magnitude and spatial distribution of the load. Such a condition results in solving the worst-case static problem in which the biggest load at some time, t, is applied. This is termed a quasistatic case. However, if the dynamic load oscillatory component is close to one or more natural frequencies, then the structural response can be much larger than the value obtained from a quasistatic calculation, and that increase can be represented by a dynamic load factor.

All of the above are discussed in much more detail in texts treating vibrations and in papers covered in the review by Noor, Burton, and Bert [34].

Historically, the linear equations of motion and the vibration of a sandwich plate were presented by Yu [1] in 1959. A text by Yu [2] in 1996 discusses both the linear as well as the nonlinear dynamics for sandwich plates, and most recently Yu [35] has included the effects of thin piezoelectric face layers.

In what follows natural frequencies are treated first, then forced linear vibrations, and finally nonlinear large-amplitude vibrations are discussed.

7.2 NATURAL FLEXURAL VIBRATIONS OF SANDWICH PLATES: CLASSICAL THEORY

Consider a rectangular sandwich plate that is midplane symmetric such that $B_{ij} = 0$. If this plate is isotropic i.e., $D_1 = D_2 = D_3 = D$, then the governing differential equation is given by Equation (3.30) for the classical theory (i.e., no transverse-shear deformation) and repeated here as

$$D\left[\frac{\partial^4 w}{\partial x^4} + 2\frac{\partial^4 w}{\partial x^2 \partial y^2} + \frac{\partial^4 w}{\partial y^4}\right] = p(x,y) \tag{7.1}$$

For dynamic loads, using D'Alembert's Principle, the equation is written as

$$D\left[\frac{\partial^4 w}{\partial x^4} + 2\frac{\partial^4 w}{\partial x^2 \partial y^2} + \frac{\partial^4 w}{\partial y^4}\right] = p(x,y,t) - \rho_m h\frac{\partial^2 w}{\partial t^2} \tag{7.2}$$

where the last term is the mass per unit planform area times the acceleration [see Equation (7.12)]. For natural vibrations the load (the forcing function) is zero and the resulting homogeneous equation is

$$D\left[\frac{\partial^2 w}{\partial x^4} + 2\frac{\partial^4 w}{\partial x^2 \partial y^2} + \frac{\partial^4 w}{\partial y^4}\right] + \rho_m h\frac{\partial^2 w}{\partial t^2} = 0 \tag{7.3}$$

This equation yields all of the natural frequencies for plates with any boundary conditions. For the easiest case, consider the sandwich plate to be simply supported on all four edges. In that case, one may guess the following form for the deflection function because it satisfies all boundary conditions:

$$w(x,y,t) = A_{mn}\sin\left(\frac{m\pi x}{a}\right)\sin\left(\frac{n\pi y}{b}\right)\cos\left(\omega_{mn}t\right) \tag{7.4}$$

where ω_{mn} is the natural circular frequency in radians per unit time.

Substituting Equation (7.4) into (7.3) one sees that the nontrivial solution exists only when

$$\omega_{mn} = \left[\frac{\pi^4 D}{\rho_m h}\left(\frac{m^2}{a^2} + \frac{n^2}{b^2}\right)^2\right]^{1/2} \tag{7.5}$$

In the above m and n are integers only. The lowest natural frequency occurs when $m = n = 1$, and this lowest frequency is always termed the fundamental natural frequency. To obtain the natural frequency in terms of cycles per second (Hz)

$$f_{mn} = \frac{\omega_{mn}}{2\pi} \tag{7.6}$$

Equation (7.5) is more accurate at low values of m and n, but as these integers increase the calculated values increasingly exceed measured values. This is because transverse shear deformation effects increase with increased values the ratio of sandwich plate thickness to the wavelength of the *m*-*n*th mode of vibration.

If the sandwich plate is specially orthotropic and midplane symmetric, then for the natural vibration problem the governing differential equation is written as in Equation (3.30),

$$D_1 \frac{\partial^4 w}{\partial x^4} + 2D_3 \frac{\partial^4 w}{\partial x^2 \partial y^2} + D_2 \frac{\partial^4 w}{\partial y^4} = -\rho_m h \frac{\partial^2 w}{\partial t^2} \tag{7.7}$$

Again, if all four edges are simply supported then the mode shapes are given by Equation (7.4) with the result that the natural circular frequency in radians per second is given by

$$\omega_{mn} = \frac{\pi^2}{\sqrt{\rho_m h}} \left[D_1 \left(\frac{m}{a}\right)^4 + 2D_3 \left(\frac{m}{a}\right)^2 \left(\frac{n}{b}\right)^2 + D_2 \left(\frac{n}{b}\right)^4 \right]^{1/2} \tag{7.8}$$

Again, the natural frequency in Hz is given by Equation (7.6).

Keep in mind that for Equation (7.8) and other equations for the frequencies for natural vibrations to accurately describe the motion, the maximum deflection must be a fraction of the plate thickness since the theory is linear. Above that level of motion, nonlinear effects become increasingly significant. Meyer-Piening [3] states that for isotropic plates the maximum deflection should not be greater than one half of the sandwich plate thickness for the linear behavior to accurately describe the motion. Experimentally, Bau-Madsen et al. [4] found a beginning of divergence between a linear relationship of deflection/load and the measured value at 0.4 (i.e., when the panel displacement amplitude is 0.4 of the sandwich panel thickness).

Ramachandra and Meyer-Piening [5] investigated frequencies of sandwich panels in contact with water and simply supported on all four edges. The water reduces the natural frequencies dramatically, and this paper should be studied to account for structures vibrating in water.

7.3 NATURAL FLEXURAL VIBRATIONS OF SANDWICH PLATES INCLUDING TRANSVERSE SHEAR DEFORMATION EFFECTS

The governing partial differential equations for a sandwich plate or panel that is specially orthotropic and midplane symmetric subjected to a lateral static load $p(x,y)$ are given in Equations (3.47) through (3.49). If one now wishes to find the natural frequencies of this sandwich composite plate, that has midsurface symmetry ($B_{ij} = 0$), no other couplings $(\ )_{16}, = (\ )_{26} = (\ )_{45} = 0$, but includes transverse shear deformation, $\varepsilon_{xz} \neq 0$, $\varepsilon_{yz} \neq 0$, then one sets $p(x,y) = 0$ in Equation (3.49), but adds $-\rho h(\partial^2 w/\partial t^2)$ to the right-hand side. In addition, because $\overline{\alpha}$ and $\overline{\beta}$ are both dependent variables that are independent of w, there will be an oscillatory motion of the lineal element across the plate thickness about the midsurface of the plate. This results in the last term on the left-hand side of Equations (3.47) and (3.48) becoming $I(\partial^2\overline{\alpha}/\partial t^2)$ and $I(\partial^2\overline{\beta}/\partial t^2)$ respectively, as shown below:

$$D_{11}\frac{\partial^2\overline{\alpha}}{\partial x^2} + D_{66}\frac{\partial^2\overline{\alpha}}{\partial y^2} + (D_{12} + D_{66})\frac{\partial^2\overline{\beta}}{\partial x\partial y} - \kappa A_{55}\left(\overline{\alpha} + \frac{\partial w}{\partial x}\right) - I\frac{\partial^2\overline{\alpha}}{\partial t^2} = 0 \quad (7.9)$$

$$(D_{12} + D_{66})\frac{\partial^2\overline{\alpha}}{\partial x\partial y} + D_{66}\frac{\partial^2\overline{\beta}}{\partial x^2} + D_{22}\frac{\partial^2\overline{\beta}}{\partial y^2} - \kappa A_{44}\left(\overline{\beta} + \frac{\partial w}{\partial y}\right) - I\frac{\partial^2\overline{\beta}}{\partial t^2} = 0 \quad (7.10)$$

$$\kappa A_{55}\left(\frac{\partial\overline{\alpha}}{\partial x} + \frac{\partial^2 w}{\partial x^2}\right) + \kappa A_{44}\left(\frac{\partial\overline{\beta}}{\partial y} + \frac{\partial^2 w}{\partial y^2}\right) - \rho_m h\frac{\partial^2 w}{\partial t^2} = 0 \quad (7.11)$$

where ρ_m is given by

$$\rho_m = \frac{1}{h}\sum_{k=1}^{N}\rho_k(h_k - h_{k-1}) \quad (7.12)$$

where ρ_k is the *mass* density of the kth lamina, and here I is

$$I = \sum_{k=1}^{N}\rho_m\, z_k^2\, dz_k \quad \text{or} \quad I = \rho_m t_f h_c^2/2 \quad (7.13)$$

for the usual foam, solid or honeycomb core sandwich panel. In Equations (7.9) through (7.11) the κ's are transverse shear deformation parameters, as discussed earlier.

Similar to the Navier procedure used in Section 7.2, and following Dobyns [36] for the simply supported plate let

$$w(x,y,t) = \sum_{m=1}^{\infty}\sum_{n=1}^{\infty} C'_{mn}\sin\left(\frac{m\pi x}{a}\right)\sin\left(\frac{n\pi y}{n}\right)e^{i\omega t} \quad (7.14)$$

$$\overline{\alpha}(x,y,t) = \sum_{m=1}^{\infty}\sum_{n=1}^{\infty} A'_{mn}\cos\left(\frac{m\pi x}{a}\right)\sin\left(\frac{n\pi y}{b}\right)e^{i\omega t} \quad (7.15)$$

$$\overline{\beta}(x,y,t) = \sum_{m=1}^{\infty}\sum_{n=1}^{\infty} B'_{mn} \sin\left(\frac{m\pi x}{a}\right) \cos\left(\frac{n\pi y}{b}\right) e^{i\omega t} \tag{7.16}$$

Substituting these into the dynamic governing equations above results in a set of homogeneous equations that can be solved for the natural frequencies of vibration

$$\begin{bmatrix} L'_{11} & L_{12} & L_{13} \\ L_{12} & L'_{22} & L_{23} \\ L_{13} & L_{23} & L'_{33} \end{bmatrix} \begin{Bmatrix} A'_{mn} \\ B'_{mn} \\ C'_{mn} \end{Bmatrix} = \begin{Bmatrix} 0 \\ 0 \\ 0 \end{Bmatrix} \tag{7.17}$$

where the unprimed quantities were defined in Equation (6.64) and

$$L'_{11} = L_{11} - \frac{\rho_m h^3}{12} \omega_{mn}^2$$

$$L'_{22} = L_{22} - \frac{\rho_m h^3}{12\ \omega_{mn}^2}$$

$$L'_{33} = L_{33} - \rho_m h \omega_{mn}^2$$

As shown before [6,37], three eigenvalues (natural frequencies) result from solving (7.17) for each value of m and n. However, two of the frequencies are significantly higher than the other because they are associated with the rotatory inertia terms, which are the last terms on the left-hand sides of Equations (7.9) and (7.10) and are very seldom important in structural responses. If they are neglected then $L'_{11} = L_{11}$ and $L'_{22} = L_{22}$ above, and the square of the remaining natural frequency can be easily found to be

$$\omega_{mn}^2 = [QL_{33} + 2L_{12}L_{23}L_{13} - L_{22}L_{13}^2 - L_{11}L_{23}^2]/\rho_m hQ \tag{7.18}$$

where, here, $Q = L_{11}L_{22} - L_{12}^2$. Also,

$$A'_{mn} = \frac{L_{12}L_{23} - L_{22}L_{13}}{Q} C'_{mn}$$

$$B'_{mn} = \frac{L_{12}L_{13} - L_{11}L_{23}}{Q} C'_{mn}$$

If transverse-shear deformation effects were neglected, Equation (7.18) would be identical to the square of Equation (7.8).

7.4 FORCED-VIBRATION RESPONSE OF A SANDWICH PLATE SUBJECTED TO A DYNAMIC LATERAL LOAD

Dobyns [36] then goes on to develop the solutions for the simply-supported laminated composite plate subjected to a dynamic lateral load $p(x,y,t)$, neglecting the rotatory inertia terms discussed above, utilizing a convolution integral $P(t)$ as seen below. Incidentally, the convolution integral is also known as the superposition integral and the Duhamel integral.

His solutions are also applicable to specially orthotropic sandwich plates using the proper stiffness matrix quantities. The solutions to Equations (7.9) through (7.11), modified to include a dynamic distributed lateral load $p(x,y,t)$ and neglecting the rotatory inertia terms are given by

$$w(x,y,t) = \frac{1}{\rho_m h} \sum_{m=1}^{\infty} \sum_{n=1}^{\infty} \left(\frac{q_{mn}}{\omega_{mn}}\right) \sin\left(\frac{m\pi x}{a}\right) \sin\left(\frac{n\pi y}{b}\right) P(t) \tag{7.19}$$

$$\overline{\alpha}(x,y,t) = \frac{1}{\rho_m h} \sum_{m=1}^{\infty} \sum_{n=1}^{\infty} \left(\frac{q_{mn}}{\omega_{mn}}\right) \frac{(L_{12}L_{23} - L_{22}L_{13})}{Q} \cos\left(\frac{m\pi x}{a}\right) \sin\left(\frac{n\pi y}{b}\right) P(t) \tag{7.20}$$

$$\overline{\beta}(x,y,t) = \frac{1}{\rho_m h} \sum_{m=1}^{\infty} \sum_{n=1}^{\infty} \left(\frac{q_{mn}}{\omega_{mn}}\right) \frac{(L_{12}L_{13} - L_{11}L_{23})}{Q} \sin\left(\frac{m\pi x}{a}\right) \cos\left(\frac{n\pi y}{b}\right) P(t) \tag{7.21}$$

where

$$P(t) = \int_0^t F(\tau) \sin\left[\omega_{mn}(t - \tau)\right] d\tau \tag{7.22}$$

and q_{mn} is the coefficient of the lateral-load function expanded in series form [see Equation (6.70)].

So for a given lateral distributed load $p(x,y,t)$, if a solution of the form given by Equations (7.19) through (7.21) is applicable, then the curvatures κ_x, κ_y, and κ_{xy} for the plate can be found from Equation (2.50), and the stresses in the faces of the sandwich plate found from Equation (6.68). The function

$P(t)$ has been solved analytically for several representative forcing functions shown in Figure 7.1.

For the sine pulse, the forcing function $F(t)$ and the convolution integral $P(t)$ are

$$F(t) = F_0 \sin(\pi t/t_1) \qquad 0 \le t \le t_1$$

$$F(t) = 0 \qquad t > t_1$$

$$P(t) = \int_0^t F(t) \sin[\omega_{mn}(t - \tau)]d\tau$$

$$= \frac{F_0 t_1 [\pi \sin \omega_{mn} t - \omega_{mn} t_1 \sin(\pi t/t_1)]}{(\pi^2 - t_1^2 \omega_{mn}^2)} \text{ for } 0 \le t \le t_1 \tag{7.23a}$$

$$P(t) = \frac{F_0 \pi t_1 [\sin \omega_{mn} t + \sin \omega_{mn}(t - t_1)]}{\pi^2 - t_1^2 \omega_{mn}^2} \text{ for } t \ge t_1 \tag{7.23b}$$

For the stepped pulse the forcing functions $F(t)$ and the convolution integral $P(t)$ are given

$$F(t) = F_0 \qquad 0 \le t \le t_1$$

$$F(t) = 0 \qquad t > t_1$$

$$P(t) = \int_0^t F(\tau) \sin[\omega_{mn}(t - \tau)]d\tau = \frac{F_0}{\omega_{mn}}[1 - \cos(\omega_{mn} t)]$$

$$\text{for } 0 \le t \le t_1 \tag{7.24}$$

$$P(t) = \frac{F_0}{\omega_{mn}}\{\cos[\omega_{mn}(t - t_1)] - \cos \omega_{mn} t\} \text{ for } t > t_1 \tag{7.25}$$

For a triangular pulse:

$$F(t) = F_0(1 - t/t_1) \qquad 0 \le t \le t_1$$

$$F(t) = 0 \qquad t > t_1$$

$$P(t) = \int_0^t F(\tau) \sin[\omega_{mn}(t - \tau)]d\tau$$

$$= \frac{F_0}{\omega_{mn}}\left[1 - \cos(\omega_{mn} t) + \frac{1}{\omega_{mn} t_1}\sin(\omega_{mn} t) - \frac{t}{t_1}\right] \text{ for } 0 \le t \le t_1 \tag{7.26}$$

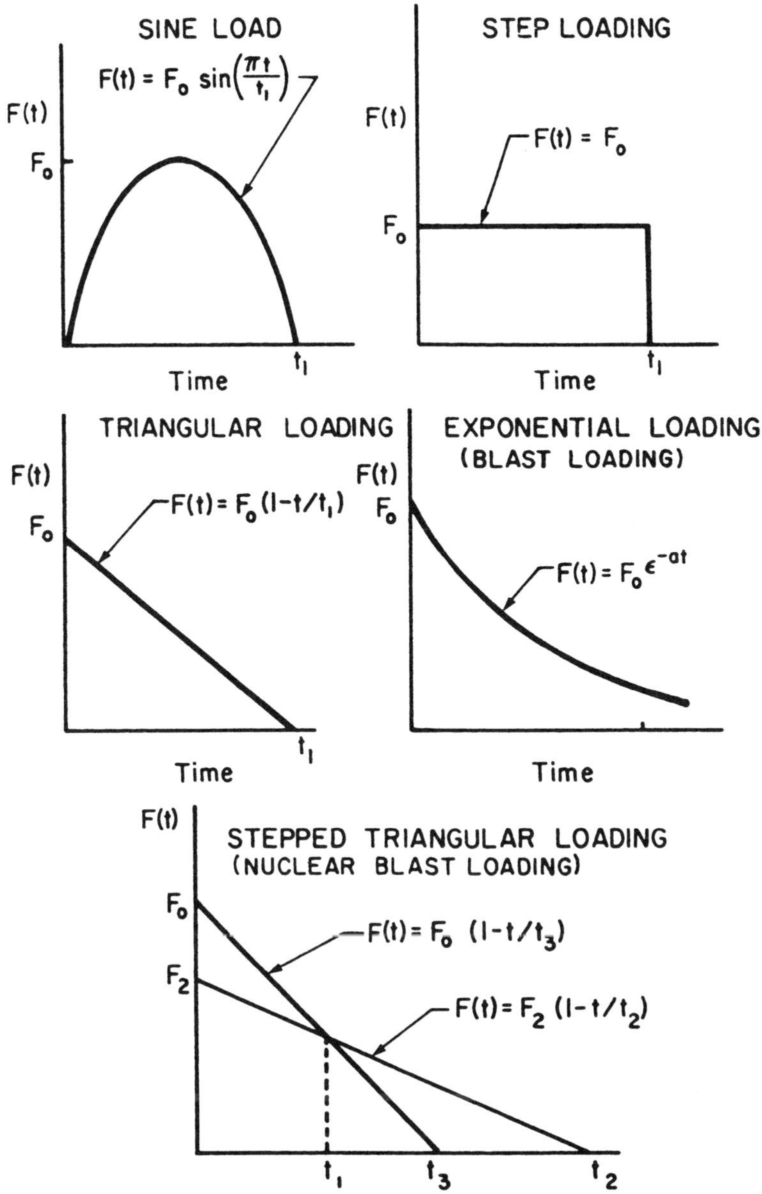

FIGURE 7.1. Representative forcing functions. (Reprinted from Reference [37].)

$$P(t) = F_0 \left\{ -\frac{1}{\omega_{mn}} \cos(\omega_{mn} t) + \frac{2}{\omega_{mn}^2 t_1} \cos\left[\omega_{mn}\left(t - \frac{t_1}{2}\right) \sin \omega_{mn}\left(\frac{t_1}{2}\right)\right\} \text{ for } t > t_1 \tag{7.27}$$

The stepped triangular pulse of Figure 7.1 simulates a nuclear-blast loading [36] where the pressure pulse consists of a long-duration phase of several seconds due to the overpressure and a short-duration phase of a few milliseconds due to the shock wave reflection. The short-duration phase has twice the pressure of the long-duration phase.

$$F(t) = F_0 (1 - t/t_3) \qquad 0 \leq t \leq t_1$$

$$F(t) = F_2 (1 - t/t_2) \qquad t \leq t \leq t_2$$

$$F(t) = 0 \qquad t > t_2$$

$$P(t) = \int_0^t F(\tau) \sin[\omega_{mn}(t - \tau)] d\tau$$

$$= \frac{F_0}{\omega_{mn}}\left[1 - \cos(\omega_{mn} t) + \frac{1}{\omega_{mn} t_3} \sin(\omega_{mn} t) - \frac{t}{t_3}\right] \text{ for } 0 < t \leq t_1 \tag{7.28}$$

$$P(t) = F_0 \left\{ \frac{1}{\omega_{mn}}\left(1 - \frac{t}{t_3}\right) \cos \omega_{mn}(t - t_1) - \frac{1}{\omega_{mn}} \cos(\omega_{mn} t) - \frac{1}{\omega_{mn}^2 t_3} \sin[\omega_{mn}(t - t_1)] + \frac{1}{\omega_{mn}^2 t_3} \sin(\omega_{mn} t)\right\}$$
$$+ F_2 \left\{ \frac{1}{\omega_{mn}}(1 - t/t_2) - \frac{1}{\omega_{mn}}(1 - t/t_2) \cos[\omega_{mn}(t - t_1)] + \frac{1}{\omega_{mn}^2 t_2} \sin[\omega_{mn}(t - t_1)]\right\} \text{ for } t_1 \leq t \leq t_2 \tag{7.29}$$

$$P(t) = F_0 \left[\frac{1}{\omega_{mn}} (1 - t_1/t_3) \cos [\omega_{mn} (t - t_1)] - \frac{1}{\omega_{mn}} \cos (\omega_{mn} t) + \frac{1}{\omega_{mn}^2 t_3} \{\sin (\omega_{mn} t) - \sin [\omega_{mn} (t - t_1)]\} \right]$$

$$+ F_2 \left[\frac{1}{\omega_{mn}} (t_1/t_2 - 1) \cos [\omega_{mn} (t - t_1)] - \frac{1}{\omega_{mn}^2 t_2} \{\sin [\omega_{mn} (t - t_2)] - \sin [\omega_{mn} (t - t_1)]\} \right] \quad \text{for } t > t_2 \qquad (7.30)$$

Lastly, the exponential pulse of Figure 7.1 may be used to simulate a high explosive (nonnuclear) blast loading when the decay parameter α is empirically determined to fit the pressure pulse of the actual blast. The equations are

$$F(t) = F_0 e^{-\alpha t}$$

$$P(t) = \int_0^t F(\tau) \sin [\omega_{mn} (t - \tau)] d\tau$$

$$= \frac{F_0[\omega_{mn} e^{-\alpha t} + \alpha \sin (\omega_{mn} t) - \omega_{mn} \cos \omega_{mn} t]}{(\alpha^2 + \omega_{mn}^2)} \quad \text{for } t > 0 \qquad (7.31)$$

It should be noted that although the forcing-function equations given above are used herein to investigate the dynamic response of a sandwich plate, these equations are useful for many other purposes.

Dobyns [36] concludes that the equations presented in this section allow one to analyze a composite material panel subjected to dynamic loads with only a little more effort than is required for the same panel subjected to static loads. He developed a Basic language program for a desktop computer to study the various dynamic loadings discussed herein. Hence, he concludes that one does not have to rely upon approximate design curves or arbitrary dynamic-magnification load factors.

With the information presented to this point, the necessary equations for the study of a composite plate without various couplings, but that include transverse shear deformation, have been developed. The plate may be subjected to various static lateral loads and a variety of dynamic loads. These loads and solutions can be used singly—or superimposed—to describe a complex dynamic input. (Those same load functions can be used for beams, shells, and many other structural configurations.) With the solutions for $\overline{\alpha}(x,y)$, $\overline{\beta}(x,y)$, and w, maximum deflections and stresses can be determined for deflection (stiffness)-critical and strength-critical structures.

Moyer et al. [7] have analyzed the dynamic response of sandwich panels, such as ship hulls, subjected to noncontact underwater explosions (UNDEX). It is interesting to note that a good indicator of panel damage is the natural-frequency response of the sandwich panel after the explosion. No frequency reduction indicates no damage; reduced natural frequencies indicates damage.

Williamson and Lagace [8] performed an extensive study of damage to sandwich panels and compared indentations caused by static and impact loads.

Olsson [9] studied the effects of impact on sandwich panels with laminated face sheets and found that impact loads result in strength reductions of 50%. These studies therefore indicate a dynamic load factor of 2.

7.5 DYNAMIC RESPONSE OF SANDWICH PLATES SUBJECTED TO LOCALIZED LOADS

One special type of lateral loading that can be deleterious to sandwich construction is a localized lateral impact to one of the faces. The indentation resulting from such an impact often negates the global approximation of constant thickness, constant flexural stiffness, and a fairly unstressed core. Here, the struck face may deflect considerably more than the other face, and the core therefore becomes compressed and sheared locally.

As Thomsen [38] states, sandwich panels are notoriously sensitive to failure by strongly localized point and line loads, resulting in premature failures. For instance Sanker [10] states that where a rectangular sandwich plate is subjected to a central concentrated lateral load, compression of the core can be 60% of the deflection of the loaded face, and thus it is incorrect to assume that compression of the core can be neglected where there are localized loads. Incidentally, Sankar finds that the core compression does not change considerably with the size of the sandwich plate. Various researchers, including Ferri and Sankar [11], have found that low-velocity impacts can be approximated by considering a static indentation.

Analytically, the Theorem of Minimum Potential Energy can be used in which the strain energy in the struck face, the core, and the other face are written out explicitly. Here the two faces, whether identical or not, are modeled as classical laminated plates. The core must be modeled as a three-dimensional orthotropic or isotropic solid. However, one can probably neglect in-plane normal and shear-stress terms. Then trial displacement functions are assumed in which the displacements are matched between each face and the core, and solutions obtained.

Some recent research in this area includes that of Nemes and Simmonds [12], and Nilsen, Moan, and Gustafson [13]. Wu, Weeks, and Sun [14] and Wu and Sun [15] have also studied the effects of localized dynamic loads on various sandwich structures. Recently, Sun, Hasebe, and Hua [39] have found that from the impact tests they conducted, the Rohacell foam-reinforced web-core construction decreases penetration damage, even though surface damage may increase compared to other sandwich architectures.

7.6 LARGE-AMPLITUDE NONLINEAR OSCILLATIONS OF SANDWICH PLATES SIMPLY-SUPPORTED ON ALL EDGES

Sandwich construction is being used increasingly for plate and shell constructions in structures such as space vehicles, deep submergence vehicles, pressure vessels, and many others. These structures, however, are often subjected to severe operational conditions. Thus, many classical methods of analysis are inadequate since the plate deflections are no longer small compared to the plate thickness. For sandwich plates composed of composite materials, orthotropy must be accounted for, and in addition, the effect of transverse shear deformation must be included.

To analyze this problem, Reissner's Variational Principle [40] (discussed in Section 5.4) is used along with Hamilton's Principle (which is discussed in Section 5.6). A simply-supported plate is treated in detail as an example. The effect of transverse-shear deformation will only be considered in the lateral motion. The reduction of results possible in the special case of isotropic, as well as transversely isotropic materials, is indicated. The solutions obtained herein reduce readily to those of the small amplitude case when the amplitudes of vibrations tend to zero.

Consider the free vibration of a rectangular sandwich plate composed of an orthotropic material, with total thickness h and edge lengths a and b. Let (x,y,z) be the right-handed Cartesian coordinates of a point of the plate before deformation. The well-known tensor expression of components of strain in the Langrangian form is the following [6]

$$2\varepsilon_{jk} = u_{j,k} + u_{k,j} + u_{i,j}u_{i,k} \tag{7.32}$$

where commas denote partial differentiation with respect to the subscripted symbol following the comma, and $i,j,k = x,y,z$.

It is now assumed that the displacement components are of the usual form, as used previously

$$
\begin{aligned}
u &= u_0(x,y,t) + z\overline{\alpha}(x,y,t) \\
v &= v_0(x,y,t) + z\overline{\beta}(x,y,t) \\
w &= w(x,y,t)
\end{aligned}
\tag{7.33}
$$

the subscript "0" is used to indicate displacements of the middle surface. It is noted that the in-plane displacements take the usual form of being the sum of a translation and the rotation of a lineal element through the thickness.

The explicit forms of the strain-displacement relations are obtained by substituting Equations (7.33) into (7.32) producing the nonlinear strain-displacement relations where in the nonlinear term, only derivatives of w are retained because in plates such derivatives are much larger than the derivatives of the in-plane displacements u_0 and v_0.

$$
\begin{aligned}
\varepsilon_x &= \frac{\partial u_0}{\partial x} + z\frac{\partial \overline{\alpha}}{\partial x} + \frac{1}{2}\left(\frac{\partial w}{\partial x}\right)^2 \\
\varepsilon_y &= \frac{\partial v_0}{\partial y} + z\frac{\partial \overline{\beta}}{\partial y} + \frac{1}{2}\left(\frac{\partial w}{\partial y}\right)^2 \\
\varepsilon_{xy} &= \frac{1}{2}\left(\frac{\partial u_0}{\partial y} + \frac{\partial v_0}{\partial x} + z\frac{\partial \overline{\alpha}}{\partial y} + z\frac{\partial \overline{\beta}}{\partial x}\right) + \frac{1}{2}\left(\frac{\partial w}{\partial x}\right)\left(\frac{\partial w}{\partial y}\right) \\
\varepsilon_{yz} &= \frac{1}{2}\left(\frac{\partial w}{\partial y} + \overline{\beta}\right);\ \varepsilon_{xz} = \frac{1}{2}\left(\frac{\partial w}{\partial x} + \overline{\alpha}\right);\ \varepsilon_z = 0
\end{aligned}
\tag{7.34}
$$

Here, ε_z is zero in accordance with most plate theories and $\varepsilon_i = \varepsilon_{ii}$ for convenience.

To simplify the derivation, consider the plate to be a single-ply specially orthotropic material. Later, it is shown that the results apply to sandwich panels as well as to laminated composite plates. Therefore, the stress-strain relations are written of the form, identical to Equation (2.9) where, because $\theta = 0$, the 1, 2, 3 subscripts can be written as x, y and z respectively.

$$\begin{aligned}
\varepsilon_x &= \frac{1}{E_x}\sigma_x - \frac{\nu_{yx}}{E_y}\sigma_y - \frac{\nu_{zx}}{E_z}\sigma_z \\
\varepsilon_y &= -\frac{\nu_{xy}}{E_x}\sigma_x + \frac{1}{E_y}\sigma_y - \frac{\nu_{zy}}{E_z}\sigma_z \\
\varepsilon_z &= -\frac{\nu_{xz}}{E_x}\sigma_x - \frac{\nu_{yz}}{E_y}\sigma_y + \frac{1}{E_z}\sigma_z \\
\varepsilon_{xy} &= \frac{1}{2G_{xy}}\sigma_{xy};\ \varepsilon_{xz} = \frac{1}{2G_{xz}}\sigma_{xz};\ \varepsilon_{yz} = \frac{1}{2G_{yz}}\sigma_{yz}
\end{aligned} \tag{7.35}$$

Because of the symmetry of the matrix of elastic coefficients, there exist the additional relations of Equation (2.14), namely

$$E_j\nu_{ij} = E_i\nu_{ji} \qquad (i,j = x,y,z) \tag{7.36}$$

which reduce the number of independent elastic constants to six.

For the case of the free vibration of a plate, the effect of transverse-normal stress can be neglected since there is no applied lateral load. We may therefore eliminate the terms containing σ_z from Equation (7.35), which reduces to

$$\begin{aligned}
\varepsilon_x &= \frac{1}{E_x}\sigma_x - \frac{\nu_{yz}}{E_y}\sigma_y \qquad \varepsilon_y = -\frac{\nu_{xy}}{E_x}\sigma_x + \frac{1}{E_y}\sigma_y \\
\varepsilon_{xy} &= \frac{1}{2G_{xy}}\sigma_{xy};\ \varepsilon_{xz} = \frac{1}{2G_{xz}}\sigma_{xz} \\
\varepsilon_{yz} &= \frac{1}{2G_{yz}}\sigma_{yz};\ \varepsilon_z = 0
\end{aligned} \tag{7.37}$$

The following expressions for the stresses are assumed in accordance with classical theory.

$$\begin{aligned}
(\sigma_x,\sigma_y,\sigma_{xy}) &= \frac{1}{h}(N_x,N_y,N_{xy}) + \frac{12z}{h^3}(M_x,M_y,M_{xy}) \\
(\sigma_{xz},\sigma_{yz}) &= \frac{3}{2h}\left[1 - \left(\frac{z}{h/2}\right)^2\right](Q_x,Q_y);\ \sigma_z = 0
\end{aligned} \tag{7.38}$$

The membrane stresses N_x/h, N_y/h and N_{xy}/h, involved in the expressions of σ_x, σ_y, and σ_{xy} respectively, that are neglected in the linear theory, can no longer be disregarded in the analysis of the large-deflection nonlinear problem.

The membrane stress resultants in terms of strains are obtained by substituting Equations (7.34) and (7.37) in Equation (7.38) and integrating across the thickness of the plate. They are

$$N_x = \frac{E_x h}{1 - \nu_{xy}\,\nu_{yx}}(\varepsilon_{x_0} + \nu_{yx}\varepsilon_{y_0})$$

$$N_y = \frac{E_y h}{1 - \nu_{xy}\,\nu_{yx}}(\varepsilon_{y_0} + \nu_{xy}\varepsilon_{x_0}) \tag{7.39}$$

$$N_{xy} = 2G_{xy}h\varepsilon_{x_0y_0}$$

where ε_{x_0} and ε_{y_0} are extensional strains of the middle surface in the x- and y-direction respectively, and $\varepsilon_{x_0y_0}$ the middle surface shear strain. From Equation (7.34) it is seen that

$$\varepsilon_{x_0} = \frac{\partial u_0}{\partial x} + \frac{1}{2}\left(\frac{\partial w}{\partial x}\right)^2, \qquad \varepsilon_{y_0} = \frac{\partial v_0}{\partial x} + \frac{1}{2}\left(\frac{\partial w}{\partial y}\right)^2$$

$$\varepsilon_{x_0y_0} = \frac{1}{2}\left(\frac{\partial u_0}{\partial y} + \frac{\partial v_0}{\partial x}\right) + \frac{1}{2}\left(\frac{\partial w}{\partial x}\right)\left(\frac{\partial w}{\partial y}\right)$$

Now Reissner's Variational Theorem [40] can be applied to this problem involving a specially orthotropic material. The function H was defined in Equation (5.15) and repeated here:

$$H = \sigma_x\varepsilon_x + \sigma_y\varepsilon_y + \sigma_z\varepsilon_z + 2\sigma_{xy}\varepsilon_{xy} + 2\sigma_{xz}\varepsilon_{xz} + 2\sigma_{yz}\varepsilon_{yz} - W(\sigma_{ij}) \tag{7.40}$$

where here

$$W(\sigma_{ij}) = \frac{1}{2}\left[\frac{1}{E_x}\sigma_x^2 + \frac{1}{E_y}\sigma_y^2 + \frac{1}{E_z}\sigma_z^2 - 2\left(\frac{\nu_{xy}}{E_x}\sigma_x\sigma_y + \frac{\nu_{yz}}{E_y}\sigma_y\sigma_z\right.\right.$$

$$\left.\left. + \frac{\nu_{xz}}{E_x}\sigma_z\sigma_x\right) + \frac{1}{G_{yz}}\sigma_{yz}^2 + \frac{1}{G_{xz}}\sigma_{xz}^2 + \frac{1}{G_{xy}}\sigma_{xy}^2\right] \tag{7.41}$$

The Reissner Functional is written, as in Equation (5.14), to be

$$\psi = \int_R HdR - \int_R F_i u_i dR - \int_{S_T} T_i u_i dS \tag{7.42}$$

The last two integrals of Equation (7.42), associated with work done by body forces and surface tractions, respectively, are eliminated in this natural oscillation problem. The strain energy is obtained by substituting Equations (7.34), (7.38), and (7.39) into Equation (7.42), setting $F_i = T_i = 0$ and integrating across the thickness of plate. It is seen that

$$\Psi = \int_0^a \int_0^b \left\{ \frac{E_x h}{2(1 - \nu_{xy}\nu_{yx})} [\bar{I}_e^2 - 2(\gamma - \nu_{yx})\overline{II}_e^2] + M_x \frac{\partial \bar{\alpha}}{\partial x} + M_y \frac{\alpha\bar{\beta}}{\partial y} \right.$$

$$- M_{xy}\left(\frac{\partial \bar{\alpha}}{\partial y} + \frac{\partial \bar{\beta}}{\partial x}\right) + Q_x\left(\frac{\partial w}{\partial x} + \bar{\alpha}\right) + Q_y\left(\frac{\partial w}{\partial y} + \bar{\beta}\right) - \frac{6M_x^2}{E_x h^3} - \frac{6M_y^2}{E_y h^3}$$

$$\left. - \frac{6M_{xy}^2}{G_{xy}h^3} + \frac{12\nu_{xy}M_x M_y}{E_x h^3} - \frac{3Q_x^2}{5G_{xz}h} - \frac{3Q_y^2}{5G_{yz}h} \right\} dxdy \tag{7.43}$$

where

$$\gamma^2 = D_y / D_x$$

$$\bar{I}_e = \varepsilon_{x_0} + \gamma \varepsilon_{y_0} \tag{7.44}$$

$$\overline{II}_e = \varepsilon_{x_0}\varepsilon_{y_0} - \frac{2G_{xy}(1 - \nu_{xy}\nu_{yx})}{E_x(\gamma - \nu_{yx})} \varepsilon_{x_0 y_0}^2$$

$\bar{I}_e$ and $\overline{II}_e$ play the role of the first and second invariants of the middle-surface strains of a plate composed of an orthotropic material [6].

The kinetic energy of the plate, after integrating across the thickness, is given by

$$T = \int_0^a \int_0^b \left\{ \frac{\rho_m h}{2} \left[\left(\frac{\partial u_0}{\partial t}\right)^2 \right.\right.$$

$$\left.\left. + \left(\frac{\partial v_0}{\partial t}\right)^2 + \left(\frac{\partial w}{\partial t}\right)^2 \right] + \frac{\rho_m h^3}{24}\left[\left(\frac{\partial \bar{\alpha}}{\partial t}\right)^2 + \left(\frac{\partial \bar{\beta}}{\partial t}\right)^2\right] \right\} dxdy \tag{7.45}$$

The kinetic energy for a simple beam is given by Equation (5.47) for comparison. According to Hamilton's principle [see Equations (5.45) and (5.46)],

$$\delta\Phi = \delta \int_{t_1}^{t_2} (T - \psi)dt = 0 \tag{7.46}$$

By virtue of the hypothesis of Berger [16], which suggests that the second invariant of the middle-surface strain of the plate can be neglected, the substitution of Equations (7.43) and (7.45) into (7.46) with $\overline{II}_e = 0$ yields

$$\delta\Phi = 0 = \int_0^a \int_0^b \left[\frac{\rho_m h^3}{12} - \left(\frac{\partial \overline{\alpha}}{\partial t} \delta\overline{\alpha} + \frac{\partial \overline{\beta}}{\partial t} \delta\overline{\beta} \right) \right.$$

$$\left. + \rho_m h \left(\frac{\partial u_0}{\partial t} \delta u_0 + \frac{\partial v_0}{\partial t} \delta v_0 + \frac{\partial w}{\partial t} \delta w \right) \right]_{t_1}^{t_2} dydx$$

$$- \int_{t_1}^{t_2} \int_0^b \left[M_x \delta\overline{\alpha} + M_{xy} \delta\overline{\beta} + \left(Q_x + \frac{E_x h}{1 - \nu_{xy}\nu_{yx}} \overline{I}_e \frac{\partial w}{\partial x} \right) \delta w \right.$$

$$\left. + \frac{E_x h}{1 - \nu_{xy}\nu_{yx}} \overline{I}_e \delta u_0 \right]_0^a dydt - \int_{t_1}^{t_2} \int_0^a \left[M_{xy} \delta\overline{\alpha} + M_y \delta\overline{\beta} \right.$$

$$\left. + \left(Q_y + \frac{E_x h}{1 - \nu_{xy}\nu_{yx}} \gamma \overline{I}_e \frac{\partial w}{\partial y} \right) \delta w + \frac{E_x h}{1 - \nu_{xy}\nu_{yx}} \gamma \overline{I}_e \delta v_0 \right]_0^b dxdt$$

$$- \int_{t_1}^{t_2} \int_0^a \int_0^b \left\{ \left[\frac{\partial \overline{\alpha}}{\partial x} - \frac{12}{E_x h^3} M_x + \frac{12\nu_{xy}}{E_x h^3} M_y \right] \delta M_x \right.$$

$$+ \left[\frac{\partial \overline{\beta}}{\partial y} - \frac{12}{E_y h^3} M_y + \frac{12\nu_{yx}}{E_y h^3} M_x \right] \delta M_y + \left[\frac{\partial \overline{\alpha}}{\partial y} + \frac{\partial \overline{\beta}}{\partial x} - \frac{12}{G_{xy} h^3} M_{xy} \right] \delta M_{xy}$$

$$+ \left[\frac{\partial M_x}{\partial x} + \frac{\partial M_{xy}}{\partial y} - Q_x - \frac{\rho_m h^3}{12} \frac{\partial^2 \overline{\alpha}}{\partial t^2} \right] \delta\overline{\alpha}$$

$$+ \left[\frac{\partial M_{xy}}{\partial x} + \frac{\partial M_y}{\partial y} - Q_y - \frac{\rho_m h^3}{12} \frac{\partial^2 \overline{\beta}}{\partial t^2} \right] \delta\overline{\beta}$$

$$+\left[\frac{\partial Q_x}{\partial x}+\frac{\partial Q_y}{\partial y}+\frac{E_x h}{1-\nu_{xy}\nu_{yx}}\frac{\partial}{\partial x}\left(\bar{I}_e\frac{\partial w}{\partial x}\right)+\frac{E_x h}{1-\nu_{xy}\nu_{yx}}\frac{\partial}{\partial y}\left(\bar{I}_e\gamma\frac{\partial w}{\partial y}\right)\right.$$

$$\left.-\rho_m h\frac{\partial^2 w}{\partial t^2}\right]\delta_w+\left[\frac{\partial w}{\partial x}+\bar{\alpha}-\frac{6}{5G_{xz}h}Q_x\right]\delta Q_x$$

$$+\left[\frac{\partial w}{\partial y}+\bar{\beta}-\frac{6}{5G_{yz}h}Q_y\right]\delta Q_y$$

$$+\left[\frac{E_x h}{1-\nu_{xy}\nu_{yx}}\frac{\partial \bar{I}_e}{\partial x}-\rho_m h\frac{\partial^2 u_0}{\partial t^2}\right]\delta u_0$$

$$+\left[\frac{\gamma E_x h}{1-\nu_{xy}\nu_{yx}}\frac{\partial \bar{I}_e}{\partial y}-\rho_m h\frac{\partial^2 v_0}{\partial t^2}\right]\delta v_0\Bigg\}\,dydxdt \tag{7.47}$$

It is to the merit of the variational principle that we can find the natural-boundary conditions explicitly from the second and third integrals in Equation (7.47) above. The first integral is used to specify the initial conditions. Further, the contents of every bracket in Equation (7.47) must vanish; the Euler-Lagrangian equations of motion are obtained as the following:

$$\frac{\partial M_x}{\partial x}+\frac{\partial M_{xy}}{\partial y}-Q_x-\frac{\rho_m h^3 f}{12}\frac{\partial^2\bar{\alpha}}{\partial t^2}=0 \tag{7.48}$$

$$\frac{\partial M_{xy}}{\partial x}+\frac{\partial M_y}{\partial y}-Q_y-\frac{\rho_m h^3 f}{12}\frac{\partial^2\bar{\beta}}{\partial t^2}=0 \tag{7.49}$$

$$\frac{\partial Q_x}{\partial x}+\frac{\partial Q_y}{\partial y}+\frac{E_x h}{1-\nu_{xy}\nu_{yx}}\left[\frac{\partial}{\partial x}\left(\bar{I}_e\frac{\partial w}{\partial x}\right)+\gamma\frac{\partial}{\partial y}\left(\bar{I}_e\frac{\partial w}{\partial y}\right)\right]-\rho_m h\frac{\partial^2 w}{\partial t^2}=0 \tag{7.50}$$

$$\frac{\partial\bar{\alpha}}{\partial x}-\frac{12}{E_x h^3}M_x+\frac{12\nu_{xy}}{E_x h^3}M_y=0 \tag{7.51}$$

$$\frac{\partial\bar{\beta}}{\partial y}-\frac{12}{E_y h^3}M_y+\frac{12\nu_{yx}}{E_y h^3}M_x=0 \tag{7.52}$$

$$\frac{\partial\bar{\alpha}}{\partial y}+\frac{\partial\bar{\beta}}{\partial x}-\frac{12}{G_{xy}h^3}M_{xy}=0 \tag{7.53}$$

$$\frac{\partial w}{\partial x} + \overline{\alpha} - \frac{6k}{5G_{xz}h} Q_x = 0 \tag{7.54}$$

$$\frac{\partial w}{\partial y} + \overline{\beta} - \frac{6k}{5G_{yz}h} Q_y = 0 \tag{7.55}$$

$$\frac{E_x h}{1 - \nu_{xy} \nu_{yx}} \frac{\partial \overline{I}_e}{\partial x} = \rho_m h \frac{\partial^2 u_0}{\partial t^2} \tag{7.56}$$

$$\frac{\gamma E_x h}{1 - \nu_{xy} \nu_{yx}} \frac{\partial \overline{I}_e}{\partial y} = \rho_m h \frac{\partial^2 v_0}{\partial t^2} \tag{7.57}$$

Two tracing constants k and f are introduced, as before, for the purpose of identifying terms which characterize the effects of transverse shear deformation and rotatory inertia respectively. It is seen that Equations (7.48) through (7.55), except (7.50), which involves the membrane effects as well as nonlinear behavior of strains, are the same form as those of the linear theory. Also, Equations (7.54) and (7.55) which characterize the effect of transverse shear deformation appear identical to equations of Eringen [17] if one changes the viscoelastic shear modulus to the elastic transverse shear modulus. Equations (7.56) and (7.57) are the simplified equations for in-plane force equilibrium by means of Berger's hypothesis.

It is hypothesized that the inertia effect of the in-plane motion can be neglected. This assumption is tantamount to following the procedure of previous investigators, see Reference [6] for references, in which they eliminated u_0 and v_0 from the equations at the outset. Hence, the right-hand side of Equations (7.56) and (7.57) are set equal to zero. Thus it is seen that,

$$\frac{\partial \overline{I}_e}{\partial x} = \frac{\partial \overline{I}}{\partial y} = 0 \tag{7.58}$$

Equation (7.58) implies that $\overline{I}_e$ is independent of the space coordinates x and y. Hence one may assume

$$\overline{I}_e = \frac{\partial u_0}{\partial x} + \frac{\partial v_0}{\partial y} + \frac{1}{2}\left(\frac{\partial w}{\partial x}\right)^2 + \frac{\gamma}{2}\left(\frac{\partial w}{\partial y}\right)^2 = \frac{\Delta^2 h^2}{12} \tag{7.59}$$

Here, Δ is a coupling parameter of the system to be defined later. From Equations (7.48) to (7.55), using Equation (7.59) and eliminating M_x, M_y, M_{xy},

Q_x, Q_y and $\overline{\alpha}$ and $\overline{\beta}$ one obtains the following in terms of the stiffness quantities for a laminated composite or sandwich panel:

$$L_1(w,\overline{\alpha},\overline{\beta}) = A_{55}\left(\frac{\partial w}{\partial x} + \overline{\alpha}\right) - k\left[D_{11}\frac{\partial^2\overline{\alpha}}{\partial x^2} + D_{12}\frac{\partial^2\overline{\alpha}}{\partial y^2} - If\frac{\partial^2\overline{\alpha}}{\partial t^2} + (\overline{H} - D_{66})\frac{\partial^2\overline{\beta}}{\partial x\partial y}\right] = 0 \tag{7.60}$$

$$L_2(w,\overline{\alpha},\overline{\beta}) = A_{44}\left(\frac{\partial w}{\partial x} + \overline{\beta}\right) - k\left[D_{12}\frac{\partial^2\overline{\beta}}{\partial x^2} + D_{22}\frac{\partial^2\overline{\beta}}{\partial y^2} - If\frac{\partial^2\overline{\beta}}{\partial t^2} + (\overline{H} - D_{66})\frac{\partial^2\overline{\beta}}{\partial x\partial y}\right] = 0 \tag{7.61}$$

$$L_3(w,\alpha,\beta) = A_{55}\left(\frac{\partial^2 w}{\partial x^2} + \frac{\partial\overline{\alpha}}{\partial x}\right) + A_{44}\left(\frac{\partial^2 w}{\partial y^2} + \frac{\partial\overline{\beta}}{\partial y}\right) + \frac{kA_{11}h^2\Delta^2}{12}\left(\frac{\partial^2 w}{\partial x^2} + \gamma\frac{\partial^2 w}{\partial y^2}\right) - \overline{M}k\frac{\partial^2 w}{\partial t^2} = 0 \tag{7.62}$$

$$\frac{\partial u_0}{\partial x} + \gamma\frac{\partial v_0}{\partial y} + \frac{1}{2}\left(\frac{\partial w}{\partial x}\right)^2 + \frac{\gamma}{2}\left(\frac{\partial w}{\partial y}\right)^2 = \frac{\Delta^2 h^2}{12} \tag{7.63}$$

where Δ (to be defined later) is a coupling parameter that is a time-dependent function whose coefficient is dependent upon the boundary conditions, and

$$(\overline{M},I) = \sum_{k=1}^{N}\int_{-h/2}^{h/2}\rho_{m_k}(1,z^2)dz$$

$$(A_{ij}, D_{ij}) = \sum_{k=1}^{N}\int_{-h/2}^{h/2}\overline{Q}_{ij_k}(1,z^2)dz,\ i,j = 1,2,6$$

$$(A_{55}, A_{44}) = \sum_{k=1}^{N}\int_{-h/2}^{h/2}(\overline{Q}_{55_k}, \overline{Q}_{44})_k\, f(z)dz$$

$$\gamma^2 = A_{22}/A_{11}$$

$$\overline{H} = D_{12} + 2D_{66}$$

$$f(z) = (5/4)[1 - 4(z/h)^2]$$

Equations (7.60) through (7.63) are the approximate equations that dominate the motion of plates. This motion consists of three independent families of modes of vibration, flexural, thickness-shear, and thickness-twist. In this work, the emphasis is on the flexural motion. As pointed out by previous works [18,41], the effect of rotatory inertia can be neglected when the plate vibrates primarily laterally in the small-deflection case. We hypothesize that the effect of rotatory inertia is also small when considering the plate undergoing large-amplitude lateral vibrations. Hence one can set $f = 0$ in Equations (7.60) and (7.61).

Consider a simply-supported plate as an example. The boundary conditions are

$$\begin{aligned} \overline{\beta} = u_0 = w = M_x = 0 \quad \text{along } x = 0,a \\ \overline{\alpha} = v_0 = w = M_y = 0 \quad \text{along } y = 0,b \end{aligned} \tag{7.64}$$

Using these conditions for u_0 and v_0, the integration of Equation (7.59) over the plate gives

$$\Delta^2 = \frac{6}{h^2ab}\int_0^a\int_0^b\left[\left(\frac{\partial w}{\partial x}\right)^2 + \gamma\left(\frac{\partial w}{\partial x}\right)^2\right]dydx \tag{7.65}$$

To satisfy the boundary conditions, solutions are assumed of the form

$$\begin{aligned} w &= \sum_m^{\alpha}\sum_n^{\alpha} W_{mn}\sin\left(\frac{n\pi x}{a}\right)\sin\left(\frac{m\pi y}{b}\right)\tau(t) \\ \overline{\alpha} &= \sum_m^{\alpha}\sum_n^{\alpha} \Gamma_{mn}\cos\left(\frac{n\pi x}{a}\right)\sin\left(\frac{m\pi y}{b}\right)\tau(t) \\ \overline{\beta} &= \sum_m^{\alpha}\sum_n^{\alpha} \Lambda_{mn}\sin\left(\frac{n\pi x}{a}\right)\cos\left(\frac{m\pi y}{b}\right)\tau(t) \end{aligned} \tag{7.66}$$

where at this point $\tau(t)$ is unknown.

The normal-mode solution is exploited since the system of governing differential equations is linear in the space functions and nonlinear in the time-dependent function left undefined in Equation (7.66). If one considers the fundamental mode of oscillation, one may use only the first term of each expression given by Equation (7.66) as an approximation to the true solution.

With this in mind, the substitution of the expression for w into Equation (7.65) yields

$$\Delta^2 = \frac{3W_{11}^2}{2h^2}(\lambda_1^2 + \gamma\delta_1^2)\tau^2 \tag{7.67}$$

in which $\lambda_1 = \pi/a$, $\delta_1 = \pi/b$. If one substitutes Equation (7.66) with $m = n = 1$ into Equations (7.60) through Equation (7.62) with $f = 0$ and $k = 1$, one obtains

$$\lambda_1 W_{11} + \left[1 + \left(\frac{6}{5G_{xz}h}\right)(D_x\lambda_1^2 + D_{xy}\delta_1^2)\right]\Gamma_{11}$$

$$+ \left(\frac{6}{5G_{xz}h}\right)(H - D_{xy})\tau_1\delta_1\Lambda_{11} = 0$$

$$\delta_1 W_{11} + \left[1 + \left(\frac{6}{5G_{yz}h}\right)(D_y\delta_1^2 + D_{xy}\lambda_1^2)\right]\Lambda_{11}$$

$$+ \left(\frac{6}{5G_{yz}h}\right)(\overline{H} - D_{xy})\,\lambda_1\delta_1\Gamma_{11} = 0$$

$$\left\{\left[G_{xz} + \Delta^2 D_x\left(\frac{6}{5h}\right)\right]\lambda_1^2\tau + \left[G_{yz} + \Delta^2\gamma D_y\left(\frac{6}{5h}\right)\right]\delta_1^2\tau\right.$$

$$\left. + \left(\frac{6\rho_m}{5}\right)\ddot{\tau}\right\}W_{11} + G_{xz}\lambda_1\tau\Gamma_{11} + G_{yz}\delta_1\tau\Lambda_{11} = 0 \tag{7.68}$$

For the nontrivial solution, the determinant of coefficients of W_{11}, Γ_{11}, and Λ_{11} in Equation (7.68) must vanish. With Equation (7.67) it gives

$$\alpha^o\ddot{\tau} + \beta^o\tau + \delta^o\tau^3 = 0 \tag{7.69}$$

where

$$\alpha^o = \frac{6\rho_m}{5}\left\{1 + \left(\frac{6}{5G_{xz}h}\right)(D_x\gamma_1^2 + D_{xy}\delta_1^2) + \left(\frac{6}{5G_{yz}h}\right)(D_{xy}\gamma_1^2 + D_y\delta_1^2)\right.$$

$$+ \left(\frac{6}{5G_{xz}h}\right)\left(\frac{6}{5G_{yz}h}\right)\{(D_xD_{xy}\lambda_1^4 + D_yD_{xy}\delta_1^4)$$

$$\left. + (D_xD_y - \overline{H}^2 + 2\overline{H}D_{xy})\,\lambda_1^2\delta_1^2\right\} \tag{7.70}$$

$$\beta^o = \left(\frac{6}{5h}\right)(D_x\lambda_1^4 + 2\overline{H}\lambda_1^2\delta_1^2 + D_y\delta_1^4) + \left(\frac{6}{5G_{xz}h}\right)\left(\frac{6}{5G_{yz}h}\right)(D_xD_{xy}G_{xz}\lambda_1^6$$

$$+ D_yD_{xy}G_{yz}\delta_1^6) + \left(\frac{6}{5G_{xz}h}\right)\left(\frac{6}{5G_{yz}h}\right)(D_xD_{xy}G_{yz} + D_xD_yG_{xz} - \overline{H}^2G_{xz}$$

$$+ 2\overline{H}D_{xy}G_{xz})\,\lambda_1^4\delta_1^2 + \left(\frac{6}{5G_{xz}h}\right)\left(\frac{6}{5G_{yz}h}\right)(D_yD_{xy}G_{xz} + D_xD_yG_{yz}$$

$$- \overline{H}^2G_{yz} + 2\overline{H}D_{xy}G_{yz})\,\lambda_1^2\delta_1^4 \tag{7.71}$$

$$\delta^o = \frac{3}{2\rho_m h}\frac{W_{11}^2}{h^2}D_x(\lambda_1^4 + 2\gamma\lambda_1^2\delta_1^2 + \gamma^2\delta_1^4)\alpha^o \tag{7.72}$$

If the plate is subjected to the initial conditions

$$\tau = 1,\ \dot{\tau} = 0 \text{ at } t = 0 \tag{7.73}$$

the solution of Equation (7.69) is

$$\tau = cn(\omega^o t, p) \tag{7.74}$$

where cn is an elliptic cosine with

$$\omega^{o2} = \frac{1 + \frac{3}{2}\hat{\beta}^2\lambda_1\left[1 + \left(\frac{6}{5h}\right)\left(\frac{D_x}{G_{xz}}\lambda_1^2 + \frac{D_y}{G_{yz}}\delta_1^2 + B\right)\right]}{1 + \left(\frac{6}{5h}\right)\left(\frac{D_x}{G_{xz}}\lambda_1^2 + \frac{D_y}{G_{yz}}\delta_1^2 + B\right)}$$

$$p^2 = \frac{\frac{3}{4}\hat{\beta}^2\gamma_1\left[1 + \left(\frac{6}{5h}\right)\left(\frac{D_x}{G_{xz}}\lambda_1^2 + \frac{D_y}{G_{yz}}\delta_1^2 + B\right)\right]}{1 + \frac{3}{2}\hat{\beta}^2\gamma_1\left[1 + \left(\frac{6}{5h}\right)\left(\frac{D_x}{G_{xz}}\lambda_1^2 + \frac{D_y}{G_{yz}}\delta_1^2 + B\right)\right]} \tag{7.75}$$

The corresponding square of the natural circular frequency for the small amplitude linear vibration is

$$\overline{\omega}^2 = \frac{1}{\rho_m h}(D_x\lambda_1^4 + 2\overline{H}\lambda_1^2\delta_1^2 + D_y\delta_1^4) \tag{7.76}$$

Note that Equation (7.76) is identical to (7.8). Also,

$$\hat{\beta} = \frac{W_{11}}{h},\ \gamma_1 = \frac{D_x\lambda_1^4 + 2D_x\gamma\lambda_1^2\delta_1^2 + D_y\delta_1^4}{D_x\lambda_1^4 + 2H\,\lambda_1^2\delta_1^2 + D_y\delta_1^4} \tag{7.77}$$

$$\begin{aligned} B = {} & \left[1 + \left(\frac{6}{5h}\right)\left(\frac{D_{xy}}{G_{yx}}\lambda_1^2 + \frac{D_{xy}}{G_{xz}}\delta_1^2 + A\right)\right]^{-1}\left\{\left(\frac{6}{5h}\right)\left(\frac{D_xD_y}{G_{xz}G_{yz}}\right.\right. \\ & \left. - \frac{\overline{H}^2}{G_{xz}G_{yz}} + 2\frac{\overline{H}D_{xy}}{G_{xz}G_{yz}} - \frac{D_xD_{xy}}{G_{xz}^2} - \frac{D_yD_{xy}}{G_{yz}^2}\right)\lambda_1^2\delta_1^2 \\ & \left. - A\left[1 + \left(\frac{6}{5h}\right)\left(\frac{D_x}{G_{xz}}\lambda_1^2 + \frac{D_y}{G_{yz}}\delta_1^2\right)\right]\right\} \end{aligned} \tag{7.78}$$

$$A = (D_x\lambda_1^4 + 2\overline{H}\lambda_1^2\delta_1^2 + D_y\delta_1^4)^{-1}(D_xD_y - H^2)\left(\frac{1}{G_{xz}}\lambda_1^2 + \frac{1}{G_{yz}}\delta_1^2\right)\lambda_1^2\delta_1^2 \tag{7.79}$$

The period T^o of $cn(\omega^0,t,p)$ is

$$T^o = \frac{4K(p)}{\omega^0} \tag{7.80}$$

where $K(p)$ is the complete elliptic integral of the first kind with parameter p. The corresponding linear period for a classical orthotropic plate not including transverse shear deformation is

$$\overline{T} = \frac{2\pi}{\overline{\omega}} \tag{7.81}$$

The frequency ratio here is defined as the ratio of the frequency determined by Equation (7.80) to that obtained for a classical plate by Equation (7.81), and is

$$R^* = \frac{f^0}{\overline{f}} = \frac{\pi}{2K(p)}\left\{\frac{1 + \frac{3}{2}\hat{\beta}^2\gamma_1\left[1 + \left(\frac{6}{5h}\right)\left(\frac{D_x}{G_{xz}}\lambda_1^2 + \frac{D_y}{G_{yz}}\delta_1^2 + B\right)\right]}{1 + \left(\frac{6}{5h}\right)\left(\frac{D_x}{G_{xz}}\lambda_1^2 + \frac{D_y}{G_{yz}}\delta_1^2 + B\right)}\right\}^{\frac{1}{2}} \tag{7.82}$$

where $f^0 = \frac{1}{T^0}$, $\overline{f} = \frac{1}{\overline{T}}$ are the frequencies of vibrations.

Numerical calculations have been made for the large-amplitude lateral oscillations of simply-supported plates composed of a typical boron/epoxy unidirectional composite material and published in Reference [6].

The results are plotted in the referenced paper, where the ordinate R^*, defined by Equation (7.82), is the ratio of the fundamental frequency using the methods developed herein to the fundamental frequency for the same orthotropic plate using classical linear theory with no provision for transverse shear deformation. The abscissa is a geometric ratio of length to thickness, varying from a moderately thick plate to a relatively thin plate. The ratio b/a is a width to length ratio and Δ is the ratio of amplitude of vibration to plate thickness.

Three separate and distinct effects are obvious. First, the deviation of all curves from the horizontal lucidly reflects the effect of transverse-shear deformation on the natural frequencies. The largest percentage effect occurs at lower $\overline{\beta}$ indicating that the further the plate is vibrating into the large amplitude, nonlinear regime, the smaller the effects of transverse shear deformation. That is expected; however, it is clear that for many fiber-reinforced composite materials transverse shear deformation effects must be included.

Second, it is also clear that the nonlinearity that occurs for the orthotropic plate is of the hardening spring type, and that the nonlinear effects are very important. It is seen that for a thin plate ($a/h = 40$), classical methods (linear, small deformation orthotropic plate theory, neglecting transverse-shear deformation) can predict a fundamental frequency differing from the correct solution by a factor of 1.7 for amplitudes of vibration only 1.5 times the plate thickness.

Third, we see that when the plate is longer in the stiffer direction (the direction of the fibers), the natural frequencies are significantly higher than when the reverse is true. This, of course, is also expected. Yu [19] found results qualitatively comparable to the first two conclusions above for the nonlinear flexural vibrations of sandwich plates. To utilize the above for anisotropic sandwich plates, it is necessary only to substitute in the proper values for the flexural stiffness, the transverse shear stiffness quantities, and the densities. Obviously, the equations given herein can be used for isotropic material monocoque and sandwich panels as well.

7.7 LINEAR AND NONLINEAR OSCILLATIONS OF SPECIALLY ORTHOTROPIC SANDWICH PANELS WITH VARIOUS BOUNDARY CONDITIONS

In all of the previous sections of this chapter, only the vibrations of plates simply supported on all four edges has been studied. For plates with other boundary conditions, Wu and Vinson [20] studied several other combinations of simply supported and clamped edges.

Consider an orthotropic sandwich plate in which the elastic properties are symmetric with respect to the middle surface. The lateral deflection is assumed to be of the same order of magnitude as the plate thickness. Analogous to the previous work [6], by using Berger's hypothesis [16] and eliminating in-plane inertia forces, the equations of motion are again given by Equations (7.59) through (7.63).

It is noted that according to Pagano [42], the parabolic distribution of transverse shear stress across the plate thickness is obvious for a laminated monocoque plate. A factor of 5/4 multiplies the distribution function used by Whitney [43], so that it can be consistent with the established shear factor from the previous work by Reissner [40] and Mindlin [41] for the homogeneous case.

As in the previous research of Yu [19] and Ray and Bert [21], it is assumed that the effect of coupling among individual vibration modes is not quantitatively significant, which was confirmed through the comparison of their analytical predictions with experimental results. Therefore, to apply the Galerkin's method, a set of solutions can be assumed as follows:

$$
\begin{aligned}
w &= W\phi_{wm}(x)\phi_{wn}(y)\tau(t) \\
\overline{\alpha} &= \Gamma\phi_{\alpha m}(x)\phi_{\alpha n}(y)\tau(t) \\
\overline{\beta} &= \Lambda\phi_{\beta m}(x)\phi_{\beta n}(y)\tau(t)
\end{aligned}
\tag{7.83}
$$

in which the ϕ's are the characteristic or modified beam functions. These are analogous to those used by Warburton [3], that satisfy the boundary conditions for clamped and simply-supported plates. The case of free edges is not considered here. Since the in-plane displacements along the boundary vanish for the boundary conditions studied, the coupling parameter can be found by integrating Equations (7.59) over the plate:

$$
\Delta^2 = \frac{6\hat{\beta}^2\tau^2(t)}{ab}\int_0^a\int_0^b [\phi'^2_{wm}(x)\phi^2_{wn}(y) + \gamma\phi^2_{wm}(x)\phi'^2_{wn}(y)]dydx \tag{7.84}
$$

For a plate of dimensions a by b, $\hat{\beta}$ is the ratio of the vibration amplitude to the plate thickness

$$
\left[= w\left(\frac{1}{2}a, \frac{1}{2}b\right)/h \right] \text{ [See Equation (7.77).]}
$$

Substituting Equation (7.84) into (7.62), using Galerkin's method ([44 Section 3.11] permits Equations. (7.60) through (7.62) to be written as follows:

$$\int_0^a \int_0^b L_1(w,\overline{\alpha},\overline{\beta})\phi_{\alpha m}(x)\phi_{\alpha n}(y)dxdy = 0$$

$$\int_0^a \int_0^b L_2(w,\overline{\alpha},\overline{\beta})\phi_{\beta m}(x)\phi_{\beta n}(y)dxdy = 0 \qquad (7.85)$$

$$\int_0^a \int_0^b L_3(w,\overline{\alpha},\overline{\beta})\phi_{wm}(x)\phi_{\beta n}(y)dxdy = 0$$

The result of using Galerkin's method shown by Equation (7.85) is a set of simultaneous homogeneous algebraic equations involving the three coefficients W, Γ, and Λ. To find the nontrivial solution, the determinant of the coefficients must vanish. This leads to an ordinary differential equation for the natural frequency. In order to obtain an explicit solution consider a fully-clamped plate as an example. It is subsequently shown that the solution has far more generality. The boundary conditions for the clamped plate are

$$u_0 = v_0 = w = \overline{\alpha} = \overline{\beta} = 0 \quad \text{on} \quad x = 0,a \quad \text{and} \quad y = 0,b \qquad (7.86)$$

The solution satisfying these boundary conditions in the x-direction are assumed in the following two forms:

(1) Odd wave number, $m = 1, 3, 5 \ldots$:

$$\phi_{wm}(x) = \cos \mu_m \left(\frac{x}{a} - \frac{1}{2}\right) + \eta_m \cosh \mu_m \left(\frac{x}{a} - \frac{1}{2}\right)$$

$$\phi_{\alpha m}(x) + \sin \mu_m \left(\frac{x}{a} - \frac{1}{2}\right) - \eta_m \sinh \mu_m \left(\frac{x}{a} - \frac{1}{2}\right) \qquad (7.87)$$

$$\phi_{\beta m}(x) = \cos \mu_m \left(\frac{x}{a} - \frac{1}{2}\right) + \eta_m \cosh \mu_m \left(\frac{x}{a} - \frac{1}{2}\right)$$

where

$$\eta_m = \frac{\sin (\mu_m/2)}{\sinh (\mu_m/2)}$$

and

$$\tan (\mu_m/2) + \tanh (\mu_m/2) = 0$$

(2) Even wave number, $m = 2, 4, 6 \ldots$:

$$\phi_{wm}(x) = \sin \mu_m \left(\frac{x}{a} - \frac{1}{2}\right) + \eta_m \sinh \mu_m \left(\frac{x}{a} - \frac{1}{2}\right)$$

$$\phi_{\alpha m}(x) = \cos \mu_m \left(\frac{x}{a} - \frac{1}{2}\right) + \eta_m \cosh \mu_m \left(\frac{x}{a} - \frac{1}{2}\right) \tag{7.88}$$

$$\phi_{\beta m}(x) = \sin \mu_m \left(\frac{x}{a} - \frac{1}{2}\right) + \eta_m \sinh \mu_m \left(\frac{x}{a} - \frac{1}{2}\right)$$

where

$$\eta_m = -\frac{\sin (\mu_m/2)}{\sinh (\mu_m/2)}$$

and

$$\tan (\mu_m/2) - \tanh (\mu_m/2) = 0$$

If one interchanges a and b and changes x to y and m to n, one obtains the proper mode functions in the y-direction. Substituting Equations (7.87) and (7.88) and the corresponding functions of y into Equations (7.84) and (7.85) one obtains the following:

$$[-\lambda_m d_{21} A_{55} \tau] W + [D_{12} + D_{66}] \lambda_m \delta_n d_{22} \tau \Lambda$$
$$+ [A_{55} d_{21} \tau + D_{11} \lambda_m^2 d_{11} \tau$$
$$+ D_{66} \delta_n^2 D_{22} \tau + I d_{21} \ddot{\tau}] \Gamma = 0$$

$$[-\delta_n d_{12} A_{44} \tau] W + [D_{12} + D_{66}] \lambda_m \delta_n d_{22} \tau \Gamma$$
$$+ [A_{44} d_{12} \tau + D_{22} \delta_n^2 d_{21} \tau$$
$$+ D_{66} \lambda_m^2 d_{22} \tau + I d_{12} \ddot{\tau}] \Lambda = 0$$

$$-\left[\left(A_{55} + \frac{A_{11} h^2 \Delta^2}{12}\right) \lambda_m^2 d_{21} \tau + \left(A_{44} + \frac{A_{11} h^2 \Delta^2 \gamma}{12}\right) \delta_n^2 d_{12} \tau\right.$$
$$\left. + \overline{M} d_{11} \ddot{\tau}\right] W + A_{55} \lambda_m d_{21} \tau \, \Gamma + A_{44} \delta_n d_{12} \tau \Lambda = 0 \tag{7.89}$$

and

$$\Delta^2 = 1.5\hat{\beta}^2(d_{21}\lambda_m^2 + d_{12}\delta_n^2)\tau^2(t) \tag{7.90}$$

where

$$\begin{aligned}
&\lambda_m = \frac{\mu_m}{a}, \ \delta_n = \frac{\mu_n}{b}, \ d_{ij} = \theta_{im}\theta_{jn} \ (i,j = 1,2) \\
&\theta_{1k} = 1 - (-1)^k \tau_{1k}^2 \\
&\theta_{2k} = 1 + (-1)^k[\eta_k^2 - (2/\mu_k) \sin \mu_k], \ k = 1, 2, 3, \ldots
\end{aligned} \tag{7.91}$$

with

$$\eta_k = (-1)^{k+1} \frac{\sin (\mu_k/2)}{\sinh (\mu_k/2)}$$

and

$$\mu_1 = 1.506\pi; \ \mu_k \cong \left(k + \frac{1}{2}\right)\pi, \text{ as } k \geq 2$$

With the expression for given by Equation (7.90) and neglecting rotatory inertia in Equation (7.89) through the equation derived from setting the determinant of the coefficients of W, Λ, and Γ equal to zero is given by Equation (7.69). Here the coefficients are defined by the following where the starred quantities below replace the coefficients in Equation (7.69):

$$\begin{aligned}
\alpha^* = {} & 1 + \left(\frac{D_{11}}{A_{55}}\frac{d_{11}}{d_{21}} + \frac{D_{66}}{A_{44}}\frac{d_{22}}{d_{12}}\right)\lambda_m^2 + \left(\frac{D_{66}}{A_{55}}\frac{d_{22}}{d_{21}} + \frac{D_{22}}{A_{44}}\frac{d_{11}}{d_{12}}\right)\delta_n^2 \\
& + \frac{D_{11}}{A_{55}}\frac{D_{66}}{A_{44}}\lambda_m^4 + \frac{D_{22}}{A_{44}}\frac{D_{66}}{A_{55}}\delta_n^4 \\
& + \left(\frac{D_{11}}{A_{55}}\frac{D_{22}}{A_{44}}\frac{d_{11}}{d_{22}} - \frac{D_{12}^2 + 2D_{12}D_{66}}{A_{55}A_{44}}\frac{d_{22}}{d_{11}}\right)\lambda_m^2\delta_n^2
\end{aligned} \tag{7.92}$$

$$\beta^* = \frac{1}{M}\Bigg[D_{11}\lambda_m^4 + 2(D_{12} + 2D_{66})\frac{d_{22}}{d_{11}}\lambda_m^2\delta_n^2 + D_{22}\delta_n^4$$

$$+ \frac{D_{11}D_{66}}{A_{44}}\frac{d_{22}}{d_{12}}\lambda_m^6 + \frac{D_{22}D_{66}}{A_{55}}\frac{d_{22}}{d_{21}}\delta_n^6$$

$$+ \left(\frac{D_{11}D_{22}}{A_{44}}\frac{d_{11}}{d_{12}} + \frac{D_{11}D_{66}}{A_{55}}\frac{d_{22}}{d_{21}} - \frac{D_{12}^2 + 2D_{12}D_{66}}{A_{44}}\frac{d_{22}^2}{d_{11}d_{21}}\right)\lambda_m^4\delta_n^2$$

$$+ \left(\frac{D_{11}D_{22}}{A_{44}}\frac{d_{11}}{d_{21}} + \frac{D_{11}D_{66}}{A_{44}}\frac{d_{22}}{d_{12}} - \frac{D_{12}^2 + 2D_{12}D_{66}}{A_{55}}\frac{d_{22}^2}{d_{11}d_{21}}\right)\lambda_m^2\delta_n^4\Bigg] \quad (7.93)$$

$$\delta^* = \frac{\hat{\beta}^2 A_{11}h^2}{8M}(d_{21}\lambda_m^2 + \gamma d_{12}\delta_n^2)\alpha^* \quad (7.94)$$

For the initial conditions, $\tau(0) = 1$ and $\dot{\tau}(0) = 0$, the solution to Equation (7.69) is an elliptic cosine function given by Equation (7.74) where here

$$\omega^* = \overline{\omega}\left[\frac{3}{2}\hat{\beta}^2\gamma_1 + \left(1 + \frac{D_{11}}{A_{55}}\frac{d_{11}}{d_{21}}\lambda_m^2 + \frac{D_{22}}{A_{44}}\frac{d_{11}}{d_{12}}\delta_n^2 + B\right)\right]^{-1} \quad (7.95)$$

$$p = \left[\frac{3}{4}\hat{\beta}^2\gamma_1(\overline{\omega}/\omega^*)\right]^{1/2} \quad (7.96)$$

and

$$\overline{\omega} = \left\{\frac{1}{M}\left[D_{11}\lambda_m^4 + 2(D_{12} + 2D_{66})\frac{d_{22}}{d_{11}}\lambda_m^2\delta_n^2 + \frac{D_{22}}{D_{11}}\delta_n^4\right]\right\}^{1/2} \quad (7.97)$$

$$\gamma_1 = (A_{11}h^2/12\overline{M}\overline{\omega}^2)(d_{21}\lambda_m^2 + \gamma d_{12}\delta_n^2)^2 \quad (7.98)$$

$$B = \left(1 + \frac{D_{66}}{A_{44}}\lambda_m^2 + \frac{D_{66}}{A_{55}}\delta_n^2 + A\right)^{-1}$$

$$\times\left[\left(\frac{D_{11}D_{22}}{A_{44}A_{55}}\frac{d_{11}}{d_{22}} - \frac{D_{12}^2 + 2D_{12}D_{66}}{A_{55}A_{44}}\frac{d_{22}}{d_{11}}\right.\right.$$

$$\left. - \frac{D_{22}D_{66}}{A_{44}^2}\frac{d_{21}}{d_{12}} - \frac{D_{11}D_{66}}{A_{55}^2}\frac{d_{12}}{d_{21}}\right)\lambda_m^2\delta_n^2$$

$$\left. - A\left(1 + \frac{D_{11}}{A_{55}}\frac{d_{11}}{d_{21}}\lambda_m^2 + \frac{D_{22}}{A_{44}}\frac{d_{11}}{d_{12}}\delta_n^2\right)\right] \qquad (7.99)$$

$$A = \left[\lambda_m^4 + \frac{D_{22}}{D_{11}}\delta_n^4 + \frac{2(D_{12} + 2D_{66})}{D_{11}}\frac{d_{22}}{d_{11}}\lambda_m^2\delta_n^2\right]^{-1}$$

$$\times\left[\left(\frac{D_{22}}{A_{44}}\frac{d_{11}}{d_{12}} - \frac{H^2}{D_{11}A_{44}}\frac{d_{22}^2}{d_{11}d_{12}}\right)\lambda_m^2\right.$$

$$\left. + \left(\frac{D_{22}}{A_{55}}\frac{d_{11}}{d_{21}} - \frac{\overline{H}^2}{D_{11}A_{55}}\frac{d_{22}^2}{d_{11}d_{21}}\right)\delta_n^2\right]\lambda_m^2\delta_n^2 \qquad (7.100)$$

Notice that in Equation (7.75) of the previous section, ω^* should agree with Equation (7.95) when the d_{ij} are unity.

Finally, the frequency ratio defined as the inverse ratio of the period due to the present nonlinear theory to that from the classical linear theory is

$$R^* = \frac{\pi}{2K(p)}\left[\frac{3}{2}\hat{\beta}^2\gamma_1 + \left(1 + \frac{D_{11}}{A_{55}}\frac{d_{11}}{d_{21}}\lambda_m^2 + \frac{D_{22}}{A_{44}}\frac{d_{11}}{d_{12}}\delta_n^2 + B\right)^{-1}\right]^{1/2} \qquad (7.101)$$

where $K(p)$ is the complete eliptical integral of the first kind with parameter p.

For the case of small amplitude, $\hat{\beta} = 0$ and $K(p) = \pi/2$. Equation (7.101) immediately reduces to the small amplitude ratio R:

$$R = \left(1 + \frac{D_{11}}{A_{55}}\frac{d_{11}}{d_{21}}\lambda_m^2 + \frac{D_{22}}{A_{44}}\frac{d_{11}}{d_{12}}\delta_n^2 + B\right)^{-1/2} \qquad (7.102)$$

which is the ratio of the frequency, including the effect of transverse shear deformation and initial in-plane stress resultants, to that obtained for the classical plate.

- - - - - - simply supported edge
/////// clamped edge

boundary conditions	d_{ij}	λ_m	δ_n
1	$\theta_{im}\theta_{jn}$	$\frac{\mu_m}{a}$	$\frac{\mu_n}{b}$
2	$\theta_{im}\,\theta_{j(2n)}$	$\frac{\mu_m}{a}$	$\frac{\mu_{2n}}{2b}$
3	θ_{im}	$\frac{\mu_m}{a}$	$\frac{n\pi}{b}$
4	$\theta_{i(2m)}\theta_{j(2n)}$	$\frac{\mu_{2m}}{2a}$	$\frac{\mu_{2n}}{2b}$
5	$\theta_{i(2m)}$	$\frac{\mu_{2m}}{2a}$	$\frac{n\pi}{b}$
6	1	$\frac{m\pi}{a}$	$\frac{n\pi}{b}$

FIGURE 7.2 Constants based on boundary conditions. (Reprinted from Reference [20].)

It should be noted that in this solution, the equations pertain explicitly to a plate fully clamped on all four edges. For other boundary conditions (i.e., combinations of clamped and simple supports, following the same procedures) solutions can be obtained by selecting assumed functions analogous to those of Warburton [3]. It is found that one derives a solution in each case of the same form as Equations (7.101) and (7.102), simply having different values of the boundary constants d_{ij}. For six combinations of boundary conditions often found, the appropriate values of d_{ij} are given in Figure 7.2, and values of θ are given in Equation (7.91). Thus, in Equations (7.101) and (7.102), the frequency ratios are defined as the ratio of that obtained from the present theory to that obtained from classical plate theory for the same plate with the same boundary conditions.

Notice that, for unsymmetric laminates, one simply replaces D_{ij} by D_{ij}^*, the reduced bending rigidity defined by

$$[D^*] = [D] - [B][A]^{-1}[B] \tag{7.103}$$

From parametric studies performed on laminated composite plates [20], applicable as well to sandwich plates, it is found that the effect of transverse shear deformation, in terms of the percentage reduction of the corresponding natural frequency determined by classical plate theory, is greater for a fully clamped plate than for a fully simply-supported plate.

In the last two sections, examples of the large amplitude nonlinear vibration of orthotropic and isotropic sandwich panels for simply supported and other boundary conditions have been presented for study. However, further reading is in order including that of Bert [22], Leissa [23], Meyer-Piening [24], Whitney and Leissa [25], Chia [26], Leissa [23], Sathyamoorthy [28], Yu [29], Nayfeh and Mook [30], and Bai and Sun [31].

7.8 VIBRATION DAMPING

Damping of sandwich structures is clearly beyond the scope of this basic textbook. However, it would be a mistake not to mention that sandwich structures provide an excellent means to incorporate significant damping over composite laminates and metallic structures through the motions of the core. With foam core especially, because it is largely subjected to transverse shear deformation, any viscoelastic characteristics of the core can result in significant vibration damping. For the study of vibration damping, the text by Nashif, Jones, and Henderson [32] is excellent.

Also, the text by Inman [33] concerning vibrations, vibration damping, control, measurement, and stability provides much-needed information useful to the study of sandwich structural vibrations.

7.9 PROBLEMS

7.1 Consider the sandwich panel of Problem 2.1 in which the dimensions are $a = 30$ in. and $b = 20$ in. The plate is simply supported on all four edges: What is the fundamental natural frequency in Hz

a. using classical plate theory?

b. including transverse-shear deformation effects?

7.2 Consider a rectangular sandwich panel, 24 in. by 30 in. in planform dimensions, simply supported on all four edges. There are two identical aluminum faces ($E = 10 \times 10^6$ psi, $\nu = 0.3$, $\rho_w = 0.10$ lb/in.^{3}, $t_f = 0.063$ in.). The core is rigid klegecell foam ($G_c = 15{,}000$ psi, $\nu = 0.4$, $\rho_w = 15$ lb/ft^3, $h_c = 1.0$ in.). What are the three lowest lateral vibration natural frequencies in Hz using:

a. Classical plate theory?

b. Plate theory including transverse-shear deformation?

c. Considering one of the faces as a plate on an elastic foundation (i.e., the core) using $k = E_c = 2G_c(1 + \nu_c) = 42{,}000$ lb/in.^{2}/in. utilizing classical plate theory, determine the fundamental natural frequency (in Hz) of the facing.

7.3 Consider the sandwich beam of Problem 4.1 except for the facing and the core thickness. This beam must be designed such that the fundamental natural frequency is 50 Hz or above. Assuming $h_c = 20t_f$, what face thickness is necessary if both ends are simply supported?

7.4 If the beam of Problem 7.3 were cantilevered, what is the fundamental natural frequency in Hz?

7.5 If the beam of Problem 7.3 were clamped at each end, what is the fundamental natural frequency in Hz?

7.10 REFERENCES

1. Yu, Y. Y., "A New Theory of Elastic Sandwich Plates - One Dimensional Core," *Journal of Applied Mechanics*, Vol. 26, pp. 415–421, 1959.
2. Yu, Y. Y., *Vibrations of Elastic Plates: Linear and Nonlinear Dynamic Modeling of Sandwiches, Laminated Composites and Piezoelectric Layers*, Springer-Verlag, New York, 1996.
3. Meyer-Piening, H-R., "Nonlinear Investigation of Deflection, Stress and Frequencies of a Rectangular Thin Sandwich Plate," in *Sandwich Constructions 2—Proceedings of the Second International Conference on Sandwich Construction*, Gainesville, Florida, Editors, D. Weissman-Berman and K-A. Olsson, EMAS Publication, United Kingdom, 1992, pp. 189–202.
4. Bau-Madsen, N. K., K-H. Svendsen and A. Kildegaard, "Finite Deformation of Sandwich Plates," in *Sandwich Constructions 2—Proceedings of the Second International Conference on Sandwich Construction*, Gainesville, Florida, Editors, D. Weissman-Berman and K-A. Olsson, EMAS Publication, United Kingdom, 1992, pp. 189–202.
5. Ramachandra, L. S. and H. R. Meyer-Piening, "Dynamic Analysis of Sandwich Structures in Contact with Water," in *Sandwich Construction 3—Proceedings of the Third International Conference on Sandwich Construction*, Southampton, Great Britain, Editor, H. G. Allen, publ. 1996, pp. 363–375.
6. Wu, C-I. and Vinson, J. R., "On the Nonlinear Oscillations of Plates Composed of Composite Materials, *Journal of Composite Materials,*" Vol. 3, pp. 548–561, July 1969.
7. Moyer, Jr., E. T. G. G. Amir, K-A. Olsson and S. E. Hellbratt, "Response of GRP Sandwich Structures Subjected to Shock Loading," in *Sandwich Constructions 2—Proceedings of the Second International Conference on Sandwich Construction*, Gainesville, Florida, Editors, D. Weissman-Berman and K-A. Olsson, EMAS Publication, United Kingdom, 1992, pp. 49–65, 1992.
8. Williamson, J. E. and P. A. Lagace, "Response Mechanisms in the Impact of Graphite/ Epoxy Honeycomb Sandwich Panels," *Proceedings of the Eighth Technical Conference of the American Society for Composites*, pp. 287–297, 1993.
9. Olsson, R., "Prediction of Impact Damage in Sandwich Panels," *Sandwich Constructions 2—Proceedings of the Second International Conference on Sandwich Construction*, Gaines-

ville, Florida, Editors, D. Weissman-Berman and K-A. Olsson, EMAS Publication, United Kingdom, 1992, pp. 659–668.

10. Ericsson, A. and B. V. Sankar, "Contact Stiffness of Sandwich Plates and Applications to Impact Problems," in *Sandwich Constructions 2—Proceedings of the Second International Conference on Sandwich Construction*, Gainesville, Florida, Editors, D. Weissman-Berman and K-A. Olsson, EMAS Publication, United Kingdom, 1992, pp. 139–159.
11. Ferri, R. and B. V. Sankar, "Static Indentation and Low Velocity Impact Tests on Sandwich Plates," ASME, Aerospace Division Publication. AD-55, *Analysis and Design Issues for Modern Aerospace Vehicles, Proceedings of the ASME/MECE*, November 1997, Dallas, TX, pp. 485–490.
12. Nemes, J. A. and K. E. Simmonds, "Low Velocity Impact Response of Foam-Core Sandwich Composites," *Journal of Composite Materials*, Vol. 26, No. 4, pp. 500–519, 1992.
13. Nilsen, P. E., T. Moan, and C. G. Gustafson, "Dynamic and Quasi-Static Indentations of PVC/GRP Sandwich Plates," in *Sandwich Constructions 2—Proceedings of the Second International Conference on Sandwich Construction*, Gainesville, Florida, Editors, D. Weissman-Berman and K-A. Olsson, EMAS Publication, United Kingdom, 1992, pp. 139–159, 1992.
14. Wu, C. L., C. A. Weeks and C. T. Sun, "Improving Honeycomb Core Sandwich Structures for Impact Resistance," *Journal of Applied Materials*, pp. 41–47, July 1995.
15. Wu, C. L. and C. T. Sun, "Low-Velocity Impact Damage in Composite Sandwich Beams," *Composite Structures*, Vol. 34, pp. 21–27, 1996.
16. Berger, H. M., "A New Approach to the Analysis of Layer Deflections of Plates," *Journal of Applied Mechanics*, Vol. 22, 1955, p. 465.
17. Eringen, A. C., "On the Nonlinear Oscillations of Viscoelastic Plates," *Journal of Applied Mechanics*, Vol. 33, 1955, p. 563.
18. Wu, C.I., "On the Lateral Vibrations of Rectangular Plates of Pyrolytic Graphite Type Materials," Master's Thesis, Department of Mechanical and Aerospace Engineering, University of Delaware, 1968.
19. Yu, Y. Y., "Nonlinear Flexural Vibrations of Sandwich Plates," *Journal of the Acoustical Society of America*, Vol. 34, 1962, p. 1176.
20. Wu, C-I and Vinson, J. R., "Nonlinear Oscillations of Laminated Specially Orthotropic Plates with Clamped and Simply Supported Edges," *Journal of the Acoustical Society of America*, Vol. 49, No. 5 (Part 2), May 1971, pp. 1561–1567.
21. Ray, J. D. and C. W. Bert, "Nonlinear Vibrations of a Beam with Pinned Ends," *J. Eng. Ind.*, p. 997, 1969.
22. Bert, C. W., "Research on Dynamics of Composite and Sandwich Plates, 1979–81," *Shock and Vibration Digest*, Vol. 14, pp. 17–34, 1982.
23. Leissa, A. W., "Advances in Vibration, Buckling and Postbuckling Studies on Composite Plates," *Proceedings of the First Conference on Composite Structures*, ed. I. M. Marshall, Applied Science, Englewood, NJ, 1981, pp. 312–334.
24. Meyer-Piening, H-R, "Remarks on Higher Order Sandwich Stress and Deflection Analysis," in *Sandwich Construction I—Proceedings of the First International Conference on Sandwich Construction*, Stockholm, Sweden, Editors, K-A. Olsson and R. P. Reichard, publ. 1989.
25. Whitney, J. M. and A. W. Leissa, *Journal of Applied Mechanics*, Vol. 36, pp. 261–266, 1969.
26. Chia, C. Y., *Nonlinear Analysis of Plates*, McGraw-Hill Publishing Company, New York, 1980.
27. Leissa, A. W., "Nonlinear Analysis of Plate & Shell Vibrations," *Proceedings of the Second*

International Conference on Recent Advances in Structural Dynamics, Southampton, England, April 1984, University of Southampton, Southampton, UK, 1984.

28. Sathyamoorthy, M. J., *Sound and Vibration*, Vol. 96, pp. 353–361, 1984.
29. Yu, Y. Y., "Nonlinear Flexural Vibrations of Sandwich Plates," *Journal of the Acoustical Society of America*, Vol. 34, pp. 1176–1183, 1962.
30. Nayfeh, A. H. and D. T. Mook, *Nonlinear Oscillations*, John Wiley and Son, New York, 1979.
31. Bai, J. M. and C. T. Sun, "Nonlinear Flexural Vibration of Sandwich Plates Subjected to In-Plane Forces," in *Sandwich Constructions 2—Proceedings of the Second International Conference on Sandwich Construction*, Gainesville, Florida, Editors, D. Weissman-Berman and K-A. Olsson, EMAS Publication, United Kingdom, 1992, pp. 203–213.
32. Nashif, A. D., D.I.G. Jones and J. P. Henderson, *Vibraton Damping*, Wiley Interscience, 1985.
33. Inman, D. J., *Vibration with Control Measurement and Stability*, Prentice Hall, Englewood Cliffs, N.J. 07632, 1989.
34. Noor, A. K., W. Scott Burton and C. W. Bert, "Computational Models for Sandwich Panels and Shells," *Applied Mechanics Reviews*, Vol. 49, No. 3, March 1996, pp. 155–199.
35. Yu, Y. Y., "Dynamics for Large Deflections of a Sandwich Plate with Thin Piezoelectric Face Layers," *Analysis and Design Issues for Modern Aerospace Vehicles*, ASME AD-Vol. 55, Ed. G. S. Simitses, pp. 285–292.
36. Dobyns, A. L., "The Analysis of Simply-Supported Orthotropic Plates Subjected to State and Dynamic Loads," *AIAA Journal*, pp. 642–650, May 1981.
37. Vinson, J. R. and R. L. Sierakowski, *The Behavior of Structures Composed of Composite Materials*, Martinus-Nijhoff Publishers (now Kluwer Academic Publishers) Dordrecht, The Netherlands, 1986.
38. Thomsen, O. T., "Analysis of Local Bending Effects in Sandwich Panels Subjected to Concentrated Loads," in *Sandwich Constructions 2—Proceedings of the Second International Conference on Sandwich Construction*, Gainesville, Florida, Editors, D. Weissman-Berman and K-A. Olsson, EMAS Publication, United Kingdom, 1992, pp.
39. Sun, C. T., R. S. Hasebe and Y. Hua, "Properties of Sandwich Structures with Reinforced Cure," in *Proceedings of the ASME Aerospace Division, Structures and Materials Committee*, ASME-AD-Vol. 55, New York, 1997.
40. Reissner, E., "On a Variational Theorem in Elasticity," *J. Math. Phys.*, Vol. 29, 1950, p. 90.
41. Mindlin, R. D., "Influence of Rotatory Inertia and Shear on Flexural Motions of Isotropic Elastic Plates," *Journal of Applied Mechanics*, 73, 1951.
42. Pagano, N. J., "Exact Solutions for Rectangular Bidirectional Composites and Sandwich Plates," *Journal of Composite Materials*, Vol. 4, pp. 20–34, 1970.
43. Whitney, J. M., *Structural Analysis of Laminated Anisotropic Plates*, Technomic Publishing Co., Inc., Lancaster, PA, 1987.
43. Warburton, G., "The Vibration of Rectangular Plates," *Proceedings of the Institute of Mechanical Engineers*, p. 371.
44. Vinson, J. R., *The Behavior of Thin Walled Structures: Beams, Plates and Shells*, Kluwer Academic Publishers, Dordrecht, 1989.

CHAPTER 8

Thermal and Moisture Effects on Sandwich Structures

8.1 GENERAL CONSIDERATIONS

All practical structures are subjected to changes in temperature and to thermal gradients. Therefore, it is necessary to understand, plan for, and include thermal effects in the design and analysis of sandwich structures. In particular, during the processing of most composites they are subjected to a thermal environment due to both the manufacturing process and the exothermal chemical reaction in curing the resin.

In addition, most polymeric materials are affected by moisture. Because the moisture effects vary significantly with temperature, the term hygrothermal effects is used [1,2]. Fortunately, there is a direct analog between thermal effects and hygrothermal effects such that if one accounts for thermal effects, one can account for moisture effects with little extra effort.

Elpass [3] points out that composite-sandwich structures must be carefully analyzed and designed to meet very close thermal-distortion tolerances for structures like communication satellite antennas or reflectors of terrestrial systems. This of course means that one must have complete thermoelastic material property data for all components of the structure, and must determine the in-plane and the normal deformations and stresses of the core, as well as the faces.

Incidentally, one can see the advantage of employing piezoelectric components as well to eradicate errors emanating from unexpected thermal/load conditions or slightly incorrect material property data. See Section 2.11 for some detail and several references for further reading.

Consider any elastic isotropic body with a constant coefficient of thermal expansion, α, in the units of in./in./°F, or equivalent, at a uniform temperature where the body is assumed to be free of thermal stresses and strains. If the body is free to deform, and the temperature is raised slowly to a temperature of ΔT degrees from the stress free temperature, the thermal strains produced at any material point can be written as

$$\varepsilon_{ij_{\mathrm{th}}} = \alpha \Delta T(x_i) \delta_{ij} \tag{8.1}$$

where x_i are the coordinate directions, and δ_{ij} is the Kronecker delta ($\delta_{ij} = 1$ for $i = j$, $\delta_{ij} = 0$ for $i \neq j$). It should be noted that thermal strains are purely dilatational ($i = j$); thermal shear strains do not exist.

In Equation (8.1), ΔT is positive when the temperature of the material point is above the stress free temperature. The coefficient of thermal expansion α is positive, i.e., the body expands when it is heated, for all isotropic engineering materials, and it is positive for most composite materials. Graphite or carbon fibers are one exception, because many of them have a negative thermal coefficient, i.e., they shrink when they are heated.

In many thermoelastic bodies, the changes in temperature within the body tend to result in strains which do not satisfy the compatibility equations. In that case isothermal strains, $\varepsilon_{ij_{\mathrm{iso}}}$, the mechanical or isothermal strains discussed in Chapter 2, are induced in order to make the total strain, $\varepsilon_{ij_{\mathrm{tot}}}$, to satisfy compatibility.

$$\varepsilon_{ij_{\mathrm{tot}}} = \varepsilon_{ij_{\mathrm{iso}}} + \varepsilon_{ij_{\mathrm{th}}} \tag{8.2}$$

In that case the "thermal stresses" are induced due to the isothermal strains induced to insure compatibility. This can occur, for example, in "thermal shock" in which there is very rapid or localized heating.

A second way that thermal stresses occur is through displacement restrictions i.e., boundary conditions on the elastic body. One simple example of this occurs when a bar is placed between immovable end grips and subsequently heated. In this case, compressive thermal stresses result.

Hence, thermal stresses are caused by two mechanisms: one by displacement restrictions, the other through induced isothermal strains to maintain compatibility. In this latter case, where thermal stresses are induced to insure compatibility the stresses are self equilibrating.

Next, consider an unrestricted thin rod at a uniform temperature. If the rod is slowly heated uniformly, such that the thermal strains satisfy the compatibility equations, the heated rod has thermal strains but no thermal stresses. Now if the unheated thin rod is placed in immovable end grips such that the rod cannot increase in length, slowly heating the rod uniformly will result in thermal stresses and no thermal strains.

In the latter case, if the compressive axial thermal stresses reach a value equal to the Euler buckling load (discussed later in Chapter 9), the rod will buckle. This is called thermal buckling. If the compressive thermal stresses exceed the compressive strength of the material at that temperature, failure will occur.

8.2 DERIVATION OF THE GOVERNING EQUATIONS FOR A THERMOELASTIC ISOTROPIC PLATE

To investigate the thermal effects on a structure, it is convenient to first look at a single layer (monocoque) rectangular plate of an isotropic material. In deriving the governing equations for this thermoelastic plate, the equilibrium equations and the strain-displacement equations are not altered from those of the isothermal plate of Chapter 2, because in the former the equations involve force balances, and in the latter, they are purely kinematic relationships involving total strains. However, the stress-strain relations are modified in accordance with Equations (8.2) and (8.1), and are given below for a simple one-ply isotropic material

$$\varepsilon_{x_{\text{iso}}} = \varepsilon_{x_{\text{tot}}} - \alpha\Delta T = \frac{1}{E}[\sigma_x - \nu\sigma_y]$$

$$\varepsilon_{y_{\text{iso}}} = \varepsilon_{y_{\text{tot}}} - \alpha\Delta T = \frac{1}{E}[\sigma_y - \nu\sigma_x]$$

or

$$\varepsilon_x = \frac{1}{E}[\sigma_x - \nu\sigma_y] + \alpha\Delta T \tag{8.3}$$

$$\varepsilon_y = \frac{1}{E}[\sigma_y - \nu\sigma_x] + \alpha\Delta T \tag{8.4}$$

In Equations (8.3) and (8.4) and in all that follows the subscript for total strains is dropped, and all strains noted explicitly are those that satisfy compatibility and appear in the strain-displacement relations. Hence, in Equations (8.3) and (8.4) the first terms on the right-hand side are really the isothermal strains, and the second terms on the right-hand side are thermal strains.

Proceeding as in Chapter 3, employing the strain-displacement relations, Equations (8.3) and (8.4) after multiplying by E become

$$\sigma_x - \nu\sigma_y + E\alpha\Delta T = E\frac{\partial u_0}{\partial x} - Ez\frac{\partial^2 w}{\partial x^2} \tag{8.5}$$

$$\sigma_y - \nu\sigma_x + E\alpha\Delta T = E\frac{\partial v_0}{\partial y} - Ez\frac{\partial^2 w}{\partial y^2} \tag{8.6}$$

Now, two quantities N^T and M^T, known as the thermal-stress resultant and the thermal-stress couple, respectively, are defined as

$$N^T = \int_{-h/2}^{+h/2} E\alpha\Delta T dz, \; M^T = \int_{-h/2}^{+h/2} E\alpha\Delta T z dz \tag{8.7}$$

Multiplying Equations (8.5) and (8.6) by dz and integrating across the thickness of the plate, then multiplying them by zdz and integrating them, provides the integrated stress-strain relations for a thermoelastic plate. It should be remembered from the discussion of Section 8.1 that the shear stress-strain relations are not altered by the inclusion of thermoelastic effects.

$$N_x - \nu N_y + N^T = Eh\frac{\partial u_0}{\partial x}$$

$$N_y - \nu N_x + N^T = Eh\frac{\partial v_0}{\partial y}$$

$$M_x - \nu M_y + M^T = \frac{-Eh^3}{12}\frac{\partial^2 w}{\partial x^2}$$

$$M_y - \nu M_x + M^T = \frac{-Eh^3}{12}\frac{\partial^2 w}{\partial y^2}$$

$$N_{xy} = \frac{K(1-\nu)}{2}\left[\frac{\partial u_0}{\partial y} + \frac{\partial v_0}{\partial x}\right] \tag{8.8}$$

$$M_{xy} = -D(1-\nu)\frac{\partial^2 w}{\partial x \partial y} \tag{8.9}$$

where K is the in-plane stiffness and D is the flexural stiffness of the isotropic structure. Rearranging the first four of the above results in

$$N_x = K\left[\frac{\partial u_0}{\partial x} + \nu\frac{\partial v_0}{\partial y}\right] - \frac{N^T}{(1-\nu)} \tag{8.10}$$

$$N_y = K\left[\frac{\partial v_0}{\partial y} + \nu\frac{\partial u_0}{\partial x}\right] - \frac{N^T}{(1-\nu)} \tag{8.11}$$

$$M_x = -D\left[\frac{\partial^2 w}{\partial x^2} + \nu\frac{\partial^2 w}{\partial y^2}\right] - \frac{M^T}{(1-\nu)} \tag{8.12}$$

$$M_y = -D\left[\frac{\partial^2 w}{\partial y^2} + \nu\frac{\partial^2 w}{\partial x^2}\right] - \frac{M^T}{(1-\nu)} \tag{8.13}$$

Introducing these thermoelastic stress-strain relations into the equilibrium equations (3.14) through (3.18) and (3.9) and (3.10), the governing differential equations for a thermoelastic plate are determined, for the case of no surface shear stresses:

$$D\nabla^4 w = p(x,y) - \frac{1}{(1-\nu)}\nabla^2 M^T \tag{8.14}$$

$$K\nabla^4 u_0 = \frac{1}{(1-\nu)}\frac{\partial}{\partial x}(\nabla^2 N^T) \tag{8.15}$$

$$K\nabla^4 v_0 = \frac{1}{(1-\nu)}\frac{\partial}{\partial y}(\vartheta^2 N^T) \tag{8.16}$$

where here the biharmonic operator $\nabla^4(\,)$ is $\nabla^2(\nabla^2(\,)) = \dfrac{\partial^4(\,)}{\partial x^4} + \dfrac{2\partial^4(\,)}{\partial x^2 \partial y^2} + \dfrac{\partial^4(\,)}{\partial y^4}$.

Also for completeness, other useful relationships are catalogued below:

$$\begin{aligned} Q_x &= D\frac{\partial}{\partial x}(\nabla^2 w) - \frac{1}{(1-\nu)}\frac{\partial M^T}{\partial x} \\ Q_y &= D\frac{\partial}{\partial y}(\nabla^2 w) - \frac{1}{(1-\nu)}\frac{\partial M^T}{\partial y} \end{aligned} \tag{8.17}$$

$$\begin{aligned} V_x &= -D\left[\frac{\partial^3 w}{\partial x^3} + (2-\nu)\frac{\partial^3 w}{\partial x \partial y^2}\right] - \frac{1}{(1-\nu)}\frac{\partial M^T}{\partial x} \\ V_y &= -D\left[\frac{\partial^3 w}{\partial y^3} + (2-\nu)\frac{\partial^3 w}{\partial y \partial x^2}\right] - \frac{1}{(1-\nu)}\frac{\partial M^T}{\partial y} \end{aligned} \tag{8.18}$$

Due to the inclusion of thermal quantities, the expressions for various normal stresses in the plate are modified as follows, as shown by Boley and Weiner [4]

$$\sigma_x = \frac{1}{h}\left[N_x + \frac{N^T}{(1-\nu)}\right] + \frac{z}{h^3/12}\left[M_x + \frac{M_T}{(1-\nu)}\right] - \frac{E\alpha\Delta T}{(1-\nu)}$$

$$\sigma_y = \frac{1}{h}\left[N_y + \frac{N^T}{(1-\nu)}\right] + \frac{z}{h^3/12}\left[M_y + \frac{M^T}{(1-\nu)}\right] - \frac{E\alpha\Delta T}{(1-\nu)} \quad (8.19)$$

The inclusion of the N^T and M^T terms in Equation (8.19) are easy to visualize since they are thermal stress resultants and couples analogous to N_x, N_y, M_x, and M_y, that are caused by lateral and in-plane "mechanical" loads. The last terms in Equation (8.19) can be visualized by the following example in which it is assumed that the first two terms do not contribute. Suppose at some value of (x,y) in a plate, the upper surface is heated while the lower surface is cooled. Thus the value of ΔT in the upper portion of the plate is positive and becomes negative in the lower-plate portion, as shown in Figure 8.1. The last term of Equation (8.19) shows that there are tensile stresses in the upper portion of the plate, while in the lower portion compressive stresses exist due to the last term of Equation (8.19). Physically, the material points in the upper portion of the plate want to expand considerably but are being restrained by those in the cooler areas of the plate, hence, tending to cause high compressive stresses. Likewise, in the cooler portion of the plate the material points wish to contract, but are being extended by the hotter portions of the plate, hence, thrown into tension. Such thermal stresses can result in material failure just as surely as stresses caused by mechanical loads.

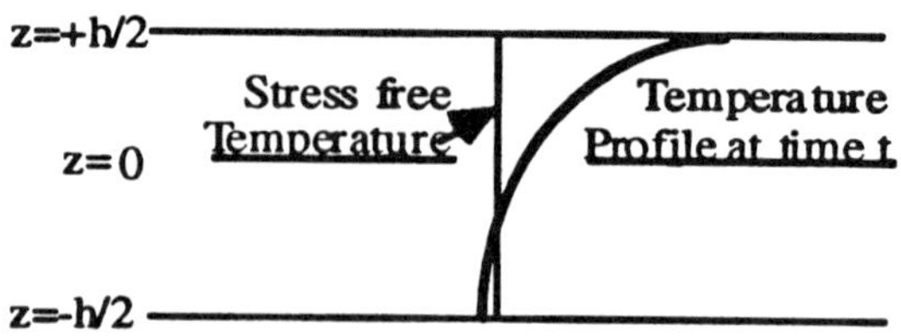

FIGURE 8.1. (Reprinted from Reference [18].)

As discussed before, shear stresses and strains are not affected by thermal effects, hence they would be as follows for a single-ply isotropic plate.

$$\sigma_{xy} = \frac{N_{xy}}{h} + \frac{M_{xy}z}{h^3/12}$$

$$\sigma_{xz} = \frac{3Q_x}{2h}\left[1 - \left(\frac{z}{h/2}\right)^2\right] \tag{8.20}$$

$$\sigma_{yz} = \frac{3Qy}{2h}\left[1 - \left(\frac{z}{h/2}\right)^2\right]$$

Of course if there exist shear stresses applied to the upper or lower surfaces of the plate, the latter two expressions must be modified as in Chapter 3. Keep in mind that the equations for stresses of (8.19) and (8.20) pertain only to a monocoque plate of an isotropic material. Other expressions must be used for composite laminated plates and sandwich plates.

To proceed with solutions of thermoelastic plates using Equations (8.14) through (8.16), one now employs the same techniques that were introduced in Chapter 6. However, the additions of thermal effects do introduce certain problems with nonhomogeneous boundary conditions that cause some analytical difficulties. These are discussed in the next section.

8.3 BOUNDARY CONDITIONS

Looking now at the boundary conditions associated with a thermoelastic plate, comparisons are made with the boundary condition for an isothermal plate, where again n denotes the direction normal to the edge and t denotes the direction parallel to the edge.

8.3.1 Simply-Supported Edge

$$w = 0, \qquad M_n = 0 \tag{8.21}$$

From Equations (8.12) and (8.13), the latter equation above is in fact

$$M_n = 0 = -D\left[\frac{\partial^2 w}{\partial n^2} + \nu\frac{\partial^2 w}{\partial t^2}\right] - \frac{M^T}{(1 - \nu)} \tag{8.22}$$

Since in the case of a simply-supported edge there is no curvature along the edge (i.e., $\partial^2 w/\partial t^2 = 0$), this equation becomes

$$\frac{\partial^2 w}{\partial n^2} = -\frac{M^T}{D(1 - \nu)} \tag{8.23}$$

Hence, the boundary conditions for a simply-supported edge of a thermoelastic plate are nonhomogeneous.

8.3.2 Clamped Edge

$$w = 0, \qquad \frac{\partial w}{\partial n} = 0 \tag{8.24}$$

These remain the same as those for the isothermal plate.

8.3.3 Free Edge

The boundary conditions are

$$M_n = 0 \text{ and } V_n = 0$$

Hence, the first condition is given by Equation (8.22) and the latter is seen to be from Equation (8.18),

$$\frac{\partial^3 w}{\partial n^3} + (2 - \nu)\frac{\partial^3 w}{\partial n \partial t^2} = -\frac{1}{D(1 - \nu)}\frac{\partial M^T}{\partial n} \tag{8.25}$$

Here the boundary conditions are seen to be nonhomogeneous also.

8.3.4 General

In many problems involving thermoelastic plates, it is seen that the boundary conditions are nonhomogeneous. Why is this important? In solving linear partial differential equations, separation of variables cannot be used with nonhomogeneous boundary conditions. Fortunately, methods are available to transform either homogeneous or nonhomogeneous partial differential equations with nonhomogeneous boundary conditions to nonhomogeneous partial differential equations with homogeneous boundary conditions, so that separation of variables may be used. A generalized method is presented in the next section.

8.4 GENERAL TREATMENT OF PLATE NONHOMOGENEOUS BOUNDARY CONDITIONS

Consider an isotropic sandwich (or monocoque) plate with the $y = 0$, b edges simply-supported. If the plate is orthotropic the same procedures would be used. The governing equation for the lateral deflection is given by Equation (8.14). From Equations (8.21) and (8.23), the boundary conditions are

$$w(x,0) = w(x,b) = 0 \tag{8.26}$$

$$\frac{\partial^2 w}{\partial y^2}(x,0) = -\frac{M^T(x,0)}{D(1-\nu)} = -\frac{M_1^T(x)}{D(1-\nu)} \tag{8.27}$$

$$\frac{\partial^2 w}{\partial y^2}(x,b) = -\frac{M^T(x,b)}{D(1-\nu)} = -\frac{M_2^T(x)}{D(1-\nu)} \tag{8.28}$$

where $M_1^T(x) = M^T(x,0)$ and $M_2^T = M^T(x,b)$.

Now introduce a function $\Psi(x,y)$ which satisfies homogeneous boundary conditions on the $y = 0$ and $y = b$ edges, by letting

$$w(x,y) = \Psi(x,y) + f_1(y)M_1^T(x) + f_2(y)M_2^T(x). \tag{8.29}$$

For this example, take $\Psi(x,y)$ to be of the Levy form [see Equation (6.11)]:

$$\Psi(x,y) = \sum_{n=1}^{\infty} \phi_n(x)\sin\left(\frac{n\pi y}{b}\right) \tag{8.30}$$

Also in Equation (8.29) $f_1(y)$ and $f_2(y)$ are to be determined later to satisfy the boundary conditions (8.26) through (8.28).

Substituting Equation (8.29) into Equations (8.26) through (8.28) results in

$$w(x,0) = \Psi(x,0) + f_1(0)M_1^T(x) + f_2(0)M_2^T(x) = 0$$

$$w(x,b) = \Psi(x,b) + f_1(b)M_1^T(x) + f_2(b)M_2^T(x) = 0$$

$$\frac{\partial^2 w(x,0)}{\partial y^2} = \frac{\partial^2 \Psi(x,0)}{\partial y^2} + f_1''(0)M_1^T(x) + f_2''(0)M_2^T(x) = \frac{M_1^T(x)}{D(1-\nu)}$$

$$\frac{\partial^2 w(x,b)}{\partial y^2} = \frac{\partial^2 \Psi(x,b)}{\partial y^2} + f_1''(b)M_1^T(x) + f_2''(b)M_2^T(x) = -\frac{M_2^T(x)}{D(1-\nu)}$$

It is required that $\Psi(x,y)$ satisfy homogeneous boundary condition at $y = 0$ and $y = b$, as stated above, therefore

$$\Psi(x,0) = \Psi(x,b) = \frac{\partial^2\Psi(x,0)}{\partial y^2} = \frac{\partial^2\Psi(x,b)}{\partial y^2} = 0 \tag{8.31}$$

Hence, from the above; it is required that

$$\begin{aligned} &f_1(0) = 0 && f_2(0) = 0 \\ &f_1(b) = 0 && f_2(b) = 0 \\ &f_1''(0) = -1/[D(1-\nu)] && f_2''(0) = 0 \\ &f_1''(b) = 0 && f_2''(b) = -1/[D(1-\nu)] \end{aligned} \tag{8.32}$$

These are the only requirements on $f_2(y)$ and $f_2(y)$. Since there are four conditions on each function, each can be assumed to be a third-order polynomial, such as

$$f_1(y) = C_0 + C_1 y + C_2 y^2 + C_3 y^3 \tag{8.33}$$

and

$$f_2(y) = k_0 + k_1 y + k_2 y^2 + k_3 y^3 \tag{8.34}$$

Substituting Equations (8.33) and (8.34) into (8.32), the result is

$$f_1(y) = \frac{1}{6bD(1-\nu)}(2b^2y - 3by^2 + y^3) \tag{8.35}$$

$$f_2(y) = \frac{1}{6bD(1-\nu)}(b^2y - y^3) \tag{8.36}$$

Using Equations (8.35) and (8.36), the substitution of Equation (8.29) into (8.14) results in the following:

$$\begin{aligned} D\nabla^4\Psi = p(x,y) &- \frac{1}{(1-\nu)}\nabla^2 M^T \\ &- \nabla^4\left\{\frac{1}{6(1-\nu)b}(2b^2y - 3by^2 + y^3)M_1^T(x)\right\} \\ &- \nabla^4\left\{\frac{b^2y - y^3}{6(1-\nu)b}M_2^T(x)\right\} \end{aligned} \tag{8.37}$$

Looking at Equation (8.37) it is seen that the original problem, which was Equation (8.14), with nonhomogeneous boundary conditions [Equations (8.26) through (8.28)] has been transformed into a problem involving a "lateral deflection" Ψ, with homogeneous boundary conditions (8.31) and an "altered loading," given by the right-hand side of Equation (8.37), which one can now simply write as $H(x,y)$. Hence,

$$D\nabla^4\Psi = H(x,y) \tag{8.38}$$

Here $\Psi(x,y)$ is given by Equation (8.30), and therefore, $H(x,y)$ must be expanded correspondingly into a Fourier series as

$$H(x,y) = \sum_{n=1}^{\infty} h_n\,(x) \sin\,(\lambda_n y) \text{ where } \lambda_n = n\pi/b \tag{8.39}$$

Substituting Equations (8.30) and (8.39) into Equation (8.38) gives

$$D\sum_{n=1}^{\infty}\left\{\phi\frac{IV}{n} - 2\lambda_n^2\phi_n'' + \lambda_n^4\phi_n\right\}\sin\,(\lambda_n y) = \sum_{n=1}^{\infty} h_n\,(x)\sin\,(\lambda_n y)$$

Hence,

$$\phi\frac{IV}{n} - 2\lambda_n^2\phi_n'' + \phi_n\lambda_n^4 = \frac{h_n\,(x)}{D} \tag{8.40}$$

It is seen that this has the same form of the ordinary differential equation in the Chapter 6 discussion of the Levy method. Now the boundary conditions at $x = 0$ and $x = a$ can be considered. As a specific example, consider them to be simply supported also. Then,

$$w(0,y) = w(a,y) = 0$$

$$M_x(0,y) = M_x(a,y) = 0$$

or

$$\begin{aligned} \frac{\partial^2 w(a,y)}{\partial x^2} &= -\frac{M^T(a,y)}{D(1-\nu)} \\ \frac{\partial^2 w(a,y)}{\partial x^2} &= -\frac{M^T(a,y)}{D(1-\nu)} \end{aligned} \tag{8.41}$$

Substituting Equation (8.29) and its derivatives into the above results in

$$w(0,y) = \Psi(0,y) + f_1(y)M_1^T(0) + f_2(y)M_2^T(0) = 0$$

$$w(a,y) = \Psi(a,y) + f_1(y)M_1^T(a) + f_2(y)M_2^T(a) = 0$$

$$\frac{\partial^2 w(0,y)}{\partial x^2} = \frac{\partial^2 \Psi(0,y)}{\partial x^2} + f_1(y)M_1^{T''}(0) + f_2(y)M_2^{T''}(0) = -\frac{M^T(0,y)}{D(1-\nu)}$$

$$\frac{\partial^2 w(a,y)}{\partial x^2} = \frac{\partial^2 \Psi(a,y)}{\partial x^2} + f_1(y)M_1^{T''}(0) + f_2(y)M_2^{T''}(a) = -\frac{M^T(a,y)}{D(1-\nu)}$$

Rearranging the above produces

$$\Psi(0,y) = -[f_1(y)M_1^T(0) + f_2(y)M_2^T(0)]$$

$$\Psi(a,y) = -[f_1(y)M_1^T(a) + f_2(y)M_2^T(a)]$$

$$\frac{\partial^2 \Psi(0,y)}{\partial x^2} = -\left[\frac{M^T(0,y)}{D(1-\nu)} + f_1(y)M_1^{T''}(0) + f_2(y)M_2^{T''}(0)\right] \tag{8.42}$$

$$\frac{\partial^2 \Psi(a,y)}{\partial x^2} = -\left[\frac{M^T(a,y)}{D(1-\nu)} + f_1(y)M_1^{T''}(a) + f_2(y)M_2^{T''}(a)\right]$$

Remembering that $\Psi(x,y)$ is given by Equation (8.30), it is logical to make the following expansions:

$$f_1(y) = \sum_{n=1}^{\infty} A_n \sin(\lambda_n y), \qquad M^T(0,y) = \sum_{n=1}^{\infty} C_n \sin(\lambda_n y)$$

$$f_2(y) = \sum_{n=1}^{\infty} B_n \sin(\lambda_n y), \qquad M^T(a,y) = \sum_{n=1}^{\infty} E_n \sin(\lambda_n y) \tag{8.43}$$

Here, A_n, B_n, C_n, and E_n are easily found by substituting Equations (8.43) and (8.30) into (8.42) and equating all coefficients. That results in the following:

$$\begin{aligned}
\phi_n(0) &= -[A_n M_1^T(0) + B_n M_2^T(0)] \\
\phi_n(a) &= -[A_n M_1^T(a) + B_n M_2^T(a)] \\
\phi_n''(0) &= -\left[\frac{C_n}{D(1-\nu)} + A_n M_1^{T''}(0) + B_n M_2^{T''}(0)\right] \\
\phi_n''(a) &= -\left[\frac{E_n}{D(1-\nu)} + A_n M_1^{T''}(a) + B_n M_2^{T''}(a)\right]
\end{aligned} \tag{8.44}$$

Hence, these four boundary values provide the necessary information to determine the constants in the solution of Equation (8.40), which is

$$\phi_n(x) = (K_1 + K_2 x)\cosh(\lambda_n x) + (K_3 + K_4 x)\sinh(\lambda_n x) + \eta_n(x) \tag{8.45}$$

Here, $\eta_n(x)$ is the particular solution of Equation (8.40). Using this, Equation (8.30) is completely solved, and in turn Equation (8.29) is solved. Subsequently, stress couples, shear resultants, and stresses can be determined everywhere using Equations (8.19) and (8.20). This general approach can be used to solve any plate problem involving nonhomogeneous boundary conditions, isotropic or orthotropic, monocoque or sandwich.

8.5 THERMOELASTIC EFFECTS ON BEAMS

Turning now to the much simpler problem of the thermoelastic isotropic beam one can modify the previous equations. For the thermoelastic beam, it can be easily shown from Chapters 3 and 4, and Section 8.2 that the governing differential equations are as follows:

$$P + P^T = EA\frac{du_0}{dx} \tag{8.46}$$

$$M_b + M_b^T = -EI\frac{d^2w}{dx^2} \tag{8.47}$$

$$V_b = -EI\frac{d^3w}{dx^3} - \frac{dM^T}{dx} \tag{8.48}$$

$$EI\frac{d^4w}{dx^4} = q(x) - \frac{d^2M^T}{dx^2} \tag{8.49}$$

$$\sigma_x = \frac{1}{A}[P + P^T] + \frac{z}{I}[M_b + M^T] - E\alpha\Delta T \tag{8.50}$$

$$P^T = \int_{-h/2}^{+h/2} Eb\alpha\Delta T dz,\ M^T = \int_{-h/2}^{+h/2} Eb\alpha\Delta T z dz \tag{8.51}$$

In the above equations, A is the cross-sectional area of the beam, and I is its area moment of inertia. Of course the stresses expressed in Equation (8.50) only apply when the beam is composed of one ply of (in this case) an isotropic material, or a unidirectional composite material.

Using these equations, solutions are easily obtained, analogous to those obtained in Chapter 4. As for thermoelastic plates, the expressions for boundary conditions of simple supported and free edges are nonhomogeneous, but in the case of beams (unlike plates) this causes no particular problem, because ordinary differential equations are involved, not partial differential equations, hence, separation of variables is not needed to obtain solutions.

With this simple beam, it is convenient to further investigate the qualitative statements made at the end of Section 8.1 regarding thermal stresses and thermal strains. Consider a rod that is free to expand or contract on the slow application of heat or cold, such that with time there is a uniform change in temperature throughout the rod. There is no beam bending, hence $M_b = 0$ and because of uniform heating $M^T = 0$. Looking at Equation (8.46) $P = 0$, because no axial load is applied. Therefore

$$\frac{du_0}{dx} = \frac{P^T}{EA} = \varepsilon_x \tag{8.52}$$

Therefore, there is a thermal strain given by Equation (8.52). From Equation (8.51) $P^T = EA\alpha\Delta T$ where A is the cross-sectional area of the beam. Therefore substituting this into Equation (8.50) results in

$$\sigma_x = 0 \tag{8.53}$$

It is, therefore, seen that although there is a thermal strain, there is no thermal stress.

Now consider the same rod of length L rigidly held in place by end fixtures such that $u_0(0) = u_0(L) = 0$, i.e., the rod cannot expand on heating. From Equation (8.46) $u_0(x) = 0$, everywhere and therefore,

$$\varepsilon_x = \frac{du_0}{dx} = 0 \tag{8.54}$$

However, from Equation (8.46)

$$P = -P^T \tag{8.55}$$

and therefore from Equation (8.50)

$$\sigma_x = -E\alpha\Delta T \tag{8.56}$$

Here there is no thermal strain, but there is a compressive thermal stress. If this thermal stress reaches the value of the Euler buckling load for the column at that temperature with those boundary conditions the column will thermally buckle. If the compressive thermal stress reaches a value of the material's compressive strength at that temperature, the column will fail.

Note: Thermal stress analysis of a structure is usually very involved, because of the complications of the thermal terms and the nonhomogeneous boundary conditions that often occur. However, the following generalization can be made. Structurally, from above it is seen that if all parts of a structure were free to expand there are no thermal stresses. Also, the worst case occurs when the structure cannot expand or contract at all, in which case from above the thermal stress is $E\alpha\Delta T$. If the structure is two dimensional such that Poisson ratio effects should be considered, then the worst case is $E\alpha\Delta T/(1 - \nu^2)$. If this is the worst case then when considering whether thermal stress effects are important in an analysis or design, simply calculate $E\alpha\Delta T/(1 - \nu^2)$ for the highest ΔT, using the largest value of E, over the particular temperature range involved, and see what the maximum value of thermal stress can be. If it is insignificant compared to the mechanical stresses and if the sum of the maximum mechanical stress plus the maximum thermal stress at the location of highest stress is less than the strength of the material at the highest temperature, then from this simple calculation one can determine that thermal stresses are not important. Conversely, they could be found easily and early to be of major importance. In either case the simple procedure can save major efforts in solutions or can prevent major structural failure caused by not including thermoelastic effects. The third major characteristic of thermal effects, self equilibration, is discussed in the following section.

8.6 SELF-EQUILIBRATION OF THERMAL STRESSES

In Section 8.1 the two mechanisms by which thermal stresses are introduced into a thermoelastic solid body were discussed. First, by displacement restrictions caused by the boundary conditions; second, by the introduction of the isothermal strains in Equation (8.2) so the total strain, satisfy compatibility (i.e., when the thermal strains, $\varepsilon_{ij_{\text{th}}}$ do not satisfy compatibility).

One other physical phenomenon occurs that is very important in the structural mechanics of planar bodies such as beams and plates; namely, if the planar body is not restricted by the boundary conditions, the thermal stresses in the body are self-equilibrating: i.e., the average stress across the thickness is zero,

$$\sigma_{i_{\text{avg}}} = \frac{1}{h}\int_{-h/2}^{+h/2} \sigma_i dz = 0 \quad (i = x,y) \tag{8.57}$$

This can be exemplified in the easiest and shortest way by considering a load-free beam lying on a friction-free surface and heated or cooled such that at any time, t, the temperature change across the thickness only is given by:

$$\Delta T(z) = \sum_{n=0}^{\infty} a_n \left(\frac{z}{h}\right)^n = a_0 + a_1\frac{z}{h} + a_2\frac{z^2}{h^2} + \ldots \tag{8.58}$$

It is seen that the term, a_0, is merely a uniform heating or cooling of the entire beam; $(a_0 + a_1 z)$ is a steady-state heat transfer situation. The entire expression represents a temperature gradient that involves additional terms occuring during transient heating or cooling and/or internal heat generation such as an exothermic reaction during cure. In the following example, it is sufficient to consider only the first three terms of Equation (8.58) to illustrate the point. From Equations (8.58) and (8.51),

$$P^T = EA\alpha\left[a_0 + \frac{a_2}{12}\right], \qquad M^T = \frac{E\alpha bh^2 a_1}{12} \tag{8.59}$$

It is important to note that the thermal resultant, P^T, is a function of the uniform temperature change, (a_0), and the transient temperature situation, (a_2). Likewise the thermal moment term, M^T, emanates only from a thermal gradient through the thickness (a_1).

For this illustrative problem, the lateral deflection is

$$w(x) = C_0 + C_1 x + C_2 x^2 + C_3 x^3 \tag{8.60}$$

where this form is the solution to Equation (8.49) for this case. The boundary conditions for this example are seen to be simply supported, i.e.,

$$w(0,L) = M_b(0,L) = 0 \tag{8.61}$$

With these boundary conditions, the constants of Equation (8.60) are found to be

$$C_0 = 0,\ C_1 = \frac{\alpha a_1 L}{2A},\ C_2 = -\frac{\alpha a_1}{2A},\ C_3 = 0 \tag{8.62}$$

and the lateral deflection is found to be

$$w(x) = \frac{\alpha a_1}{2A}\quad x(L - x) \tag{8.63}$$

It is seen that only if a_1 is nonzero i.e., a thermal gradient through the thickness, will the beam deflect at all. Also from Equation (8.50), it is seen that

$$\sigma_x = E\alpha\frac{a_2}{12}\left[1 - \frac{12_z^2}{h^2}\right] \qquad (i = x,y) \tag{8.64}$$

It is extremely important to see that stresses will occur in the beam if and only if a_2 is nonzero. Therefore, it is seen that for a steady-state temperature distribution (a_n, $n \geq 2 = 0$) the beam is stress free, whether a deflection occurs or not. Substituting Equation (8.64) into (8.57) results in, for all cases,

$$\sigma_{i_{\text{avg}}} = 0 \tag{8.65}$$

Therefore, in those cases in which boundary conditions do not constrain the thin beam or plate, the average stress across the thickness is zero, i.e., the stresses are self-equilibrating.

8.7 RECTANGULAR COMPOSITE-MATERIAL PLATE SUBJECTED TO LATERAL AND HYGROTHERMAL LOADS

Numerous publications are devoted to the thermal response of laminated anisotropic plates and shells, reviewed in 1992 by Noor and Burton [5]. Relatively few studies have been done regarding heat transfer in sandwich plates and shells. The studies that have been conducted up to 1996 are contained in the recent review of Noor, Burton and Bert [6]. Among the major studies are those of Chamis, Aiello, and Murthy [7] in 1988. Among recent research published in 1997 are papers of Noor, Starnes, and Peters [8] and Noor and Peters [9].

Whenever a plate or panel structure is composed of a composite material such that the plate is orthotropic or generally anisotropic, and if transverse-shear deformation effects are important, and if there is a thermal environment present, then obtaining a solution through solving the partial differential equations with the attendant nonhomogeneous boundary conditions is a major challenge. Even if the solution is obtained it requires a major effort. Far more promising is the approach utilizing the Theorem of Minimum Potential Energy.

8.7.1 Minimum Potential Energy Expressions

Sloan [10] performed such a study for a monocoque composite-material laminated plate, and because the results are quite applicable to anisotropic sandwich plates, simply by using the appropriate stiffness quantities, the following is given in some detail because the results are so general.

The stress-strain relations to be used are given by Equation (2.40), wherein the $\overline{Q}_{ij}$ stiffness matrix quantities are given by Equation (2.41). The strain-displacement relations are given by Equation (2.46), where the form of the panel displacements are given by Equation (2.47).

Employing the Theorem of Minimum Potential Energy [Equations (5.1) and (5.2)] for this plate, it is seen that on summing all of the strain-energy density functions for each lamina across N laminae the potential energy for this plate subjected to a laterally-distributed loading $p(x,y)$ is

$$
\begin{aligned}
V = \frac{1}{2}\sum_{k=1}^{N}\int_A\int_{h_{k-1}}^{h_k} & \{\sigma_x\,[\varepsilon_x - \alpha_x\Delta T - \beta_x\Delta m] \\
& + \sigma_y\,[\varepsilon_y - \alpha_y\Delta T - \beta_y\Delta m] + \sigma_{xz}[2\varepsilon_{xz}] \\
& + \sigma_{yz}\,[2\varepsilon_{yz}] + \sigma_{xy}\,[2\varepsilon_{xy} - \alpha_{xy}\Delta T - \beta_{xy}\Delta m]\}_k\,dzdA \\
& - \int_A p(x,y)w(x,y)dA
\end{aligned}
\qquad (8.66)
$$

Here A refers to the planform area of the plate whose dimensions are $0 \leq x \leq a$, $0 \leq y \leq b$ and $-h/2 \leq z \leq h/2$. It is noted that the strains used in the strain-energy relations are the isothermal strains, hence one sees the differences between total strain and the thermal and hygrothermal strains in Equation (8.66).

Now, the result of substituting the constitutive equations from Equation (2.64) and the strain-displacement relations from Equation (2.46) into (8.66) is presented here again for continuity in this chapter, although it is a repeat of Equation (5.70).

$$\begin{aligned}
V = \int_A \Bigg\{ & \frac{A_{11}}{2}\left(\frac{\partial u_0}{\partial x}\right)^2 + B_{11}\frac{\partial u_0}{\partial x}\frac{2\overline{\alpha}}{\partial x} + \frac{D_{11}}{2}\left(\frac{\partial \overline{\alpha}}{\partial x}\right)^2 + A_{12}\frac{\partial u_0}{\partial x}\frac{\partial v_0}{\partial y} \\
& + B_{12}\left[\frac{\partial u_0}{\partial x}\frac{\partial \overline{\beta}}{\partial y} + \frac{\partial v_0}{\partial y}\frac{\partial \overline{\alpha}}{\partial x}\right] + D_{12}\frac{\partial \overline{\beta}}{\partial y}\frac{\partial \overline{\alpha}}{\partial x} + A_{16}\left[\frac{\partial u_0}{\partial x}\frac{\partial u_0}{\partial y} + \frac{\partial u_0}{\partial x}\frac{\partial v_0}{\partial x}\right] \\
& + D_{16}\left[\frac{\partial \overline{\alpha}}{\partial x}\frac{\partial \overline{\alpha}}{\partial y} + \frac{\partial \overline{\alpha}}{\partial x}\frac{\partial \overline{\beta}}{\partial x}\right] + B_{16}\left[\frac{\partial u_0}{\partial x}\frac{\partial \overline{\alpha}}{\partial y} + \frac{\partial u_0}{\partial x}\frac{\partial \overline{\beta}}{\partial x} + \frac{\partial u_0}{\partial y}\frac{\partial \overline{\alpha}}{\partial x} + \frac{\partial v_0}{\partial x} + \frac{\partial \overline{\alpha}}{\partial x}\right] \\
& + \frac{A_{22}}{2}\left(\frac{\partial v_0}{\partial y}\right)^2 + B_{22}\frac{\partial v_0}{\partial y}\frac{\partial \overline{\beta}}{\partial y} + \frac{D_{22}}{2}\left(\frac{\partial \overline{\beta}}{\partial y}\right)^2 + A_{26}\left[\frac{\partial v_0}{\partial y}\frac{\partial u_0}{\partial y} + \frac{\partial v_0}{\partial y}\frac{\partial v_0}{\partial x}\right] \\
& + B_{26}\left[\frac{\partial v_0}{\partial x}\frac{\partial \overline{\alpha}}{\partial y} + \frac{\partial v_0}{\partial y}\frac{\partial \overline{\beta}}{\partial x} + \frac{\partial u_0}{\partial y}\frac{\partial \overline{\beta}}{\partial y} + \frac{\partial v_0}{\partial x}\frac{\partial \overline{\beta}}{\partial y}\right] + D_{26}\left[\frac{\partial \overline{\alpha}}{\partial y}\frac{\partial \overline{\beta}}{\partial y} + \frac{\partial \overline{\beta}}{\partial x}\frac{\partial \overline{\beta}}{\partial y}\right] \\
& + A_{45}\left[\overline{\alpha}\overline{\beta} + \overline{\alpha}\frac{\partial w}{\partial y} + \overline{\beta}\frac{\partial w}{\partial x} + \frac{\partial w}{\partial x}\frac{\partial w}{\partial y}\right] + A_{55}\left[\frac{\overline{\alpha}^2}{2} + \overline{\alpha}\frac{\partial w}{\partial x} + \frac{1}{2}\left(\frac{\partial w}{\partial x}\right)^2\right] \\
& + A_{44}\left[\frac{\overline{\beta}^2}{2} + \overline{\beta}\frac{\partial w}{\partial y} + \frac{1}{2}\left(\frac{\partial w}{\partial y}\right)^2\right] + A_{66}\left[\frac{1}{2}\left(\frac{\partial u_0}{\partial y}\right)^2 + \frac{\partial u_0}{\partial y}\frac{\partial v_0}{\partial x} + \frac{1}{2}\left(\frac{\partial v_0}{\partial x}\right)^2\right] \\
& + B_{66}\left[\frac{\partial u_0}{\partial y}\frac{\partial \overline{\alpha}}{\partial y} + \frac{\partial u_0}{\partial y}\frac{\partial \overline{\beta}}{\partial x} + \frac{\partial v_0}{\partial x}\frac{\partial \overline{\alpha}}{\partial y} + \frac{\partial v_0}{\partial x}\frac{\partial \overline{\beta}}{\partial x}\right] \\
& + D_{66}\left[\frac{1}{2}\left(\frac{\partial \overline{\alpha}}{\partial y}\right)^2 + \frac{\partial \overline{\alpha}}{\partial y}\frac{\partial \overline{\beta}}{\partial x} + \frac{1}{2}\left(\frac{\partial \overline{\beta}}{\partial x}\right)^2\right] - \frac{\partial u_0}{\partial x}(N_{1j}^T + N_{1j}^m) \\
& - \frac{\partial v_0}{\partial y}(N_{2j}^T + N_{2j}^m) - \left(\frac{\partial u_0}{\partial y} + \frac{\partial v_0}{\partial x}\right)(N_{6j}^T + N_{6j}^m) - \frac{\partial \overline{\alpha}}{\partial x}(M_{1j}^T + M_{1j}^m) \\
& - \frac{\partial \overline{\beta}}{\partial y}(M_{2j}^T + M_{2j}^m) - \left(\frac{\partial \overline{\alpha}}{\partial y} + \frac{\partial \overline{\beta}}{\partial x}\right)(M_{6j}^T + M_{6j}^m) \\
& + T^* + M^* + \overline{MT^*} - p(x,y)w(x,y)\Bigg\} dA
\end{aligned} \tag{8.67}$$

where all quantities except displacements are defined by Equations (2.56) through (2.59), (2.60) through (2.63) and

$$T^* = \frac{1}{2}\sum_{k=1}^{N}\int_{h_{k-1}}^{h_k} [\overline{Q}_{ij}]_k\,[\alpha_i]_k\,[\alpha_j]_k\,[\Delta T(z,t)]^2 dz$$

$$M^* = \frac{1}{2}\sum_{k=1}^{N}\int_{h_{k-1}}^{h_k} [\overline{Q}_{ij}]_k\,[\beta_i]_k\,[\beta_j]_k\,[\Delta m(z,t)]^2 dz$$

$$\overline{MT}^* = \sum_{k=1}^{N}\int_{h_{k-1}}^{h_k} [\overline{Q}_{ij}]_k\,[\alpha_i]_k\,[\beta_j]_k\,\Delta T(z,t)\Delta m(z,t) dz$$

Equation (8.67) is the most general formulation and it is seen that without the static-load term there are 30 terms to represent the strain energy in this composite material panel. Referring to Equation (2.64) and its ensuing discussion it is seen that if the laminate has no stretching-shearing coupling ($A_{16} = A_{26} = 0$) then two terms would be dropped; if no twisting-stretching coupling ($B_{16} = B_{26} = 0$), two more would be dropped; likewise two more can be dropped if there is no bending-twisting coupling ($D_{16} = D_{26} = 0$). If the laminate were symmetric about the midplane, a very common construction, then five B_{ij} terms would be canceled out, because there is no bending-stretching coupling. Also, if the stiffness matrix terms for a solid core or honeycomb-sandwich-core sandwich panel are used, then Equation (8.67) can be used for a sandwich panel as well.

8.7.2 Moisture and Temperature Distributions

Having qualitatively discussed the hygrothermal effects earlier; it is important to be able to quantitatively determine $\Delta T(z)$ and $\Delta m(z)$ given a set of surface conditions.

For transient solutions for the thermal and moisture gradients for a sandwich structure one must account for the thermal properties of the face and core materials. Such analyses are beyond the scope of this text. For a steady-state case the solutions can be represented far more easily by simple linear functions in the z-direction such as the following for the moisture quantity $\Delta m(z)$ to use in all of the previous equations:

$$\Delta m(z) = \frac{1}{2}\left[\Delta m\left(\frac{h}{2}\right) - \Delta m\left(-\frac{h}{2}\right)\right] + \frac{z}{h}\left[\Delta m\left(\frac{h}{2}\right) - \Delta m\left(-\frac{h}{2}\right)\right] \quad (8.68)$$

Here, the quantities refer to the moisture content in the sandwich face at $z = +h/2$ and the bottom face at $z = -h/2$. There is an analogous expression for $\Delta T(z)$, found by simply replacing all Δm by ΔT.

From these solutions all of the thermal and moisture stress resultants and stress couples can be formulated straightforwardly. Obviously if the matrix of the composite structure is a metal or ceramic, the Δm, hygrothermal terms can be neglected; they only pertain to polymeric matrix composites.

8.7.3 Boundary Conditions for the Equations that Include Transverse Shear Deformation Effects

Solutions are discussed for three types of boundary conditions; all edges simply supported, all edges clamped, and two opposite edges simply supported, the other two clamped. Because of the necessity to include asymmetric lamina stacking sequences, resulting in bending-stretching coupling, these boundary conditions must be carefully specified. Following Whitney [11] and Wu and Vinson [12] two simply supported and two clamped boundary conditions are defined as

$$\begin{aligned} S1&: w_n = M_n = u_{0n} = N_{nt} = 0 \\ S2&: w_n = M_n = N_n = v_{0t} = 0 \\ C1&: w_n = \overline{\alpha}_n = u_{0n} = N_{nt} = 0 \\ C2&: w_n = \overline{\alpha}_n = N_n = v_{0t} = 0 \end{aligned} \tag{8.69}$$

In the above, w is the lateral deflection, u_0 and v_0 are the in-plane midsurface displacements, M is the stress couple, N is the in-plane stress resultant, $\overline{\alpha}$ is the rotation, and n is the direction normal to the edge, while t is the direction tangent to the edge.

8.7.4 Solution Approach

In this illustrative problem the Rayleigh-Ritz method is used, and is recommended in general. Earlier, for the beam, no assumptions were made regarding the form of the displacements for the solution, hence an Euler-Lagrange Differential Equation (5.13), the beam equation, resulted, which subsequently needs to be solved. However, the Rayleigh-Ritz procedure includes specifying a "guessed at" form of the displacement solution. In this case the resulting Euler-Lagrange equations are algebraic, and if good assumed displacements are used, provide very accurate solutions.

Good assumed displacements are more formally termed "admissible functions," defined as those that at least satisfy all geometric boundary conditions, i.e., displacement constraints and rotation constraints. Better yet are those functions which, in addition, satisfy part or all of the stress boundary conditions (that is, stress couples, in-plane and shear resultants).

In studies by Flaggs [13], Smith [14], and Linsenmann [15] that included transverse-shear deformations, the rotations $\overline{\alpha}$ and $\overline{\beta}$ were determined as functions of the lateral deflection, $w(x,y)$ through analyzing a beam in bending with transverse shear deformation effects included. However, here a functional form for these variables will be assumed following Wu and Vinson [12].

For the boundary conditions $S1$,

$$w(x,y) = \sum_{m_{odd}}^{\infty} \sum_{n_{odd}}^{\infty} W_{mn} \sin(\lambda_m x) \sin(\lambda_n y) \tag{8.70}$$

$$\overline{\alpha}(x,y) = \sum_{m_{odd}}^{\infty} \sum_{n_{odd}}^{\infty} \Gamma_{mn} \cos(\lambda_m x) \sin(\lambda_n y) \tag{8.71}$$

$$\overline{\beta}(x,y) = \sum_{m_{odd}}^{\infty} \sum_{n_{odd}}^{\infty} \Lambda_{mn} \sin(\lambda_m x) \cos(\lambda_n y) \tag{8.72}$$

$$u_0(x,y) = \sum_{m_{odd}}^{\infty} \sum_{n_{odd}}^{\infty} U_{mn} \sin(\lambda_m x) \cos(\lambda_n y) \tag{8.73}$$

$$v_0(x,y) = \sum_{m_{odd}}^{\infty} \sum_{n_{odd}}^{\infty} V_{mn} \cos(\lambda_m x) \sin(\lambda_n y) \tag{8.74}$$

The form for the $S2$ boundary condition is identical to the above for the first three variables. The last two variables become for $S2$:

$$u_0(x,y) = \sum_{m_{odd}}^{\infty} \sum_{n_{odd}}^{\infty} U_{mn} \cos(\lambda_m x) \sin(\lambda_n y) \tag{8.75}$$

$$v_0(x,y) = \sum_{m_{odd}}^{\infty} \sum_{n_{odd}}^{\infty} V_{mn} \sin(\lambda_m x) \cos(\lambda_n y) \tag{8.76}$$

The assumed forms of displacements and rotations for the $C1$ *clamped boundary conditions* are the characteristic beam functions originally formulated by Warburton [16], extended by Wu and Vinson [12], and completely characterized by Young and Felgar [17]. They are

$$w(x,y) = \sum_{m_{\text{odd}}}^{\infty} \sum_{n_{\text{odd}}}^{\infty} W_{mn} \phi_{w_m}(x)\, \phi_{w_n}(y) \tag{8.77}$$

$$\overline{\alpha}(x,y) = \sum_{m_{\text{odd}}}^{\infty} \sum_{n_{\text{odd}}}^{\infty} \Gamma_{mn} \phi_{\overline{\alpha}m}(x)\, \phi_{\overline{\alpha}n}(y) \tag{8.78}$$

$$\overline{\beta}(x,y) = \sum_{m_{\text{odd}}}^{\infty} \sum_{n_{\text{odd}}}^{\infty} \Lambda_{mn} \phi_{\overline{\beta}m}(x)\, \phi_{\overline{\beta}n}(y) \tag{8.79}$$

$$u_0(x,y) = \sum_{m_{\text{odd}}}^{\infty} \sum_{n_{\text{odd}}}^{\infty} U_{mn} \sin(\lambda_m x) \cos(\lambda_n y) \tag{8.80}$$

$$v_0(x,y) = \sum_{m_{\text{odd}}}^{\infty} \sum_{n_{\text{odd}}}^{\infty} v_{mn} \cos(\lambda_m x) \sin(\lambda_n y) \tag{8.81}$$

where, because m is odd only, the following repeats Equation (7.87):

$$\begin{aligned}
\phi_{wm}(x) &= \cos\left[\mu_m\left(\frac{x}{a} - \frac{1}{2}\right)\right] + \eta_m \cosh\left[\mu_m\left(\frac{x}{a} - \frac{1}{2}\right)\right] \\
\phi_{\alpha m}(x) &= \sin\left[\mu_m\left(\frac{x}{a} - \frac{1}{2}\right)\right] - \eta_m \sinh\left[\mu_m\left(\frac{x}{a} - \frac{1}{2}\right)\right] \\
\phi_{\beta m}(x) &= \cos\left[\mu_m\left(\frac{x}{a} - \frac{1}{2}\right)\right] + \eta_m \cosh\left[\mu_m\left(\frac{x}{a} - \frac{1}{2}\right)\right]
\end{aligned} \tag{8.82}$$

where η_m is found from

$$\eta_m = \sin(\mu_m/2)/\sinh(\mu_m/2) \tag{8.83}$$

and μ_m is found from the following expression,

$$\tan(\mu_m/2) + \tanh(\mu_m/2) = 0 \tag{8.84}$$

By interchanging $\overline{\alpha}$ and $\overline{\beta}$, substituting y for x, n for m, and b for a, Equation (8.82) yields the proper functions for the y-direction in Equations (8.77) through (8.81). The solution of the transcendental Equation (8.84) provides the necessary values of μ_m and μ_n in Equations (8.83) and (8.82). The solutions are

$$\mu_1 = 1.50562\pi$$
$$\mu_2 = 2.49975\pi, \qquad \mu_k = \left(k + \frac{1}{2}\right)\pi \text{ for } k \geq 3 \tag{8.85}$$

In the $C2$ boundary conditions u_0 and v_0 are given by Equations (8.75) and (8.76).

For two edges simply supported, and the two opposite edges clamped, the following forms are employed where $x = 0$ and a are the clamped boundary conditions and $y = 0$ and b are the simply supported edges,

$$w(x,y) = \sum_{m_{odd}}^{\infty} \sum_{n_{odd}}^{\infty} W_{mn} \phi_{wm}(x) \sin(\lambda_n y) \tag{8.86}$$

$$\overline{\alpha}(x,y) = \sum_{m_{odd}}^{\infty} \sum_{n_{odd}}^{\infty} \Gamma_{mn} \phi_{\overline{\alpha}m}(x) \cos(\lambda_n y) \tag{8.87}$$

$$\overline{\beta}(x,y) = \sum_{m_{odd}}^{\infty} \sum_{n_{odd}}^{\infty} \Lambda_{mn} \phi_{\overline{\beta}m}(x) \sin(\lambda_n y) \tag{8.88}$$

where

$$\lambda_n = n\pi/b$$

The $u_0(x,y)$ and $v_0(x,y)$ variables are given by either Equations (8.73) and (8.74) or (8.75) and (8.76) depending upon the desired boundary condition.

8.7.5 Calculation of Deflections, Rotations, and Stresses

The values of the coefficients W_{mn}, Γ_{mn}, Λ_{mn}, U_{mn}, and V_{mn} are obtained from substituting the above equations into Equation (8.67), taking variations, performing integrations, and solving the set of simultaneous equations. Following that, the above coefficients are substituted into Equations (8.70) through (8.88) to find the displacements and rotations. From these the laminate strains and curvatures are found from Equations (2.48) through (2.50). Then the stresses everywhere in each lamina are calculated using Equation (2.44), after determining in the thermal and hygroscopic contributions. For a sandwich panel the stresses are determined from Equations (1.5) and (1.7).

8.9 REFERENCES

1. Vinson, J. R. and T-W. Chou, *Composite Materials and Their Use in Structures*, Applied Science Publishers, London, 1975.
2. Vinson, J. R. and R. L. Sierakowski, *The Behavior of Structures Composed of Composite Materials*, Martinus-Nijhoff Publishers, (now Kluwer Academic Publishers) Dordrecht, The Netherlands, 1986.
3. Elpass, W.J., "Design and Analysis of High Precision Sandwich Structures," in *Sandwich Constructions 2—Proceedings of the Second International Conference on Sandwich Construction*, Gainesville, Florida, Editors D. Weissman-Berman and K-A Olsson, EMAS Publication, United Kingdom, 1992, pp. 695–717.
4. Boley, B.A. and J. H. Weiner, *Theory of Thermal Stresses*, J. Wiley and Sons, New York, 1960.
5. Noor, A. K. and W. S. Burton, "Computational Models for High Temperature Multilayered Composite Plates and Shells," *Applied Mechanics Reviews*, Vol. 45, No. 10, pp. 419–446, 1992.
6. Noor, A. K., W. Scott Burton and C. W. Bert, "Computational Models for Sandwich Panels and Shells," *Applied Mechanics Reviews*, Vol. 49, No. 3, March 1996, pp. 155–199.
7. Chamis, C. C., R.A. Aiello, R. A. and P. L. Murthy, "Fiber Composite Sandwich Behavior: Thermostructural Computational Simulation," *Journal of Composite Technology and Research*, Vol. 10, No. 3, pp. 93–99, Fall 1988.
8. Noor, A. K., J. H. Starnes, Jr. and J. M. Peters, "Analysis of Curved Sandwich Panels Subjected to Combined Temperature Gradient and Mechanical Loads," *Proceedings of the AIAA/ASME/ASCE/AHS/ASC Structures, Structural Dynamics and Materials Conference*, April 7–10, 1997, Vol. 4, AIAA, New York, 1997, pp. 2446–2470.
9. Noor, A. K. and J. M. Peters, "Analysis of Curved Sandwich Panels with Cutouts Subjected to Combined Temperature Gradient and Mechanical Loads," in *Proceedings of the ASME Aerospace Division, Structures and Materials Committee*, Editor, G. J. Simitses, ASME-AD-Vol. 55, New York, 1997.
10. Sloan, J. G., "The Behavior of Rectangular Composite Material Plates Under Lateral and Hygrothermal Loads," MMAE Thesis, University of Delaware, 1979 (also, AFOSR-TR-78-1477, July 1978).
11. Whitney, J. M., *Structured Analysis of Laminated Anisotropic Plates*, Technomic Publishing Co., Inc., Lancaster, PA, 1987.
12. Wu, C. I. and J. R. Vinson, "Nonlinear Oscillations of Laminated Specially Orthotropic Plates with Clamped and Simply Supported Edges," *Journal of the Acoustical Society of America*, Vol. 49, Part 2, May 1971, pp. 1561–1567.
13. Flaggs, D. L., "Elastic Stability of Generally Laminated Composite Plates Including Hygrothermal Effects," MMAE Thesis, University of Delaware, 1978.
14. Smith, A. P., "The Effect of Transverse Shear Deformation on the Elastic Stability of Orthotropic Plates Due to In-Plane Loads," MMAE Thesis, University of Delaware, 1973.

15. Linsenmann, D. R., "Stability of Plates of Composite Materials," MMAE Thesis, University of Delaware, 1974.

16. Warburton, G., "The Vibration of Rectangular Plates," *Proceedings of the Institute of Mechanical Engineers*, p. 371.

17. Young, D. and R. F. Felgar, Jr., "Tables of Characteristic Functions Representing Normal Modes of Vibration for a Beam," The University of Texas Engineering Research Series Report No. 44, July 1, 1949.

18. Vinson, J. R., *The Behavior of Thin Walled Structures: Beams, Plates and Shells*, Kluwer Academic Publishers, Dordrecht, The Netherlands, 1989.

CHAPTER 9

Elastic Instability (Buckling) of Sandwich Panels

9.1 GENERAL CONSIDERATIONS

As stated previously, structures usually fail in one of four ways.

- overstressing (strength-critical structure)
- over deflecting (stiffness-critical structures)
- resonant vibration
- buckling

For many cases, because sandwich structures (compared to monocoque structures) minimize stress, are extremely stiff, and have high fundamental frequencies, care must be taken to insure that unanticipated buckling does not undermine a structural design.

In monocoque structures for given plate dimensions, material, boundary conditions, and a given load type (in-plane compression, in-plane shear), only one buckling load will result in buckling. This is the lowest eigenvalue of a countable infinity of such eigenvalues. All other eigenvalues exist mathematically, but only the lowest value has physical significance. This differs from natural frequencies in which several eigenvalues can be very important.

For the simplest case, a column, a rudimentary introduction was given previously. While philosophically the simple examples cover the topic of buckling; more complex structures can have several types of buckling instabilities, any one of which can destroy the structure.

Historically, there are four major textbooks dealing primarily with elastic stability or buckling. These are authored by Bleich [1] , Timoshenko and Gere [2], Brush and Almroth [9] and Simitses [9]. Although these texts deal primarily with structures other than sandwich, the solutions can be applied by using the appropriate flexural stiffnesses.

9.2 THE BUCKLING OF AN ORTHOTROPIC SANDWICH PLATE SUBJECTED TO IN-PLANE LOADS-CLASSICAL THEORY

From previous developments, it was seen that for a plate there are five equations associated with the in-plane loads N_x, N_y, and N_{xy} and the in-plane displacements they cause, namely u_0 and v_0. The in-plane equilibrium equations are given by Equations (3.9) and (3.10). From Equation (2.64), for the case of midplane symmetry ($B_{ij} = 0$) and no thermal or moisture considerations it is seen that

$$N_x = A_{11}\varepsilon_{x_0} + A_{12}\varepsilon_{y_0} + 2A_{16}\varepsilon_{x_0y_0} \tag{9.1}$$

$$N_y = A_{12}\varepsilon_{x_0} + A_{22}\varepsilon_{y_0} + 2A_{26}\varepsilon_{x_0y_0} \tag{9.2}$$

$$N_{xy} = A_{16}\varepsilon_{x_0} + A_{26}\varepsilon_{y_0} + 2A_{66}\varepsilon_{x_0y_0} \tag{9.3}$$

Likewise, for the midplane symmetric panel, the six governing equations involving M_x, M_y, M_{xy}, Q_x, Q_y, and w, are given by Equations (3.16), (3.17), (3.18), (3.25), (3.26), and (3.27), the latter three neglecting the D_{16} and D_{26} terms. One can see there is no coupling between in-plane and lateral action for the plate with midplane symmetry. Yet it is well known and often observed that in-plane loads do cause lateral deflections through buckling, which is usually disastrous.

The answer to the paradox is that in the above discussion only *linear* elasticity theory is considered, while the physical event of buckling is a nonlinear problem. For brevity, the development of the nonlinear theory will not be included herein because it is included in so many other texts [1–4,16,17].

The results of including the terms to predict the advent or inception of buckling for the beam and plate are, modifying Equation (3.30),

$$D_1\frac{\partial^4 w}{\partial x^4} + 2D_3\frac{\partial^4 w}{\partial x^2\partial y^2} + D_2\frac{\partial^4 w}{\partial y^4} = p(x,y) + N_x\frac{\partial^2 w}{\partial x^2} + 2N_{xy}\frac{\partial^2 w}{\partial x\partial y} + N_y\frac{\partial^2 w}{\partial y^2} \tag{9.4}$$

where clearly there is a coupling between the in-plane loads and the lateral deflection. In this equation the flexural stiffness quantities can be the sandwich flexural stiffness quantities.

It should be noted that the buckling loads, like the natural frequencies, are independent of the lateral loads, which will be disregarded in what follows.

However, in actual structural analysis, the effect of lateral loads, in combination with the in-plane loads could cause overstressing and failure before the in-plane buckling load is reached. However, the buckling load is still independent of the type or magnitude of the lateral load, as are the natural frequencies. Incidentally, common sense dictates that if one is designing a structure to withstand compressive loads, with the possibility of buckling being the failure mode, one had better design the structure to be midplane symmetric, so that $B_{ij} = 0$. Otherwise the bending-stretching coupling would likely cause overstressing before the buckling load is reached.

Looking now at Equation (9.4) for the buckling of the composite plate under an axial load N_x only, and ignoring $p(x,y)$ the equation becomes

$$D_1 \frac{\partial^4 w}{\partial x^4} + 2D_3 \frac{\partial^4 w}{\partial x^2 \partial y^2} + D_2 \frac{\partial^4 w}{\partial y^4} - N_x \frac{\partial^2 w}{\partial x^2} = 0 \tag{9.5}$$

Again, one may assume the buckling mode for a sandwich plate to be that of the Navier solution for the case of the plate simply supported on all four edges:

$$w(x,y) = \sum_{m=1}^{\infty} \sum_{n=1}^{\infty} A_{mn} \sin\left(\frac{m\pi x}{a}\right) \sin\left(\frac{n\pi y}{b}\right) \tag{9.6}$$

Substituting Equation (9.6) into (9.5) it is seen that the equation is satisfied only when N_x has certain values, namely the critical values, $N_{x_{cr}}$,

$$N_{x_{cr}} = -\frac{\pi^2 a^2}{m^2}\left[D_1\left(\frac{m}{a}\right)^4 + 2D_3\left(\frac{m}{a}\right)^2\left(\frac{n}{b}\right)^2 + D_2\left(\frac{n}{b}\right)^4\right] \tag{9.7}$$

Again several things are clear: Equation (9.5) is a homogeneous equation—this is an eigenvalue problem; one cannot determine the value of A_{mn}; and again only the lowest value of $N_{x_{cr}}$ is of any importance. However, it is not clear which values of m and n result in the lowest critical buckling load. All values of n appear in the numerator, so $n = 1$ is the necessary value. But m appears in several places, and depending upon the value of the flexural stiffnesses D_1, D_2, and D_3, and the length to width ratio of the plate, a/b, it is not clear which value of m will provide the lowest value of $N_{x_{cr}}$. However, for a given plate this is easily determined computationally.

What about the buckling loads of composite material sandwich plates with boundary conditions other than simply supported? It is seen that all combinations of beam vibrational mode shapes are applicable for plates with various boundary conditions. These have been developed by Warburton [18] and all derivatives and integrals of those functions catalogued conveniently by Young and Felgar [19,20] for easy use. Likewise they can be used instead of Equation (9.6) to obtain solutions of Equation (9.5) to determine the critical buckling load per unit width, $N_{x_{cr}}$.

The buckling loads calculated in this section do not include transverse shear deformation effects, and are therefore only approximate—but they are useful for preliminary design, because of their relative simplicity. If transverse shear deformation were included, the buckling loads are lower than those calculated with classical theory. Therefore, the buckling loads calculated, neglecting transverse shear deformation, are not conservative. Transverse shear deformation effects will now be investigated.

9.3 ELASTIC STABILITY OF A COMPOSITE SANDWICH PANEL INCLUDING TRANSVERSE-SHEAR DEFORMATION AND HYGROTHERMAL EFFECTS

In this section, a general buckling theory is formulated that accounts for the hygrothermal effects as well as transverse shear deformation and all of the couplings discussed in earlier sections as performed by Flaggs [21]. Again the Theorem of Minimum Potential Energy is employed, so the strain energy of the plate is identical to that previously used, namely Equation (5.70). However, in the absence of a lateral load $p(x,y)$, the last term of Equation (5.70) is absent, but in its place are the effects of in-plane stress resultants N_x, N_{xy}, and N_y, that can cause an elastic instability. Equation (5.70) becomes

$$V = \int_R W dR[\text{See Equation (5.70)}] - \iint \left\{ N_x \left[\frac{\partial u_0}{\partial x} + \frac{1}{2}\left(\frac{\partial w}{\partial x}\right)^2 \right] \right.$$

$$\left. + N_{xy}\left[\frac{\partial u_0}{\partial y} + \frac{\partial v_0}{\partial x} + \frac{\partial w}{\partial x}\frac{\partial w}{\partial y} \right] + N_y \left[\frac{\partial v_0}{\partial y} + \frac{1}{2}\left(\frac{\partial w}{\partial y}\right)^2 \right] \right\} dA \qquad (9.8)$$

These terms are treated in detail in References [5] and [6], and will not be developed in detail here, but they are standard for an isotropic panel as well as a composite laminate.

The buckling loads, N_x, N_{xy}, or N_y, are determined by finding the value of the load at which bifurcation occurs, that is, loads at which the plate can be in equilibrium in both a straight configuration (i.e., $w = 0$) and in a slightly deformed ($w \neq 0$) configuration. This is accomplished through setting the variation of the potential energy V in Equation (9.8) equal to zero, as in Equation (5.2). This operation results in an eigenvalue problem that can be solved for nontrivial solutions that are discrete values of the applied loads.

The lowest critical load is the actual physical buckling load. Unlike solving for several natural frequencies, all of which could be important, only the lowest buckling load has any physical meaning.

In Section 8.7 u_0, v_0, w, $\overline{\alpha}$, and $\overline{\beta}$ were all considered to be unknowns to be found through the solution. In this study, simply to illustrate another approach, for the rotations $\overline{\alpha}$ and $\overline{\beta}$, the rotations are solved for in terms of the lateral deflection w. To do this, consider the laminate to be a beam, hence modifying Equation (5.70) to only have an x-dependence. Solving that problem results in three equations:

$$A_{11}\frac{d^2u_0}{dx^2} + B_{11}\frac{d^2\overline{\alpha}}{dx^2} = 0 \tag{9.9}$$

$$A_{55}\frac{d\overline{\alpha}}{dx} - A_{55}\frac{d^2w}{dx^2} - N_x\frac{d^2w}{dx^2} = 0 \tag{9.10}$$

$$-A_{55}\overline{\alpha} - A_{55}\frac{dw}{dx} + B_{11}\frac{d^2u_0}{dx^2} + D_{11}\frac{d^2\overline{\alpha}}{dx^2} = 0 \tag{9.11}$$

These equations can be solved for $\overline{\alpha}$, the result being as follows. Even though $\overline{\alpha}$ is determined for a beam, a y dependence is permitted since a plate is being considered.

$$\overline{\alpha}(x,y) = -\frac{\partial w}{\partial x} - \left(\frac{D_{11}A_{11} - B_{11}^2}{A_{11}}\right)\left(\frac{A_{55} - N_x}{A_{55}^2}\right)\frac{\partial^3 w}{\partial x^3} \tag{9.12}$$

In Equation (9.12) note that the first bracketed term is precisely the reduced flexural stiffness term discussed previously. Similarly, assuming a beam in the y-direction, it is found that

$$\overline{\beta}(x,y) = \frac{\partial w}{\partial y} - \left(\frac{D_{22}A_{22} - B_{22}^2}{A_{22}}\right)\left(\frac{A_{44} - N_y}{A_{44}^2}\right)\frac{\partial^3 w}{\partial y^3} \tag{9.13}$$

Equations (9.12) and (9.13) can now be substituted into Equation (9.8) so that the potential energy expression contains only u_0, v_0, and w as unknown functions.

From Equation (8.69) the simply supported case $S1$ and the clamped case $C1$ are chosen as a good set of examples to study to investigate various effects. Hence, the forms of displacements chosen are given by Equations (8.70), (8.73), and (8.74) for the S1 case and by Equations (8.80) and (8.81) are chosen for the $C1$ case. However, Flaggs and Vinson chose a form for w that differs from Equation (8.77). It must be remembered that an admissible function satisfies at least the geometric boundary conditions but that function is not unique, hence in this case the following is used:

$$W(x,y) = \sum_{m=1}^{\infty}\sum_{n=1}^{\infty} W_{mn}\left\{\cos\left[\frac{(m-1)\pi x}{a}\right] - \cos\left[\frac{(m+1)\pi x}{a}\right]\right\}$$
$$\times\left\{\cos\left[\frac{(n-1)\pi y}{b}\right] - \cos\left[\frac{(n+1)\pi y}{b}\right]\right\} \qquad (9.14)$$

These displacements can now be substituted into Equation (9.8). Then taking the variation with respect to the unknown amplitudes W_{mn}, Γ_{mn}, and Λ_{mn} results in the eigenvalue problem below, where one simply sets $N_y = \lambda N_x$:

$$\begin{bmatrix} E_{11} & E_{12} & E_{13} \\ E_{21} & E_{22} & E_{23} \\ E_{31} & E_{32} & E_{33} \end{bmatrix}\begin{Bmatrix} \Gamma_{mn} \\ \Lambda_{mn} \\ W_{mn} \end{Bmatrix} = N_x \begin{bmatrix} 0 & 0 & F_{13} \\ 0 & 0 & F_{23} \\ F_{31} & F_{32} & F_{33} \end{bmatrix}\begin{Bmatrix} \Gamma_{mn} \\ \Lambda_{mn} \\ W_{mn} \end{Bmatrix}$$
$$+ N_x^2 \begin{bmatrix} 0 & 0 & 0 \\ 0 & 0 & 0 \\ 0 & 0 & G_{33} \end{bmatrix}\begin{Bmatrix} \Gamma_{mn} \\ \Lambda_{mn} \\ W_{mn} \end{Bmatrix} \qquad (9.15)$$

where the E_{ij}, F_{ij} and G_{ij} quantities are symbolic but can be found through the variational process.

Equation (9.15) can be simplified by uncoupling the third set of simultaneous linear algebraic equations, by substituting for the Γ_{mn} and Λ_{mn} equations in terms of W_{mn}, resulting in

$$[E]\{W_{mn}\} = N_x[F]\{W_{mn}\} + (N_x)^2\,[G]\{W_{mn}\} \qquad (9.16)$$

For a symmetrically laminated $[0, 45, -45, 90]_{4s}$ *T*300/5208 graphite-epoxy composite plate with simply supported and clamped boundary conditions, the buckling loads using the generally laminated plate theory are calculated for both steady state and transient hygrothermal conditions. Figures 9.1 and 9.2 show the effects on the applied buckling load, N_x, of different steady-state hygrothermal environments for clamped and simply-supported boundary conditions. The effects of temperature and moisture are both quite clear. Note how along the abscissa, the combination of moisture and temperature combine to produce buckling with no axially compressive load. It is straightforward to develop analogous plots for various sandwich panels.

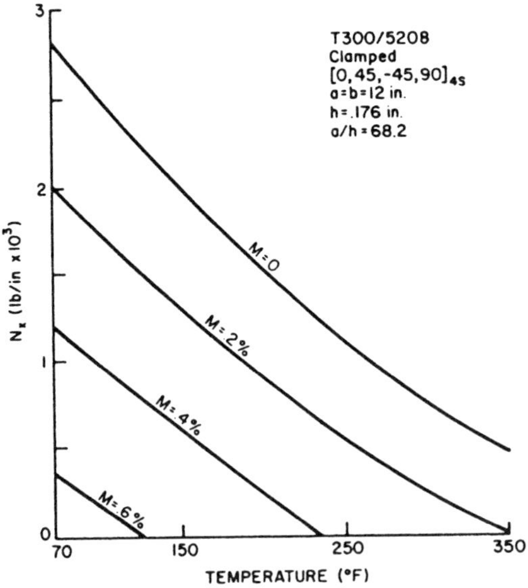

FIGURE 9.1. Buckling load per unit width as a function of temperature and moisture. (Reprinted from Reference [17].)

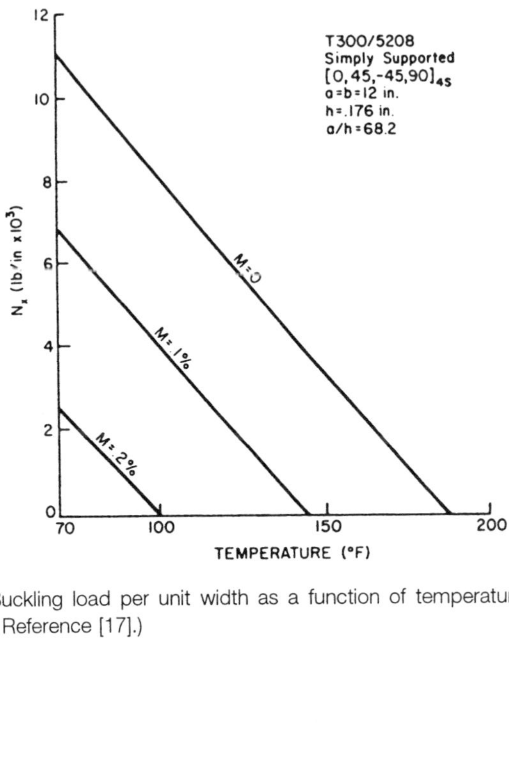

FIGURE 9.2. Buckling load per unit width as a function of temperature and moisture. (Reprinted from Reference [17].)

9.4 THE BUCKLING OF AN ISOTROPIC PLATE ON AN ELASTIC FOUNDATION SUBJECTED TO BIAXIAL IN-PLANE COMPRESSIVE LOADS

It is important to consider that besides overall buckling of the entire plate, it is possible that a sandwich face plate may buckle, due to loads applied to the face. In this case the plate can be considered to be supported on a uniform elastic foundation, namely the core. In such a case the buckling equation for this phenomenon is

$$D\nabla^4 w + kw + \overline{N}_x \frac{\partial^2 w}{\partial x^2} + \overline{N}_y \frac{\partial^2 w}{\partial y^2} = 0 \tag{9.17}$$

where D is the flexural stiffness of the face plate, w is the lateral displacement of the face plate, k is the foundation modulus in force/unit area/unit deflection, and $\overline{N}_x$, $\overline{N}_y$ are the compressive loads per unit width in the subscripted direction (i.e., $\overline{N}_x = -N_x$, etc.).

Considering this localized buckling phenomenon, it has been found that the plate boundary conditions at the outer plate edges do not affect the buckling load. Therefore, for analytical simplicity, assume simply-supported edges on all four sides. Therefore, the Navier approach may be used for the solution, with the lateral deflection assumed to be

$$w(x,y) = A_{mn} \sin\left(\frac{m\pi x}{a}\right) \sin\left(\frac{n\pi y}{b}\right) \tag{9.18}$$

where A_{mn} is the deflection amplitude, a is the plate dimension in the x-direction, and b is the plate dimension in the y direction.

For simplification, let $\phi = \overline{N}_y/\overline{N}_x$ and $r = a/b$. Substituting Equation (9.18) into (9.17) and using the above

$$\overline{N}_{x_{cr}} = \frac{(\pi^4 D/a^4)(m^2 + n^2 r^2)^2 + k}{(\pi^2/a^2)(m^2 + n^2 r^2 \phi)} \tag{9.19}$$

If $\phi = 1$, and $r = 1$, then the response is independent of direction. When the in-plane loads are caused by the cooling of a sandwich plate wherein the coefficients of thermal expansion between face and core cause the face to be compressed, then $\phi = 1$. Further because the buckling is a localized phenomenon, one can let $r = 1$. Then Equation (9.19) may be written as

$$\overline{N}_{x_{cr}} = \frac{(\pi^4 D/a^2)(m^2 + n^2)^2 + ka^2}{\pi^2(m^2 + n^2)} \tag{9.20}$$

First it is seen that the minimum value of $\overline{N}_x$ will occur when $m = n = 1$, therefore

$$\overline{N}_x = \frac{(4\pi^4 D/a^2) + ka^2}{2\pi^2} \tag{9.21}$$

To find the dimension a resulting in a minimum value of $\overline{N}_{x_{cr}}$, set $\partial \overline{N}_{x_{cr}} / \partial a = 0$, with the result that

$$a = 2^{1/2} \pi \left(\frac{D}{k}\right)^{1/4} \tag{9.22}$$

This is the half wavelength of the buckle that will occur, and it can be determined that this is a localized buckle in a reasonably sized face plate. Substituting Equation (9.22) into (9.21) results in

$$\overline{N}_{x_{cr}} = 2(kD)^{1/2}$$

As defined, $\overline{N}_{x_{cr}}$ is a compressive force per unit width and equal to $\overline{N}_{y_{cr}}$, since $\phi = 1$, or in the usual notation, where $\overline{N}_i = -N_i$,

$$N_{x_{cr}} = N_{y_{cr}} = -2(kD)^{1/2} \tag{9.23}$$

The buckling stress in the face plate is therefore

$$\sigma_{cr} = \frac{-2}{h}(kD)^{1/2} \tag{9.24}$$

It has been found that in the fabrication of some sandwich plates, because of the cooling down subsequent to joining the faces to the core in a rolling operation, differential thermal contractions caused sufficiently high compressive stresses in the faces to cause thermal buckling.

9.5 THE BUCKLING OF HONEYCOMB CORE SANDWICH PANELS SUBJECTED TO IN-PLANE COMPRESSIVE LOADS

Consider a rectangular sandwich panel of length a (the load direction), width b, face thickness t_f, core depth h_c, core cell wall thickness t_c, and diameter

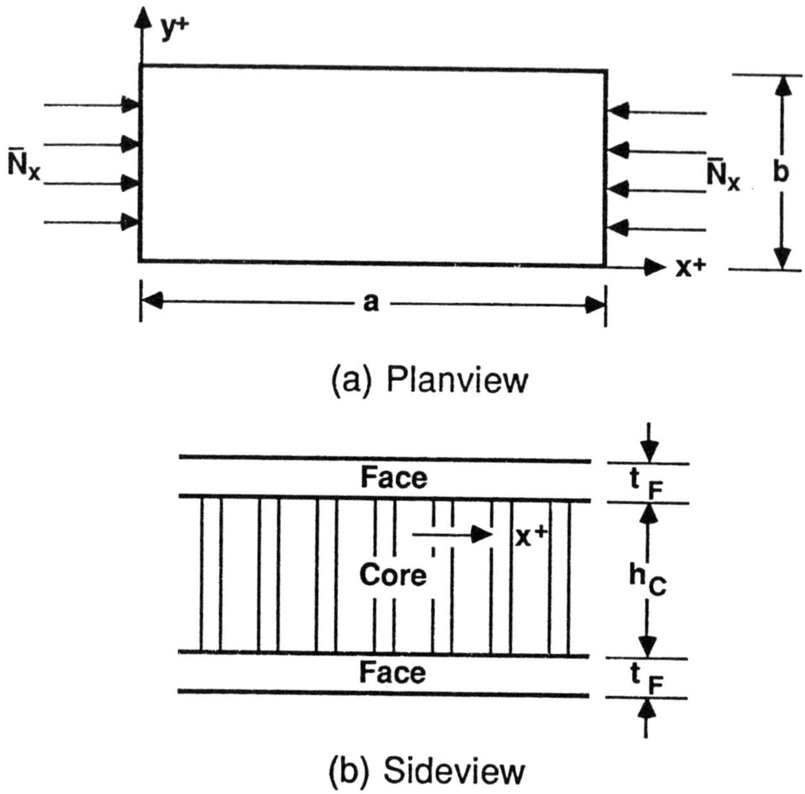

FIGURE 9.3. Sandwich panel with in-plane compressive load in the x-direction. (Reprinted from Reference [17].)

d of a circle inscribed in the cell as shown in Figures 9.3 and 9.4 for the hexagonal-cell honeycomb core. A sketch analogous to Figure 9.4 could also be drawn for the square-cell honeycomb. It is assumed for this study that the core is composed of an isotropic material of shear stiffness G_c and modulus of elasticity E_c. If the core is orthotropic, the properties normal to the plane of the panel for E_c and G_c are used. Consider the faces to be composed of identical composite materials that are balanced about their own planes of symmetry with no unwanted couplings, i.e., $B_{ij} = 0$ and all $(\ )_{16}$ and $(\ )_{26}$ terms equal to zero. It is assumed that the in-plane load in the x-direction is uniform and has the value of N_x (load/unit width) in compression, so $\overline{N}_x = -N_x$ for convenience.

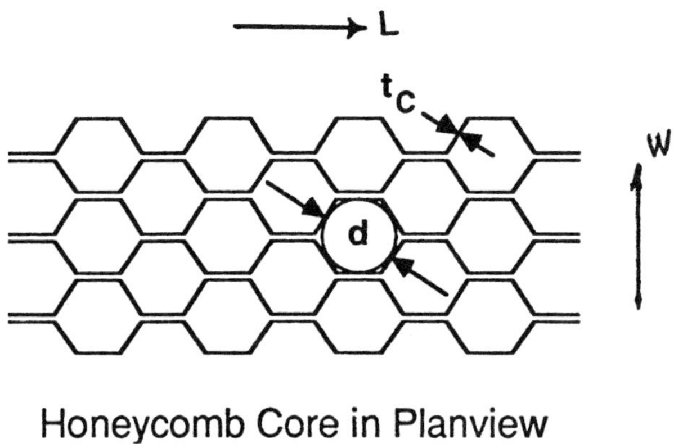

FIGURE 9.4. Honeycomb core in plan view. (Reprinted from Reference [17].)

9.5.1 Face Stresses

It is assumed that the in-plane loads are resisted only by the faces, not the honeycomb core. Therefore, the applied compressive face stress in the load direction is written as

$$\sigma_x = N_x/2t_f \tag{9.25}$$

For an applied load N_x, the face stress σ_x is, of course, restricted to some prescribed maximum value to prevent overstressing.

The honeycomb sandwich panel shown in Figure 9.3 can be overstressed according to Equation (9.25), but can also buckle in one of several modes, any one of which will render the panel to be useless. These modes of buckling are overall instability, core shear instability, face wrinkling, and face dimpling (monocell buckling). These are discussed in turn.

9.5.2 Overall Instability

The equation to use for the compressive face stress for overall buckling of the subject panel that includes the effects of transverse shear deformation is given by [22] as follows, where the bar over the modulus indicates that for stresses above the proportional limit, a plasticity reduction factor should be used.

$$\sigma_{cr} = \frac{\pi^2}{4(1 - \nu_{xy}\nu_{yx})} \sqrt{\bar{E}_{fx}E_{fy}} \frac{h_c^2}{b^2} K, \text{ for } V_x \leq k_1 B_1 r \tag{9.26}$$

In this equation, the coefficient K is given by

$$K = \frac{B_1C_1 + 2B_2C_2 + \dfrac{C_3}{B_1} + A\left(\dfrac{V_y}{C_4} + V_x\right)}{1 + (B_1C_1 + B_3C_2)\dfrac{V_y}{C_4} + \left(\dfrac{C_3}{B_1} + B_3C_2\right)V_x + \dfrac{V_yV_xA}{C_4}} \tag{9.27}$$

where

$$A = C_1C_3 - B_2C_2^2 + B_3C_2\left(B_1C_1 + 2B_2C_2 + \frac{C_3}{B_1}\right) \tag{9.28}$$

$$B_1 = \sqrt{\frac{D_y}{D_x}},\ B_2 = \frac{D_yV_{xy} + 2D_{xy}}{\sqrt{D_xD_y}} = 2B_3 + B_1\,\nu_{xy},\ B_3 = \frac{D_{xy}}{\sqrt{D_xD_y}} \tag{9.29}$$

$$V_y = \frac{\pi^2}{b^2}\frac{\sqrt{D_xD_y}}{U_{yz}},\ V_x = \frac{\pi^2}{b^2}\frac{\sqrt{D_xD_y}}{U_{xz}} \tag{9.30}$$

$$U_{xz} = G'_{cx}h_c, \quad U_{yz} = G'_{cy}h_c \tag{9.31}$$

In Equation (9.30) the V_i quantities are seen to be particular ratios of the sandwich-panel flexural stiffness D, to the transverse shear stiffness U, hence the V quantities are called the transverse-shear flexibility parameters. The D quantities relate to face properties while the U quantities are core properties. The effective core moduli, G'_{cx} and G'_{cy} are defined in Equations (9.44) and (9.45).

The constants C_1 through C_4 in Equations (9.27) and (9.28) are associated with the boundary conditions and are listed as follows where n is the number of buckling waves in the direction of the compressive loading.

(1) All edges simply supported

$$C_1 = C_4 = \frac{a^2}{n^2b^2}, \quad C_2 = 1, \quad C_3 = \frac{n^2b^2}{a^2} \tag{9.32}$$

(2) Loaded edges simply supported, other edges clamped

$$C_1 = \frac{16}{3}\frac{a^2}{n^2b^2}, \quad C_2 = \frac{4}{3}, \quad C_3 = \frac{n^2b^2}{a^2}, \quad C_4 = \frac{4}{3}\frac{a^2}{n^2b^2} \tag{9.33}$$

(3) Loaded edges clamped, other edges simply supported

$$C_1 = C_4 = \frac{3}{4}\frac{a^2}{b^2} \quad \text{for} \quad n = 1$$

$$C_1 = C_4 = \frac{1}{(n^2 + 1)}\frac{a^2}{b^2} \quad \text{for} \quad n \geq 2 \tag{9.34}$$

$$C_2 = 1,\ C_3 = \frac{(n^2 + 6n^2 + 1)}{(n^2 + 1)}\frac{b^2}{a^2}$$

(4) All edges clamped

$$C_1 = 4C_4 = \frac{4a^2}{b^2} \quad \text{for} \quad n = 1$$

$$C_1 = 4C_4 = \frac{16}{3(n^2 + 1)}\frac{a^2}{b^2} \quad \text{for} \quad n \geq 2 \tag{9.35}$$

$$C_2 = \frac{4}{3},\quad C_3 = \frac{(n^4 + 6n^2 + 1)}{n^2 + 1)}\frac{b^2}{a^2}$$

Obviously, if the critical face stress given by Equation (9.26) is higher than the allowable compressive stress of the face material, the panel will be overstressed before there is an overall buckling problem.

9.5.3 Core Shear Instability

Referring to the expressions for overall stability, if the value of V_x is increased through increasing the panel bending stiffnesses (D_x and D_y) or decreasing the core transverse shear stiffness (U_{xz}), the value of K in Equation (9.27) is decreased. There exists a value of V_x that causes K to equal $1/V_x$. This value depends both on the boundary conditions and the effective shear moduli of the core (G'_{cx}, G'_{cy}), that are defined in Equations (9.44) and (9.45). At this particular value of $K = 1/V_x$, K is independent of the length-to-width ratio a/b and n is infinite. For values of V_x greater than this value, $K = 1/V_x$, which is true for a great number of practical sandwich panels because they have high values of V_x. Under these conditions, the critical stress can be written as follows:

$$\sigma_{cr} = \frac{G'_{cx}h_c}{2t_f}, \quad \text{for} \quad V_x \geq k_1 B_1 r \tag{9.36}$$

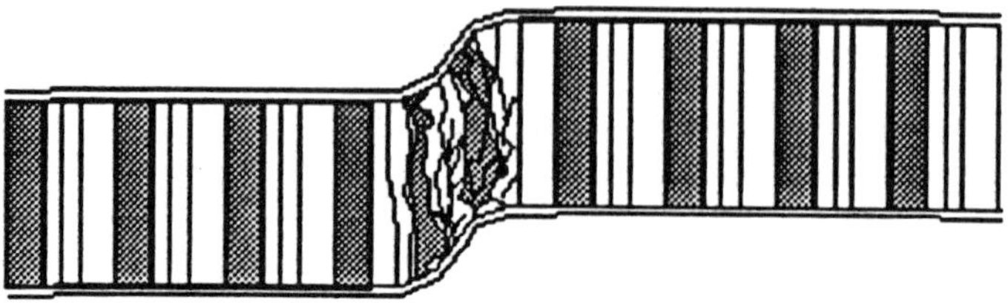

FIGURE 9.5. Core shear instability. (Reprinted from Reference [17].)

This value of critical stress is called the core-shear instability stress and cannot be exceeded for any given sandwich construction. It is seen that Equation (9.36) is independent of the panel length, width, and boundary conditions. Core shear instability is illustrated in Figure 9.5.

The particular value of V_x at which $K = 1/V_x$, at which the critical stresses for overall buckling and core shear instability are equal. This value is given in Equation (9.37) and is dependent on the boundary conditions and the effective shear moduli of the core, G'_{cx} and G'_{cy}, defined in Equations (9.44) and (9.45). The values for the boundary condition factors k_1 are listed in Table 9.1.

$$V_x = k_1 B_1 r \quad \text{where here} \quad r = G'_{cx}/G'_{cy} \tag{9.37}$$

More specifically this can be written as follows, where $\overline{E}$ refers to a modulus multiplied by a suitable plasticity reduction factor, when stresses exceed the proportional limit.

$$\frac{\pi^2}{2(1 - \nu_{xy}\nu_{yx})k_3} \frac{t_f h_c d}{b^2 t_c} \frac{\overline{E}_{fx}}{G_c} = k_1 r \tag{9.38}$$

9.5.4 Face Wrinkling Instability

Wrinkling occurs across many cells of the honeycomb core and, under the loading conditions described here, extends across the width of the plate, but

TABLE 9.1. Boundary Condition Factors k_i *for Various Edge Conditions.*

Boundary Condition	k_1
All edges simply supported	1
Loaded edges simply supported, other edges clamped	3/4
Loaded edges clamped, other edges simply supported	1
All edges clamped	3/4

FIGURE 9.6. Face wrinkling instability. (Reprinted from Reference [17].)

is localized in the direction of the axial load; that is, the wrinkle is essentially a short wavelength buckle, as shown in Figure 9.6. Heath [23] derived an expression for this mode of instability for the case of isotropic materials as

$$\sigma_{cr} = \left[\frac{2}{3}\frac{t_f}{h_c}\frac{E_c'E_f}{(1-\nu^2)}\right]^{1/2} \tag{9.39}$$

Heath defined E_c' incorrectly in his paper, but Hemp [5] clarified the point in an earlier paper. The face-wrinkling stability equation for isotropic faces given by Equation (9.39) can be modified for anisotropic materials to be:

$$\sigma_{cr} = \left[\frac{2}{3}\frac{t_f}{h_c}\frac{E_c'\sqrt{E_{fx}E_{fy}}}{(1-\nu_{xy}\nu_{yx})}\right]^{1/2} \tag{9.40}$$

An earlier equation for face wrinkling was developed by Hoff and Mautner [9], and given by

$$\sigma_{cr} = c(E_{fx}E_{cx}'G_{cxz}')^{1/3} \tag{9.41}$$

for a honeycomb sandwich panel where E_{cx}' and G_{cxz}' are given by Equations (9.44) and (9.46), and c is a constant usually 0.5, 0.6, or 0.65. Note that here the critical stress depends on material properties only. It is not clear whether one should use Equation (9.40) or (9.41). The analyst/designer can be conservative by using the equation which gives the lowest critical stress for the particular case being studied.

9.5.5 Monocell Buckling or Face Dimpling

In honeycomb core sandwiches, a fourth type of instability occurs because the faces over one cell can buckle as a small plate supported by the cell walls. Methods of analysis developed at the USFPL used an empirical equation having the form of the plate-buckling equation with the numerical coefficient determined by empirical means. The result for an isotropic face material is as follows, where the subscript T denotes the tangent modulus.

$$\sigma_{cr} = \frac{2E_{f_T}}{(1 - \nu^2)} \left(\frac{t_f}{d}\right)^2$$

For anisotropic faces, the expression can be written as

$$\sigma_{cr} = \frac{2(E_{fx_T} E_{fy})^{1/2}}{(1 - \nu_{xy}\nu_{yx})} \left(\frac{t_f}{d}\right)^2 \tag{9.42}$$

9.5.6 Core Properties

Mechanical properties of honeycomb core used in the previous equations are called "effective" and are designated with a prime, because they are properties associated with the core acting as a homogeneous material having these "effective" properties. They are functions of the core material's properties, ρ_c, E_c, G_{cx}, and G_{cy}, the core wall thickness t_c, and the cell size d. To truly optimize the structure for minimum weight, it is advantageous to relate the effective properties back to fundamental geometry and material properties. For foam or solid core sandwich panels there is no need for effective properties and the actual mechanical properties are used for E_c, G_{cx} and G_{cy} and ρ_c.

For hexagonal-cell honeycomb core having some double walls as shown in Figure 9.4, the properties were developed by Kaechele [24], and shown below. These effective core properties can be related to the geometry and actual material properties as follows, where ρ_c, G_c, and E_c are the weight density, the shear modulus and the modulus of elasticity of the core material itself:

$$\rho_c' = k_2(t_c/d)\rho_c \tag{9.43}$$

$$G_{cx}' = k_3(t_c/d)G_c \tag{9.44}$$

$$G_{cy}' = k_4(t_c/d)G_c \tag{9.45}$$

$$E_c' = k_2(t_c/d)E \tag{9.46}$$

It is seen from Equations (9.44) and (9.45) that here

$$r = G_{cx}'/G_{cy}' = k_3/k_4 \tag{9.47}$$

TABLE 9.2. Values of k_2, k_3 and k_4 for Various Honeycomb Constructions.

Type of Construction	k_2	k_3	k_4
Hexagonal (Kaechele)	8/3	5/3	1
Hexagonal (MIL HDBK-23)	8/3	4/3	8/15
Square cell (Kaechele)	2	1	1
Square cell (MIL HDBK-23)	2	1	1

For the hexagonal cell construction of Figure 9.4 as well as other types of honeycomb core, the values of k_2, k_3, and k_4 are given in Table 9.2 according to Kaechele [24] and MIL HDK-23 [22]. For other honeycomb configurations, the values of these constants can be easily derived using the methods of Kaechele [24]. A fairly inclusive listing of commercially available honeycomb core properties are given in Appendix 1.

9.5.7 Plasticity Effects

The extension of elastic buckling theory to account for the buckling of structures at stresses above the proportional limit of the material has been widely studied. Many investigations have used the elastic equations, wherein Young's modulus E has been multiplied by a plasticity reduction factor η. However, there is considerable difference of opinion about a correct form for η. These expressions range in complexity and it is not at all clear which expression has more merit. For structural optimization, if the compressive stress-strain curve of the face material has a proportional limit, then for stresses above the proportional limit, all values of E_{fx} can be replaced by $\overline{E}_{fx}$ where

$$\overline{E}_{fx} = \eta E_{fx} \tag{9.48}$$

One often uses a plasticity reduction factor in which $\overline{E}_{fx} = (E_{fx_T} E_{fx})^{1/2}$ where the subscript T denotes the tangent modulus at that face stress. This will require an iteration to match face stress with tangent modulus. It is unlikely that for the uniaxial compression in the x-direction, the stresses in the y-direction will cause deviation from the elastic value E_y, but this could also be modified.

9.5.8 Weight Relationship

To obtain a weight per unit area of the honeycomb sandwich structure, the following may be used:

$$W = 2t_f\rho_f + \rho_c' h_c + W_{ad} \tag{9.49}$$

The weight of the adhesive or other joining material, W_{ad}, cannot be easily related to the variables discussed earlier and is dependent upon the material, method of joining, fabrication techniques, and skill and temperament of the personnel. Since in many cases this is a small fraction of the weight and, because of the factors involved, it will not be specified further and need not be accounted for in the comparisons to select the optimum geometry and materials. However, care should be used to include it when comparing structures employing no adhesive or with other types of construction.

9.5.9 Analysis and Design Methods

For a honeycomb sandwich panel (such as that shown in Figure 9.3) to withstand an in-plane compressive load N_x, in force per unit width, the materials and component sizes must be sufficient to insure that overstressing [Equation (9.25)], overall buckling [Equation (9.26)], core shear instability [Equation (9.36)], face wrinkling [Equation (9.40) or (9.41)], and face dimpling [Equation (9.42)] will not occur. If any of the first four occur the panel is useless. If face dimpling occurs, it may not cause structural failure, but peeling of any coating such as paint could occur, the surface may be "unsightly," a boundary layer could be tripped from laminar to turbulent flow, or permanent core crushing in that vicinity may occur. These should be avoided.

9.6 THE BUCKLING OF SOLID- OR FOAM-CORE SANDWICH PANELS SUBJECTED TO IN-PLANE COMPRESSIVE LOADS

9.6.1 Face Stresses

For foam core sandwich panels and many other solid core panels it can be assumed that all in-plane loads, and bending loads as well, are resisted by the faces only. Therefore, in this case Equation (9.25) is used to determine the face stresses in terms of the applied load per unit width, N_x.

9.6.2 Overall Buckling

The equation for the overall buckling of an anisotropic composite sandwich panel with a solid core subjected to an in-plane compressive load is given by Equations (9.26) through (9.31). However in this case the actual mechanical properties of the core, G_{cx} and G_{cy} are used in Equation (9.31).

9.6.3 Core Shear Instability

Core shear crimping or core shear instability will occur at a face stress lower than that of overall panel buckling when

$$V_x \geq k_1 B_1 r \tag{9.50}$$

The critical stress value on core shear instability is given by

$$\sigma_{cr} = \frac{G_{cx} h_c}{2t_f} \tag{9.51}$$

It is seen that overall panel buckling and core shear instability will occur at the same face stress value when $V_x = k_1 B_1 r$, or more specifically when

$$\frac{\pi^2}{2(1 - \nu_{xy}\nu_{yx})} \frac{\overline{E}_{fx}}{G_{cx}} \frac{h_c t_f}{b^2} = k_1 r \tag{9.52}$$

Thus Equations (9.26) and (9.51) completely describe the conditions of the simultaneous buckling of overall panel and core shear instability without the complexities of using Equations (9.27) through (9.35) with the lengthy determination of K.

9.6.4 Face Wrinkling

In addition to overall panel buckling and core shear instability, a short wavelength buckling can occur if the faces are thin and can be described by

$$\sigma_{cr} = \left[\frac{2}{3} \frac{t_f}{h_c} \frac{E_c (E_{fx} E_{fy})^{1/2}}{(1 - \nu_{xy}\nu_{yx})}\right]^{1/2} \tag{9.53}$$

Of course the Hoff-Mautner equation is still used also

$$\sigma_{cr} = C[E_{fx} E_{cx} G_{cxz}]^{1/3} \tag{9.54}$$

where the constant C is 0.5, 0.6 or 0.65 by various users. Again, it is seen that with this equation the critical strain is dependent on material properties only. Plantema [25] uses 0.82 for the constant C in Equation (9.52), and Dreher [7] says that this corresponds well with the experimental data. He states emphatically that $C = 0.5$ does not correspond with his test results.

9.6.5 Weight Relationship

The weight of the solid-core sandwich panel per unit planform area, w is given by

$$W = 2t_f\rho_f + \rho_c h_c + W_{ad} \tag{9.55}$$

where ρ_i are the weight density of the materials involved and W_{ad} is the nonanalytic weight per unit planform area of the adhesive bonding the face to the core.

9.6.6 Analysis and Design Methods

For a foam- or solid-core sandwich panel to withstand an in-plane compressive load N_x, in force per unit width, the materials and component thicknesses must be sufficient to insure that overstressing [Equation (9.25)], overall buckling [Equation (9.26)], core shear instability [Equation (9.51)], and face wrinkling [Equation (9.53)] will not occur, because any one of the above will cause panel failure. With the satisfactory design, the panel weight is determined from Equation (9.55). As discussed previously, foam cores available today comprise an almost continuous value of shear modulus G and density ρ, usually a linear relationship between them.

9.7 BUCKLING OF A TRUSS-CORE SANDWICH PANEL SUBJECTED TO UNIAXIAL COMPRESSION

Another type of sandwich construction for panels subjected to uniaxial compression involves a corrugated core, sometimes referred to as single-truss core. Several promising and clever manufacturing methods have been devised to make unique use of fiber-reinforced polymer matrix composites for this type of construction. At least one of these involves weaving the filaments of the face material and core material together at the junctions, thus increasing the structural integrity of the joint and avoiding the joining problems associated with conventional construction.

The analysis methods developed herein are applicable to both metallic and composite material construction and account for material orthotropy in both the face and core. They are also applicable to panels at elevated or lowered temperature, under steady state and nearly uniform temperatures. Only the stress-strain curve is necessary for each temperature under consideration.

Consider the flat corrugated-core (truss-core) sandwich panel cross-section

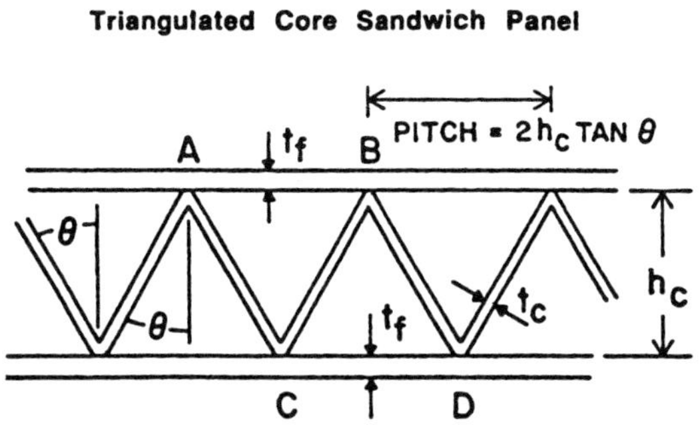

FIGURE 9.7. Cross-section of corrugated-core sandwich panel. (Reprinted from Reference [27].)

idealized in Figure 9.7 for the panel shown in Figure 9.3(a). For a given material system there are four geometric parameters to consider; namely the core depth (h_c), the web thickness (t_c), the face thickness (t_f), and the angle the web makes with a line normal to the faces (θ).

This panel will fail if any of the following five events occur: overstressing of the face, overstressing of the core, overall panel instability, local face buckling, or web buckling. It can be shown that face wrinkling, which can occur in honeycomb sandwich construction will not occur in this type of construction, because it can be shown that local face buckling and web buckling will invariably occur at lower values of the applied load. Likewise, core shear instability, in the sense of shear crimping in honeycomb sandwich construction cannot occur for the same reason. Hence, there are three modes of instability and four geometric parameters.

Unlike honeycomb and solid or foam-core sandwich construction, with a truss-core sandwich, it is intended that the core will carry or resist a portion of the in-plane compressive load. Thus, it is necessary to define some elastic and geometric quantities before proceeding.

9.7.1 Elastic and Geometric Constants

These are determined from those given in more general form by Libove and Hubka [8]. The core area per unit width and the area moment of inertia of the core per unit width are given by

$$\overline{A}_c = \frac{t_c}{\sin \theta} \tag{9.56}$$

$$\overline{I}_c = \frac{t_c h_c}{12 \sin \theta} = \frac{\overline{A}_c h_c^2}{12} \tag{9.57}$$

The extensional stiffness of the plate per unit width in the x-direction, A_{11}, is given by the following for isotropic materials,

$$A_{11} = E_c\bar{A}_c + 2E_f t_f \tag{9.58}$$

where E_c and E_f are the compressive moduli of elasticity of the core and face material respectively. Therefore Equation (9.58) is A_{11} for the truss-core panel in the language of composite theory.

The transverse shear stiffness per unit width, in the x-direction and the y-direction, respectively, are found to be, in the notation of [8],

$$D_{q_x} = \frac{G_c t_c \cos\theta}{\tan\theta} \tag{9.59}$$

$$D_{q_y} = \frac{E_c t_c}{(1 - \nu_c^2)} \cos^2\theta \sin\theta \tag{9.60}$$

The latter expression agrees with that derived by Anderson [9]. Thus, Equations (9.59) and (9.60) are A_{55} and A_{44} for the truss-core panel [see Equation (2.68)].

Lastly, the moment of inertia per unit width of the faces, considered as membranes, with respect to the sandwich middle surface, is seen to be, as before,

$$\bar{I}_f = \frac{t_f h_c^2}{2} \tag{9.61}$$

9.7.2 Overall Instability

The best expression describing the overall instability of a corrugated-core sandwich panel composed of isotropic materials under uniaxial compressive loads is derived by Seide [10] NACA TN2679 (which is used in ANC-23 in slightly modified form [22])

$$N_x = \frac{\pi^2 E_f \bar{I}_f K}{b^2} \tag{9.62}$$

where K is the buckling coefficient derived and plotted in Figures 2 and 4 of Reference [10]. It is given as a function of length to width ratio (a/b), for the cases of the unloaded edges simply-supported and clamped, for various values of the transverse shear flexibility parameter *V*, defined as

$$V = \frac{\pi^2 E_f \bar{I}_f}{b^2 D_{q_y}} + \frac{\pi^2(1 - \nu_c^2)}{2\cos^2\theta \sin\theta}\left(\frac{t_f}{t_c}\right)\left(\frac{h_c}{b}\right)^2\left(\frac{E_f}{E_c}\right) \tag{9.63}$$

The buckling coefficient *K* for this type of construction has the same general characteristics as that of a flat homogeneous plate; namely, that for $a/b > 1$, successive minima occur for increasing numbers of half sine waves in the loading direction, each minimum having the same value of *K*. Hence, this minimum value can be taken as the lower bound for all panels where $a > b$. Therefore, Figures 3 and 5 of Reference [10] can be used. These figures make use of the ratio $E_c\bar{I}_c/E_f\bar{I}_f$, which for this construction is given by

$$\frac{E_c\bar{I}_c}{E_f\bar{I}_f} = \frac{E_c}{E_f}\frac{t_c}{t_f}\frac{1}{6\sin\theta} \tag{9.64}$$

For example, for panels with the unloaded edges simply supported the buckling coefficient *K* is given as follows:

$$K = \frac{\left[\left(\frac{1}{\beta}\right) + \beta^2\right]}{(1 - \nu_f^2) + 2(1 + \nu_f)\beta^2} + \frac{E_c\bar{I}_c}{E_f\bar{I}_f\beta^2}$$

where here $\beta = a/b$.

There is no published analytical expression describing the overall instability of corrugated-core sandwich panels utilizing orthotropic materials and for uniaxial compressive loads. However, it is not difficult to deduce the form by observing the difference in the analogous expression for honeycomb-core-sandwich panels for isotropic and orthotropic materials.

From the isotropic expressions on pages 53, 82, and 96 of ANC-23 [22], it is seen that when the overall instability expressions are written for orthotropic materials E_f is replaced by $\sqrt{E_{fx}E_{fy}}$ when flexural properties are involved, while E_f is replaced by E_{fx} in extensional property expressions. It is therefore deduced that for corrugated-core panels utilizing orthotropic materials using Equations (9.61), (9.62) and the above, the critical load per unit width is

$$N_x = \frac{\pi^2\sqrt{E_{fx}E_{fy}}\, t_f h_c^2 K}{2b^2} \tag{9.65}$$

It is also hypothesized that Figures 3 and 5 of Reference [12] may be used to determine K if the following expressions are used instead of Equations (9.63) and (9.64):

$$V = \frac{\pi^2(1 - \nu_{xyc}\nu_{yxc})}{2\cos^2\theta\sin^2\theta}\left(\frac{t_f}{t_c}\right)\left(\frac{h_c}{b}\right)^2 \frac{\sqrt{E_{fx}E_{fy}}}{\sqrt{E_{cx}E_{cy}}} \tag{9.66}$$

$$\frac{E_c\bar{I}_c}{E_f\bar{I}_f} = \frac{\sqrt{E_{cx}E_{cy}}}{\sqrt{E_{fx}E_{fy}}}\frac{t_c}{t_f}\frac{1}{6\sin\theta} \tag{9.67}$$

9.7.3 Face Plate Instability

From Figure 9.7 it is seen that each plate element of the face from A to B can buckle due to the axial loading N_x. Since the support conditions at A and B, the unloaded edges, are not known precisely, it is conservative to assume a simple support. Since for almost all constructions the panel length a is greater than the distance AB, the buckling coefficient K is taken as four. Anderson [11] discusses the effect of more complex buckling modes due to the interaction between face and core elements. However, he shows that, at most, the buckling coefficient would be 4.21 for simultaneous buckling of face and core elements. Hence the value of four appears very realistic as a conservative value. The face plate instability equation can therefore be written in terms of the quantities given in Figure 9.7 as

$$\sigma_{cr} = \frac{\pi^2 E_f}{12(1 - \nu_f^2)}\frac{t_f^2}{h_c^2}\frac{1}{\tan^2\theta} \tag{9.68}$$

where σ_f is the critical stress in the face.

Utilizing the critical stress expression for an orthotropic plate given by Timoshenko and Gere [2], and using the terminology of Figure 9.7, the expression can be written as

$$\sigma_{f_{cr}} = \frac{\pi^2 E_{of} t_f^2}{12h_c^2(1 - \nu_{xyf}\nu_{yxf})\tan^2\theta} \tag{9.69}$$

where

$$2E_{oi} = \sqrt{E_{ix}E_{iy}} + \nu_{yxi}E_{ix} + 2G_{xyi}(1 - \nu_{xyi}\nu_{yxi}) \quad (i = c,f) \tag{9.70}$$

9.7.4 Web-Plate Instability

Similar to the above, the plate instability equation for a web element can be written as

$$\sigma_{w_{cr}} = \frac{\pi^2 E_c}{3(1 - \nu_c^2)} \frac{t_c^2}{h_c^2} \cos^2 \theta \tag{9.71}$$

where $\sigma_{w_{cr}}$ is the critical stress in the web element. Likewise, for orthotropic composite materials

$$\sigma_{w_{cr}} = \frac{\pi^2 E_{oc} t_c^2 \cos^2 \theta}{3h_c^2(1 - \nu_{xyc}\nu_{yxz})} \tag{9.72}$$

where E_{oc} is given by Equation (9.70).

9.7.5 Load-Stress Relationship

By enforcing equal axial strains in the core and face to insure compatibility in the overall construction, the following important relationship is easily derived for the stress in the core material as a function of the face stresses,

$$\sigma_c = \sigma_f \frac{E_{cx}}{E_{fx}} \tag{9.73}$$

Then since $N_{xf} = 2t_f\, \sigma_f$ and $N_{xc} = \sigma\, A_c$, the face stress can be written in terms of the applied load per unit width $N_x = N_{xf} + N_{xc}$:

$$\sigma_f = \frac{N_x}{\left(\frac{E_c}{E_f} \frac{t_c}{\sin \theta} + 2t_f\right)} \tag{9.74}$$

For orthotropic composite materials the modulus values to use in Equations (9.73) and (9.74) are those in the load direction.

Both Equations (9.68) and (9.69) hold only when stresses in the face and core are both below the proportional limit of each material. Above the proportional limit an iterative procedure would be needed to insure compatibility in determining an analogous relation to Equation (9.73). If both core and face materials are the same $\sigma_c = \sigma_f$ and the procedures which follow hold above the proportional limit if a suitable plasticity reduction factor is used with the modulus of elasticity.

9.7.6 Weight Relationship

From Figure 9.7 it is seen that

$$W = 2\rho_f t_f + \rho_c \overline{A}_c + W_{ad} \tag{9.75}$$

where ρ_c and ρ_f are the weight density of the core and face material respectively, and W and W_{ad} are the weight per unit plan form area of the panel and the weight of the adhesive or other material used to join face and core, respectively.

9.7.7 Analysis and Design Methods

For the truss core panel subjected to a uniform in-plane compressive load per unit width, N_x, the face stress is given by Equation (9.74) and the core stress by Equation (9.73). These stresses must be maintained at or below the allowable compressive stress for the face material and core material respectively. For orthotropic materials the properties in the load direction should be used (the x-direction). It is also necessary that, with a specific applied load N_x, neither overall buckling [Equations (9.62) or (9.65)] nor face plate buckling [Equations (9.68) or (9.69)] nor web plate buckling [Equations (9.71) or (9.72)] occur.

Once satisfied of the structural integrity, Equation (9.75) is used to determine the panel weight.

Under in-plane compressive loads, because the core in truss-core sandwich panels does resist a portion of the load, and because the construction has such a high flexural stiffness, truss-core sandwich panels can efficiently resist higher loads than honeycomb solid- or foam-core sandwich panels. As a secondary advantage, this type of construction can be used as a heat exchanger or a liquid storage container.

9.8 ELASTIC STABILITY OF A WEB-CORE SANDWICH PANEL SUBJECTED TO A UNIAXIAL COMPRESSIVE IN-PLANE LOAD

9.8.1 Introduction

Consider a flat web-core sandwich panel, generalized to include some arbitrary angle θ, as shown in Figure 9.8. The overall geometry and loading is given in Figure 9.3(a). There are five geometric variables; namely, the core depth h_c, the web thickness t_c, the face thickness t_f, the angle the web makes with a line normal to the faces (θ) , and the distance between web elements d_f.

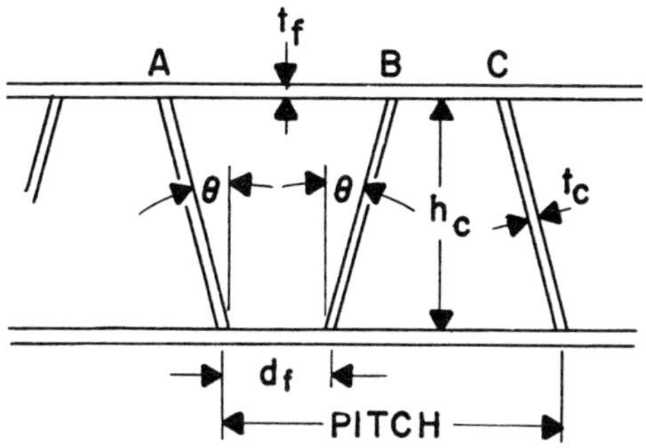

FIGURE 9.8. Generalized web-core sandwich panel. (Reprinted from Reference [28].)

The panel is considered to fail if any of the following instabilities occur: overall panel instability, local face buckling in the region from A to B, local face buckling in the region B to C, and web element buckling (see Figure 9.8). In each case, the expression used to mathematically describe the instabilities is the best available in the literature. Overstressing can occur if the stresses in either the face or core exceed established allowable stresses for the materials used. Hence, there are five geometric variables and four modes of failure.

9.8.2 Elastic and Geometric Constants

The elastic and geometric constants for the web-core construction can be determined from those given in more general form by Libove and Hubka [8]. The core area per unit width and the moment of inertia of the core about the centroidal axis per unit width are given by

$$\bar{A}_c = t_c h_c / (d_f + h_c \tan\theta)\cos\theta \tag{9.76}$$

$$\bar{I} = t_c h_c^3 / 12 \cos\theta (d_f + h_c \tan 0) = \bar{A}_c h_c^2 / 12 \tag{9.77}$$

The transverse shear stiffness of the core, per unit width in the x-direction (D_{q_y}) of an element of the sandwich panel cut by two y-z planes is seen to be negligible, due to the lack of structural continuity of the web core. Hence, following Libove and Hubka [8], as well as Seide [10].

$$D_{q_Y} \to 0 \tag{9.78}$$

Hence, the transverse shear flexibility parameter in the y-direction is given by

$$V_y = \pi^2 E_f \bar{I}_f / b^2 D_{q_Y} \to \infty \tag{9.79}$$

This is not to say that the construction as a whole has no transverse-shear stiffness, but rather that the stiffness that does exist depends upon the faces to provide the continuity. In fact Seide discusses this case and states that for the case of $D_{q_X} = \infty$ and $D_{q_Y} = 0$, the compressive buckling load is finite, rather than being equal to zero, and varies with plate aspect ratio.

It should be noted that when $\theta \neq 0$ and $d_f = 0$, the construction has a continuous core and is called corrugated core or truss core, with the result that $D_{q_Y} \neq 0$, and there is a specific value of θ for which the weight is a minimum. This construction is treated in Section 9.7.

The area moment of inertia per unit width of the faces considered as membranes, with respect to the sandwich middle surface, per unit width, is seen to be

$$\bar{I}_f = t_f h_c^2/2 \tag{9.80}$$

9.8.3 Load-Stress Relationship

For a given load N_x, the face stress and the core stress are given by Equations (9.73) and (9.74) where $\bar{A}_c$ is given by Equation (9.76).

9.8.4 Overall Panel Instability

Just as in Section 9.7.2 earlier, the best expression applicable to the overall instability of a web-core sandwich panel composed of isotropic materials under uniaxial compressive loads is derived by Seide [10] and given as follows:

$$N_x = \pi^2 E_f \bar{I}_f K/2b^2 \tag{9.81}$$

where K is the buckling coefficient derived and plotted in Figures 2 and 4 of Reference [10] for various boundary conditions. For this construction, with simply supported unloaded edges, K can be given explicitly as

$$K = \frac{[(1/\beta) + \beta^2]}{1 - \nu_f^2 + 2(1 + \nu_f)\beta^2} + \frac{E_c \bar{I}_c}{E_f \bar{I}_f \beta^2} \tag{9.82}$$

where here $\beta = a/b$.

There is no published analytical expression describing the overall panel instability of web-core sandwich panels utilizing orthotropic materials subjected to uniaxial compressive loads. However, as in previous sections, it is not difficult to deduce the form of the equation by observing the differences in the analogous expressions for honeycomb-core sandwich panels for isotropic and orthotropic materials.

The remainder of this subsection is identical to Section 9.7.2, but is repeated here for completeness. From the expressions in Reference [22], it is seen that when the overall instability expressions are written for orthotropic materials E_f is replaced by $(E_{fx}E_{fy})^{1/2}$ when flexural properties are considered, while E_f is replaced by E_{fx} when extensional properties are involved. It is therefore deduced that for web-core panels utilizing orthotropic materials, Equation (9.81) becomes, utilizing Equation (9.80),

$$N_x = \pi^2 (E_{fx}E_{fy})^{1/2} t_f h_c^2 K/2b^2 \tag{9.83}$$

It is also hypothesized that Figures 2 and 4 of Reference [12] may be extended to find the buckling coefficient K for the orthotropic construction if $E_c\bar{I}_c/E_f\bar{I}_f$ is determined by

$$\frac{E_c\bar{I}_c}{E_f\bar{I}_f} = \frac{(E_{cx}E_{cy})^{1/2} t_c h_c}{6(E_{fx}E_{fy})^{1/2} t_f(d_f + h_c \tan\theta)\cos\theta} \tag{9.84}$$

Likewise Equation (9.82) may be extended to orthotropic construction by replacing

$$(1 - \nu_f^2) \text{ by } (1 - \nu_{xyf}\nu_{yxf}), \text{ and } (1 + \nu_f) \text{ by } (1 + \nu_{xyf})$$

9.8.5 Face Plate Instability

From Figure 9.8 it is seen that the face elements *A-B* and *C-D* may each undergo an elastic instability under uniaxial compressive loads. Since the unloaded edge supports are not known precisely, it is conservative to assume a simple support. Since for almost all panels, the panel length a is much greater than the width, which is the distance *A* to *B* or *C* to *D*, the buckling coefficient is taken as 4.

For the faces made of an orthotropic material the expression for the critical face stress given by Timoshenko and Gere [2] is used. In terms of the quantities defined in Figure 9.8 the critical stress can be written as follows for the region *A* to *B*:

$$\sigma_{f_1} = \left(\frac{\pi^2 E_{of} t_f^2}{3(1 - \nu_{xyf}\nu_{yxf})(d_f + 2h_c \tan\theta)^2}\right. \tag{9.85}$$

and in the region *B* to *C*:

$$\sigma_{f_2} = \pi^2 E_{of} t_f^2 / 3(1 - \nu_{xyf}\nu_{yxf}) d_f^2 \tag{9.86}$$

where $E_{0i}(i = c,f)$ is given by Equation (9.70).

9.8.6 Web-Plate Instability

Similar to the above, the plate instability equation for a web element composed of an orthotropic material can be written as

$$\sigma_c = \pi^2 E_{oc} t_c^2 \cos^2 \theta / 3(1 - \nu_{xyc}\nu_{yxc}) h_c^2 \tag{9.87}$$

9.8.7 Load-Stress Relationship

For the construction of Figure 9.8, it is seen that the load per unit width, N_x, is related to the face and web stresses, σ_f and σ_c, by the following relationship:

$$N_x = \sigma_c \overline{A}_c + 2\sigma_f t_f \tag{9.88}$$

By equating the axial strains in the core and face to insure compatibility in the overall construction, the following relationship is easily derived:

$$\sigma_c = \sigma_f E_{cx} / E_{fx} \tag{9.89}$$

Thus from Equations (9.88), (9.89), and (9.76), the load per unit width N_x is related to the face stress as follows:

$$N_x = \sigma_f \{[E_{cx} t_c h_c / E_{fx}(d_f + h_c \tan \theta) \cos \theta] + 2t_f\} \tag{9.90}$$

Both Equations (9.89) and (9.90) hold only when stresses in the face and core are below the proportional limit of each material, where Hooke's Law applies. Above the proportional limit of either material an iterative procedure would be needed to insure compatibility in determining an analogous relationship to Equation (9.79), employing some reduced moduli $\overline{E}_{ci}$ and $\overline{E}_{fi}$ involving a plasticity reduction factor.

If both core and face materials are the same, $\sigma_c = \sigma_f$, then the same procedures apply for stresses above the proportional limit if a suitable plasticity reduction factor is used with the modulus of elasticity.

9.8.8 Weight Relation

From Figure 9.8, it is seen that

$$W = 2\rho_f t_f + \rho_c \overline{A}_c + W_{ad} \tag{9.91}$$

where ρ_c and ρ_f are the specific weight of the core and face material, respectively, and W and W_{ad} are the panel weight per unit of planform area and the weight of the adhesive or other material used to join face to core, respectively.

At the outset, independent of the material system, it is clear that from Equations (9.85) and (9.86) that for minimum weight construction of the web-core sandwich

$$\theta = 0^\circ \tag{9.92}$$

This is intuitively obvious. The result is that all other expressions used for optimization are simplified, and the construction show in Figure 9.8, with $\theta = 0^\circ$ now assumes the familiar web-core configuration, shown in Figure 1.1(c).

9.8.9 Design and Analysis

For the web-core panels subjected to an in-plane load per unit width, N_x, it is clear the components must be sized such that neither the faces nor the web elements are overstressed [Equations (9.88) and (9.89)]. Also, overall buckling [Equations (9.81) through (9.84)], face-element buckling [Equations (9.85) through (9.86)], and web-plate buckling [Equation (9.87)] cannot be allowed. Finally the panel weight can be determined from Equation (9.91).

9.9 BUCKLING OF HONEYCOMB CORE SANDWICH PANELS SUBJECTED TO IN-PLANE SHEAR LOADS

If the honeycomb core sandwich panel discussed in Section 9.5 and shown in Figure 9.3 is subjected to in-plane shear loads, the following equations in this section can be used for their design and analysis [26]. Again, the honeycomb sandwich panel faces can be overstressed, or any of four buckling modes can occur.

9.9.1 Load-Face Stress Relationship

Since the honeycomb core does not take any of the applied in-plane load, the applied load face stress relationship is

$$\frac{N_{xy}}{2t_f} = \frac{N_{yx}}{2t_f} = \sigma_{xy} = \sigma_{yx} \tag{9.93}$$

9.9.2 Overall Panel Buckling

The overall panel buckling equation is

$$N_{xy} = \frac{\pi^2 \sqrt{\overline{D}_x \overline{D}_y}}{b^2} L_{cr} \tag{9.94}$$

where following the terminology of the original source document the buckling coefficient is L_{cr}; where, as seen earlier in the text

$$\overline{D}_x = \frac{\overline{E}_{fx} h_c^2 t_f}{2(1 - \nu_{xy}\nu_{yx})}, \quad \overline{D}_y = \frac{\overline{E}_{fy} h_c^2 t_f}{2(1 - \nu_{xy}\nu_{yx})}$$

and where, if the stresses are above the proportional limit, the following plasticity-reduction factor may be used, where the subscript T refers to the tangent modulus.

$$\overline{E}_{fx} = \sqrt{E_{fx_T} E_{fx}}, \quad \overline{E}_{fy} = \sqrt{E_{fy_T} E_{fy}}$$

L_{cr} is a buckling coefficient defined by curves in Reference [12]. Overall buckling will occur when

$$V_x \leq k_1 S_{xi} \tag{9.95}$$

where V_x and k_1 have been defined previously in Section 9.5 and S_{xi} is defined as

$$S_{xi} = \frac{\left[\dfrac{1 - \overline{\nu}}{2}\right] 4r + (1 - r)^2}{\left[\dfrac{1 - \overline{\nu}}{2}\right] (1 + r^2) \left[r + \dfrac{b^2}{a^2}\right]} \tag{9.96}$$

where $\overline{\nu}$ in this expression is defined as

$$\overline{\nu} = \sqrt{\nu_{xy}\nu_{yx}}$$

and in this equation,

$$r = V_y/V_x = G'_{cx}/G'_{cy} = k_3/k_4 \tag{9.97}$$

9.9.3 Core Shear Instability

Core shear instability occurs when $V_x \geq k_1 S_{xi}$, and is described by

$$N_{xy} = h_c \sqrt{G'_{cx} G'_{cy}} \tag{9.98}$$

9.9.4 Face Wrinkling

The appropriate equation for the critical face stress for face wrinkling is:

$$\sigma_{xy} = \left[\frac{2t_f \, E'_c \, \sqrt{\overline{E}_{fx} \overline{E}_{fy}}}{3h_c \, (1 - \nu_{xy}\nu_{yx})}\right]^{1/2} \tag{9.99}$$

9.9.5 Monocell Buckling

The equation to use to determine the critical face stress for monocell buckling is:

$$\sigma_{xy} = \frac{2\sqrt{E_{fx_T} E_{fy_T}}}{(1 - \nu_{xy}\nu_{yx})} \left(\frac{t_f}{d}\right)^2 \qquad (9.100)$$

where here, the authors of Equation (9.100) have employed the tangent modulus for the use of this equation with face stresses that are above the material's proportional limit.

9.9.6 Analysis and Design

Again, the weight equation is given by Equation (9.49). At this point any honeycomb-core panel can be designed or analyzed for specified in-plane shear loads. The faces must be sized such that the allowable shear stress does not exceed Equation (9.94). Also, the panel sizes must preclude overall buckling [Equation (9.94) through (9.97)], core shear instability [Equation (9.98)], face wrinkling [Equation (9.99)], and face dimpling [Equation (9.100)]. Obviously, if the face material is isotropic then $E_{fx} = E_{fy}$.

9.10 BUCKLING OF A SOLID-CORE OR FOAM-SANDWICH PANEL SUBJECTED TO IN-PLANE SHEAR LOADS

Again the expression for overall buckling is given by Equation (9.94). This can occur when Equation (9.95) holds where for this case

$$V_x = \frac{\pi^2}{b^2} \frac{\sqrt{\overline{E}_{fx}\overline{E}_{fy}}}{G_{cx}} \frac{h_c t_f}{2(1 - \nu_{xy}\nu_{yx})} \leq k_1 S_{x1} \qquad (9.101)$$

The core shear instability equation and the face-wrinkling equations are given by Equations (9.98) and (9.99), but in this case effective core properties $(\ )'$ must be replaced by the actual core material properties for the solid-core or foam-core materials.

Thus, the applied load-face stress relation and the weight equation are

$$N_{xy} = 2t_f\sigma_{xy} \qquad (9.102)$$

$$W = 2t_f\rho_f + \rho_c h_c + W_{ad} \qquad (9.103)$$

any solid-core sandwich panel can now be analyzed and designed for in-plane shear loads.

9.11 BUCKLING OF A TRUSS-CORE SANDWICH PANEL SUBJECTED TO IN-PLANE SHEAR LOADS

9.11.1 Introduction

Consider the flat corrugated-core sandwich panel cross section of Figure 9.7. For a given material system, there are four geometric parameters with which to optimize; namely the core depth (h_c), the web thickness (t_c), the face thickness (t_f), and the angle the web makes with a line normal to the faces (θ). The overall panel to be considered is shown in planform in Figure 9.9. This panel of width b and length a is subjected to in-plane shear loads per unit edge distance N_{xy} and N_{yx} (lb/in.).

In addition to overstressing, the panel is considered to fail if any of the following instabilities occur: overall instability, shear instability of the faces, and shear instability in the web. Thus, there are three modes of instability, and four geometric variables. Since panels in which the faces and cores utilize different orthotropic materials are the most general materials system, it is convenient to derive all expressions for that situation.

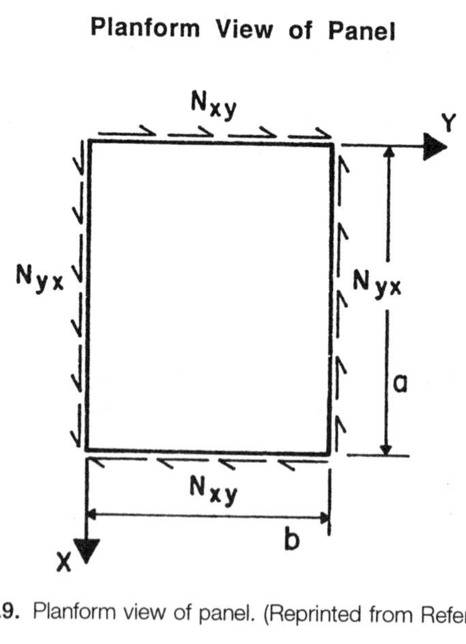

FIGURE 9.9. Planform view of panel. (Reprinted from Reference [29].)

9.11.2 Overall Stability

The overall instability of a truss-core sandwich panel under in-plane shear loading is given in ANC-23 [22] as well as other places as

$$\tau_f^{cr} = \frac{(E_{fy}^3 E_{fx})^{1/4}}{(1 - \nu_{xyf}\nu_{yxf})}\left(\frac{h_c}{b}\right)^2 \bar{j} \tag{9.104}$$

where τ_f^{cr} is the critical shear stress in the face of the sandwich and where $\bar{j}$ is a buckling coefficient given by Vinson and Shore [26], related to the buckling coefficient j.

The coefficient j, found in [22], for orthotropic panels with simply-supported edges whose axes of elastic symmetry are parallel to the edges is given in Figure 9.10. In this figure, j is plotted as a function of B_2 and l/r where

$$B_2 = \frac{2G_{xyf}(1 - \nu_{xyf}\nu_{yxf}) + E_{fy}\nu_{xyf}}{\sqrt{E_{fx}E_{fy}}}, \quad \frac{1}{r} = \frac{b}{a}\left(\frac{E_{fx}}{E_{fy}}\right)^{1/4} \tag{9.105}$$

9.11.3 Face-Plate Instability

Looking at Figure 9.7, it is seen that each plate element of the faces from A to B can buckle due to the applied shear loads N_{xy} and N_{yx}. Since the support condition of the plate element along the edges depicted by A and B are not known precisely, it is conservative to assume that they are simply-supported edges. For such a case, the governing equation is given by Timoshenko and Gere [2] and others, for an orthotropic plate whose axes of elastic symmetry are parallel to the edges, of thickness h and width b,

$$\tau_{cr} = \frac{k}{3}\frac{(E_y^3 E_x)^{1/4}}{(1 - \nu_{xy}\nu_{yx})}\left(\frac{h}{b}\right)^2 \tag{9.106}$$

In this expression k is a coefficient plotted in Reference [2] (Figure 9.42) as a function of two parameters: β and $1/\theta$. If one looks at the original figures, it is obvious that if $\beta = 1/r$ and $1/\theta = B_2$, then Figure 9-42 of Reference [2] and Figure 9.10 herein are identical. Hence, the k of Equation (9.106) is identical to j of Figure 9.10 under these conditions.

From Figure 9.7 it is seen that for the face-plate instability Equation (9.106) can be written as

$$\tau_{f_{cr}} = \frac{k_f}{12}\frac{(E_{fy}^3 E_{fx})^{1/4}}{(1 - \nu_{xyf}\nu_{yxf})}\frac{t_f^2}{h_c^2 \tan^2\theta} \tag{9.107}$$

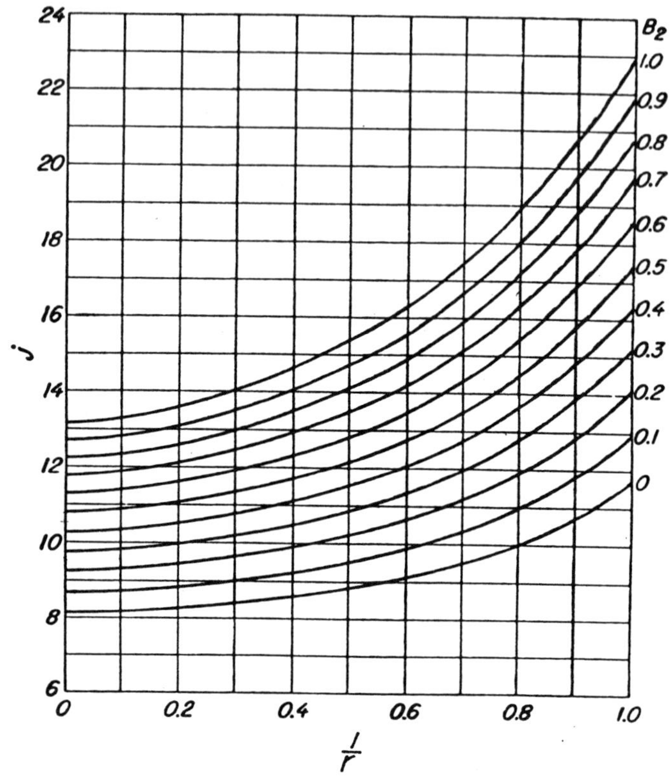

FIGURE 9.10. Buckling coefficient.

where $k_f = j$ is given by Figure 9.10, in which for this plate element B_2 is given by Equation (9.105) and

$$\frac{1}{r} = \frac{2h_c \tan \theta}{a} \left(\frac{E_{fx}}{E_{fy}} \right)^{1/4} \tag{9.108}$$

9.11.4 Web-Plate Instability

Likewise, the local plate elements of the triangulated core can become unstable due to shear stresses induced into the core by the shearing of the faces. Again, the conservative assumption is made here that the web elements are simply supported along the edges A and C depicted in Figure 9.7.

Referring to Equation (9.107) and the geometry of Figure 9.7, it is seen that the expression describing the web-plate instability can be written as

$$\tau_{f_{cr}} = \frac{k_c}{3} \frac{[E_{yc}^3 E_{xc}]^{1/4}}{(1 - \nu_{xyc}\nu_{yxc})} \frac{t_c^2}{h_c^2} \cos^2\theta \tag{9.109}$$

where $k_c = j$ is found from Figure 9.10, where here

$$B_2 = \frac{2G_{xyc}(1 - \nu_{xyc}\nu_{yxc}) + E_{cy}\nu_{xyc}}{\sqrt{E_{cx}E_{cy}}} \text{ and } \frac{1}{r} = \frac{h_c}{a\cos\theta}\left(\frac{E_{cx}}{E_{cy}}\right)^{1/4} \tag{9.110}$$

9.11.5 Load Stress Relationship

Looking at the construction shown in Figure 9.9 along the edges at $x = 0$ or $x = a$, the shear resultant N_{yx} is primarily resisted by the two faces. Even if the core elements are bonded or otherwise fastened to some edge fixture through which the shear N_{yx} is transmitted little load will be introduced into the core web plates directly. Hence, the load stress relationship is taken to be

$$\frac{N_{yx}}{2t_f} = \frac{N_{xy}}{2t_f} = \tau_f \tag{9.111}$$

This is not to imply that loads are not introduced into the core elements by the faces, as shown below.

9.11.6 Core Stress–Face Stress Relationship

Consider the repeated unit of the triangulated-core construction shown in Figure 9.11. Due to the shearing deformations of the faces, shearing deformations occur in the core since the web element and the face are bonded or otherwise connected to their junction along *db*, *gh*, etc.

The following relationships are valid in the elastic range, where $\varepsilon_{xyi}(i = c,f)$ are the shearing strains and the other symbols are given in Figures 9.7, 9.8 and 9.11.

$$\gamma_1 = 2\varepsilon_{xyi} = \frac{\tau_i}{G_{xyi}}, \quad \gamma_f = \frac{\delta_f}{h_c \tan\theta}, \quad \gamma_c = \frac{\delta_c}{h_c}\cos\theta$$

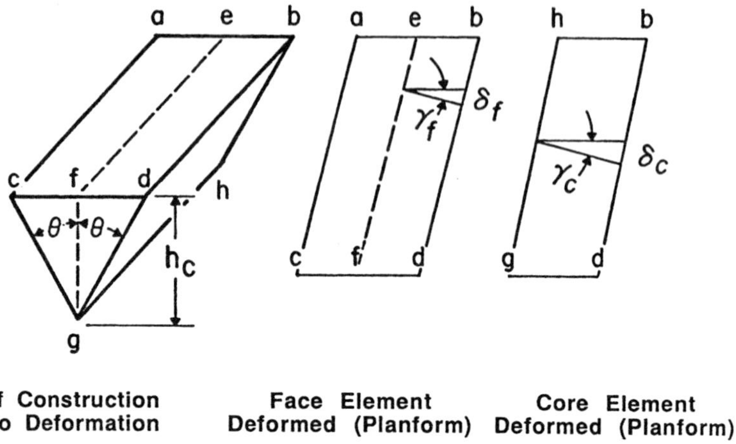

FIGURE 9.11. Triangulated-core construction. (Reprinted from Reference [29].)

For compatability of deformations, $\delta_c = \delta_f$ hence

$$\tau_c = \frac{G_{xyc}}{G_{xyf}} \tau_f \sin \theta \tag{9.112}$$

Note that $\tau_c < \tau_f$ when the same materials are used for faces and core.

9.11.7 Weight Relationship

The weight relationship from Figure 9.7 is repeated here for completeness,

$$W = 2\rho_f t_f + \rho_c \overline{A}_c + W_{ad} = 2\rho_f t_f + \frac{\rho_c t_c}{\sin \theta} + W_{ad} \tag{9.113}$$

where ρ_f and ρ_c are the weight density of the face and core materials respectively; W_{ad} is the weight in lb/in.^{2} of planform area of the adhesive or any other material used to join face and core; W is the total weight in lb/in.^{2} of planform area of the entire panel; and $\overline{A}_c = t_c/\sin \theta$.

9.12 BUCKLING OF A WEB-CORE SANDWICH PANEL SUBJECTED TO AN IN-PLANE SHEAR LOAD

9.12.1 Overall Buckling

For the web-core sandwich subjected to in-plane shear loads, it is seen that all the applied load is taken by the face components, and none of the applied shear can be induced into the web elements because they are perpendicular to the faces when $\theta = 0°$ (see Figure 9.8).

The overall panel in-plane shear instability is given as follows, following Libove and Hubka [8], and Vinson and Shore [26]

$$\tau_{f_{cr}} = \frac{(E_{fy}^3 E_{fx})^{1/4}}{1 - \nu_{xyf}\nu_{yxf}} \left(\frac{h_c}{b}\right)^2 \frac{1}{1 + 4\left|\frac{K_m(V=0)}{K_m} - 1\right|} j \tag{9.114}$$

where τ_{fcr} is the critical shear stress in the face, and where for this construction

$$K_m = \frac{A}{B_1C_1 + B_3C_2 + V_xA} \tag{9.115}$$

$$K_m(V = 0) = B_1C_1 + 2B_2C_2 + \frac{C_3}{B_1} \tag{9.116}$$

$$A = C_1C_3 - B_2^2C_2^2 + B_3C_2\left(B_1C_1 + 2B_2C_2 + \frac{C_3}{B_1}\right) \tag{9.117}$$

$$B_1 = (E_{fy}E_{fx})^{1/2} \tag{9.118}$$

$$B_2 = \frac{2G_{xyf}(1 - \nu_{yxf}\nu_{yxf}) + E_{fy}\nu_{xyf}}{(E_{fx}E_{fy})^{1/2}} \tag{9.119}$$

$$B_3 = \frac{G_{xyf}(1 - \nu_{xyf}\nu_{yxf})}{(E_{fx}E_{fy})^{1/2}} \tag{9.120}$$

C_1 through C_4 are given in Reference [22] for various boundary conditions and constructions. For the web-core construction with in-plane shear loading ($n = 1$) the constants are

All edges simply supported:

$$C_1 = C_4 = a^2/b^2,\ C_2 = 1,\ C_3 = b^2/a^2 \tag{9.121}$$

$x = 0$, a simply supported, $y = 0$, b clamped:

$$C_1 = \frac{16}{3}\frac{a^2}{b^2},\ C_2 = \frac{4}{3},\ C_3 = b^2/a^2,\ C_4 = \frac{4}{3}\frac{a^2}{b^2} \tag{9.122}$$

$x = 0$, a clamped; $y = 0$, b simply supported

$$C_1 = C_4 = \frac{3}{4}\frac{a^2}{b^2}, \; C_2 = 1, \; C_3 = 4\frac{b^2}{a^2} \tag{9.123}$$

All edges clamped:

$$C_1 = 4C_4 = 4a^2/b^2, \; C_2 = \frac{4}{3}, \; C_3 = 4b^2/a^2 \tag{9.124}$$

The coefficient j is obtained from Reference [2] for panels with simply-supported edges, given herein as Figure 9.10,where for this construction, B_2 is given by Equation (9.119) above and

$$1/r = (b/a)\,(E_{fx}/E_{fy})^{1/4} \tag{9.125}$$

9.12.2 Face-Plate Buckling

From Figure 9.7, it is seen that each face plate can buckle between each web element due to the applied shear loads N_{xy} and N_{yx}. Since the support conditions along the face element at the web element are not known precisely, it is conservative to assume simply-supported edges. For that case Timoshenko and Gere [2] provide a stability equation for an orthotropic plate, which can be written as

$$\tau_{fcr} = \frac{k_f}{3}\frac{(E_{fy}^3 E_{fx})^{1/4}}{(1 - \nu_{xyf}\nu_{yxf})}\frac{t_f^2}{d_f^2} \tag{9.126}$$

where it can be shown that $k_f = j$ of Figure 9.10, B_2 is given by Equation (9.119) and $1/r$ is

$$\frac{1}{r} = \frac{d_f}{a}\left(\frac{E_{fx}}{E_{fy}}\right)^{1/4} \tag{9.127}$$

Note that for $\theta = 0°$, when the panel is subjected to in-plane shear loads, the web plates are unloaded. If $\theta \neq 0°$, then some load is introduced into the web plates analogous to the truss-core construction discussed earlier.

9.12.3 Load-Stress Relationship

Looking at the construction of Figure 9.7, it is seen that in-plane shear resultants N_{xy} and N_{yx} are resisted by the two faces. Hence, the load-stress relationship is seen to be, again,

$$\frac{N_{xy}}{2t_f} = \frac{N_{yx}}{2t_f} = \tau_f \tag{9.128}$$

9.12.4 Weight Relationship

The weight of the web-core sandwich panel per unit planform area is given by

$$W - W_{ad} = 2\rho_f t_f + \rho_c \overline{A}_c = 2\rho_f t_f + \frac{\rho_c t_c h_c}{d_f} \tag{9.129}$$

where W_{ad} is the weight of the adhesive or other bonding material.

9.13 OTHER CONSIDERATIONS

Of course, the theory and analysis becomes far more complex if the sandwich panel is curved, has dissimilar faces, initial imperfections, and dynamic loading. Librescu, Hause, and Camarda [13] developed methods of analysis for giving accurate predictions for static and dynamic behavior subjected to complex mechanical and thermal loads in the prebuckled and postbuckled ranges, and made comparisons with available experimental data.

Smidt [14] also investigated curved sandwich panels because of their applications to high-speed boats, containers, tanks and aircraft. His experimental research was compared to finite element solutions he obtained. He comments that in Equation (9.54) the coefficient can vary from 0.5 to 0.8.

Also, Laine and Rio [15] found that for foam-core sandwich panels typical of those used in ship construction, loaded up to 60% of their elastic-buckling load, creep (that they define as more than 10% influence on deformation) is significant. They therefore advocate either the use of a factor of safety of 2 on the critical buckling load, or introduce a creep law into the structural analysis.

9.14 PROBLEMS

9.1 For the sandwich panel of Problem 7.1, using the materials of Problem 2.1,
   a. What is the overall buckling load, $N_{x_{cr}}$ (lb/in) the sandwich plate can withstand, using classical plate theory?

b. What is the overall buckling load $N_{x_{cr}}$ (lb/in.) the plate can withstand including transverse shear deformation effects?
c. What is the critical load $N_{x_{cr}}$ for core-shear instability?
d. What is the critical load $N_{x_{cr}}$ for face wrinkling?

9.2 If the sandwich panel of Problem 9.1 were clamped on all four edges, solve 9.1(b), (c) and (d) again for this case.

9.3 Consider a foam-core sandwich panel that is 16″ x 16″ in planform dimensions, simply supported on all four edges. Each face is made of cross-ply Kevlar 49/epoxy composite whose properties are given in Table 10.1. The face stacking sequence is [0°/90°/90°/0°], wherein each ply is 0.0055 in. thick. The foam core has $G_c = 15{,}000$ psi , and $\rho_c = 2$ lb/ft^3, and is 3/4 in. in thickness.

a. Determine each element of the $[A]$, $[B]$, and $[D]$ stiffness matrices for the sandwich.
b. Could the perturbation technique of Section 6.4 be used to solve problems for this plate?
c. If the sandwich plate is subjected to an in-plane compressive load in the x-direction, what is the critical buckling load per unit edge distance, $N_{x_{cr}}$ using classical-plate theory?
d. What is the mode of buckling?
e. If transverse shear deformation effects are included, what is the overall critical buckling load, $N_{x_{cr}}$?
f. What is the fundamental natural frequency in Hz, using classical plate theory?
g. What is the answer to f if transverse shear deformation effects are included?

9.4 Consider a hexagonal-cell honeycomb core sandwich panel, simply supported on all four edges, subjected to an in-plane compressive load, N_x (lb/in.). The face material is made of unidirectional T300/934 graphite epoxy, whose properties are given in Table 10.1. The core is aluminum whose properties are $E = 10 \times 10^6$ psi, and $\nu = 0.3$. The dimensions of the panel are:

$a = 48$ in.	$t_f = 0.066$ in.	$h_c = 0.50$ in.
$b = 30$ in.	$t_c = 0.001$ in.	$d = 0.50$ in.

a. At what face stress will face dimpling occur?
b. Would face dimpling ever occur in this panel under this load situation? Why?

c. At what face stress will face wrinkling occur?
d. For this panel, which will have the lower buckling stress, overall panel buckling (i.e. $V_x \leq k_1 B_1 r$) or core-shear instability (i.e., $V_x \geq k_1 B_1 r$)?
e. At what face stress will core-shear instability occur?
f. Of all modes of failure, how will this sandwich panel fail, when subjected to an increasing in-plane compressive loading?
g. Therefore what is the greatest in-plane compressive load N_x (lb/in. of width) that this sandwich panel can withstand?
h. What is the greatest in-plane compressive load, (pounds) that this sandwich panel can withstand without failing?
i. What is the total weight of the panel if the density of the faces is 0.055 lb/in.^{3} and the density of the aluminum is 0.100 lb/in.^{3}?

9.5 Regarding Problem 9.4, consider a monocoque laminated panel of the same size made of the unidirectional T300/934 graphite/epoxy, that is $2t_f = 2 \times 0.066'' = 0.132''$ thick, simply supported on all four edges. Thus this panel has the same face material, but no core.
a. What in-plane buckling load per unit width, $N_{x_{cr}}$ can this monocoque panel withstand before buckling?
b. What is the face stress at that load?
c. What is the panel weight per unit planform area?
d. What thickness would be necessary to have the same buckling load $N_{x_{cr}}$ as the sandwich panel considered in Problem 9.4 above?
e. What would the monocoque panel weigh in *d* above? What is the comparison of the weight of this panel to the sandwich panel of 9.4 above?

9.6 A designer is considering various alternatives in the design of a rectangular sandwich panel measuring 60 inches by 30 inches in planform. The panel must withstand an in-plane compressive load in the longer direction of $N_x = -600$ lb/in. Consider the foam core to weigh 15 lb/ft^3, and $h_c = 25\ t_f$.
a. For a sandwich with aluminum faces ($E = 10 \times 10^6$ psi, $\nu = 0.3$, $\rho = 0.1$ lb/in.^{3}, $\sigma_{\text{all}} = \pm 30{,}000$ psi). To simply support this panel or all four edges will require 20 lb. of support structure; to clamp it on all four edges will require 40 lb of support structure. For the plate to not be overstressed or buckle, which design will result in less system weight?
b. One alternative is to use magnesium faces ($E = 6.5 \times 10^6$ psi, $\nu = 0.3$, $\sigma_{\text{all}} = \pm 30{,}000$ psi, $\rho = 0.065$ lb/in.^{3}). Would a simply supported or clamped magnesium faced system weigh less than the better aluminum-faced sandwich system?

c. What about steel faces ($E = 30 \times 10^6$ psi, $\nu = 0.3$, $\sigma_{all} = \pm 60{,}000$ psi, $\rho = 0.283$ lb/in.^{3}); could a steel-faced sandwich result in weight savings?

9.15 REFERENCES

1. Bleich, H. H., *Buckling of Metal Structures*, McGraw-Hill Book Co., Inc., 1952.
2. Timoshenko, S. and J. Gere, *Theory of Elastic Stability*, 2nd Edition, McGraw-Hill Book Co., Inc., New York, 1961.
3. Brush, D. O. and B. O. Almroth, *Buckling of Bars, Plates and Shells*, McGraw-Hill Book Co., Inc., New York, 1975.
4. Simitses, G. J., *An Introduction to the Elastic Stability of Structures*, Prentice-Hall, Inc., Englewood Cliffs, NJ, 1976.
5. Hemp, "On a Theory of Sandwich Construction," College of Aeronautics Report No. 15, 1948.
6. Hoff, N. J. and S. E. Mautner, "Buckling of Sandwich Type Panels," *Journal of the Aeronautical Sciences*, Vol. 12, No. 3, 1945, pp. 285–297.
7. Dreher, G., "Stability Failure of Sandwich Structures," in *Sandwich Constructions 2—Proceedings of the Second International Conference on Sandwich Construction*, Gainesville, Florida, Editors, D. Weissman-Berman and K-A. Olsson, EMAS Publications, United Kingdom, 1992.
8. Libove, C. and Hubka, R. E., "Elastic Constants for Corrugated-Core Sandwich Plates," NACA TN 2289, 1951.
9. Anderson, Melvin S., "Optimum Proportions of Truss Core and Web-Core Sandwich Plates Loaded in Compression," NASA TN D-98, September 1959.
10. Seide, P., "The Stability Under Longitudinal Compression of Flat Symmetric Corrugated-Core Sandwich Plates with Simply Supported Loaded Edges and Simply Supported or Clamped Unleaded Edges," NACA TN 2679, 1952.
11. Anderson, M. S., "Local Instability of the Elements of a Truss-Core Sandwich Plate," NASA TR R-30, 1959.
12. Kuenzi, E. W., W. S. Ericksen and J. Zahan, "Shear Stability of Flat Panels of Sandwich Construction," Forest Products Laboratory Report 1560, Rev. N62-13403.
13. Librescu, L., T. Hause and C. J. Camarda, "Geometrically Nonlinear Theory of Initially Imperfect Sandwich Curved Panels Incorporating Nonclassical Effects," *AIAA Journal*, Vol. 35, No. 8, August 1997, pp. 1393–1403.
14. Smidt, S., "Testing of Curved Sandwich Panels and Comparison with Calculations Based on the Finite Element Method," in *Sandwich Construction 3—Proceedings of the Third International Conference on Sandwich Construction*, Southampton, Great Britain, Editor, H. G. Allen, EMAS Publications, United Kingdom, 1996, pp. 665–680.
15. Laine, C. and G. Rio, "Buckling of Sandwich Panels Used in Shipbuilding: Experimental and Theoretical Approach," in *Sandwich Constructions 2—Proceedings of the Second International Conference on Sandwich Construction*, Gainesville, Florida, Editors, D. Weissman-Berman and K-A. Olsson, EMAS Publications, United Kingdom, 1992, pp. 243–252.
16. Vinson, J. R. and T-W. Chou, *Composite Materials and Their Use in Structures*. Applied Science Publishers, London, 1975.

17. Vinson, J. R. and R. L. Sierakowski, *The Behavior of Structures Composed of Composite Materials*, Martinus-Nijhoff Publishers (now Kluwer Academic Publishers) Dordrecht, The Netherlands, 1986.

18. Warburton, G., "The Vibration of Rectangular Plates," *Proceedings of the Institute of Mechanical Engineers*, Institute of Mechanical Engineers, London, U.K., p. 371.

19. Young, D. and R. F. Felgar, Jr., "Tables of Characteristic Functions Representing Normal Modes of Vibration for a Beam," The University of Texas Engineering Research Series Report No. 44, July 1, 1949.

20. Felgar, R. P. Jr., "Formulas for Integrals Containing Characteristic Functions of a Vibrating Beam," The University of Texas Bureau of Engineering Research Circular No. 14, 1950.

21. Wu, C. I. and J. R. Vinson, "Nonlinear Oscillations of Laminated Specially Orthotropic Plates with Clamped and Simply Supported Edges," *Journal of the Acoustical Society of America*, Vol. 49, Part 2, May 1971, pp. 1561–1567.

22. Anon., "Materials, Properties and Design Criteria Part II, Sandwich Construction for Aircraft," *Military Handbook 23* (ANC-23), Department of the Air Force Research and Development Command; Department of the Navy, Bureau of Aeronautics; and the Department of Commerce, Civil Aeronautics Administration, 2nd Edition, 1955.

23. Heath, W. G., "Sandwich Construction, Part 2: The Optimum Design of Flat Sandwich Panels," *Aircraft Engineering*, Vol. 32, August 1960, pp. 230–235.

24. Kaechele, L. E., "Minimum Weight Design of Sandwich Panels," USAF. Project Rand Research Memorandum, RM 1895, AD-133011, March 1957.

25. Plantema, F. J., *Sandwich Construction: The Bending and Buckling of Sandwich Beams, Plates and Shells*, John Wiley and Sons, New York, 1966.

26. Vinson, J. R. and S. Shore, "Structural Optimization of Flat Corrugated Core Sandwich Panels Under In-Plane Shear Loads and Combined Uniaxial Compression and In-Plane Shear Loads," Naval Air Engineering Center Report NAEC-ASC-1110, July, 1967.

27. Vinson, J. R., and S. Shore, "Minimum Weight Corrugated Core Sandwich Panels Subjected to Uniaxial Compression," presented at the Third Annual Symposium on High Performance Composites, St. Louis, October 1967.

28. Vinson, J. R., "Minimum Weight Web-Core Composite Sandwich Panels Subjected to In-Plane Compressive Loads," presented at the International Symposium on Composite Materials and Structures, Beijing, China, June 13–16, 1986.

29. Vinson, J. R., "Minimum Weight Triangulated (Truss) Core Composite Sandwich Panels Subjected to In-Plane Shear Loads," in *Composites 86—Recent Advances in Japan and the United States*, Eds. K. Kawata, S. Umekawa and A. Kobayashi, Proceedings of the Third Japan-U.S. Conference on Composite Materials, Japan Society of Composite Materials, Publishers, Tokyo, 1986, pp. 663–671.

CHAPTER 10

Structural Optimization to Obtain Minimum-Weight Sandwich Panels

10.1 INTRODUCTION

It is sometimes sufficient to be able to design and analyze sandwich panels to insure structural integrity. However, it is very desirable to be able to design a sandwich panel that not only successfully resists the applied loads, but is also of minimum weight. A designer could proceed with a search for the minimum-weight solution by trial and error through examining all of the possible combinations of materials and thicknesses for each sandwich element. However, even with excellent insight and intuition, this random-walk approach is time consuming and is not guaranteed to provide the absolute minimum weight panel.

Fortunately, more rational methods have been developed so that the absolute minimum weight can be found analytically for honeycomb, solid, foam, truss-, and web-core rectangular sandwich panels. The methods apply to panels that are subjected to either a uniaxial compressive load, N_x, or to in-plane shear loads, N_{xy}. Each optimization uses the best available equations (established through long use) to depict the various failure modes. When, and if, new and better equations are formulated, they can be used to perform the same optimization procedures as those presented here.

In each case, the age-old principle of the "weakest link in the chain" is applied. The philosophy is that with any sandwich panel there are several failure modes, each independent of the others, and any one of which would result in the panel's failure. Each failure mode is thus a "link" in the sandwich "chain" that resists the applied load, and failure of any one "link" means failure of the whole "chain." Logically, the most efficient chain is one in which each link fails simultaneously. In the sandwich, each "link" has an associated weight varying directly with its load carrying ability. Therefore, the minimum-weight panel ("the chain") is found by insuring that each of the failure modes ("the links") occur simultaneously. Conversely, one could say that if a particular failure mode is significantly stronger than the others, then the material associated with this excess strength could either be removed or could be reallocated to bring all modes closer to equality. This extra material is truly "dead weight" (and cost) as the overall panel will fail via some other mode before ever reaching the load associated with the strongest mode.

In performing the optimization, unique expressions are obtained for each geometric variable of the minimum-weight panel. Additionally, a "universal relationship" is found that relates the applied load index to unique values of face- and core-element stresses for given materials and boundary conditions. The methods developed are also applicable to panels at any steady-state temperature if the material properties are known at that temperature. In order to select the best material systems to achieve minimum weight for a given load, material property factors of merit are also found; these factors of merit are tabulated for many of the current material systems. The optimization methods have been found to be extremely useful for the following reasons:

(1) Unique values of each geometric variable are found in order to obtain the minimum-weight panel for a specific set of conditions, including the load index, material system, panel dimensions, and boundary conditions.
(2) Various material systems can be compared for best materials selection.
(3) The best stacking sequence can be determined for laminated composite face materials.
(4) The optimum sandwich panel weight can be compared with other types of sandwich, reinforced panel, or monocoque architectures.
(5) Weight penalties for nonoptimal construction can be determined rationally.

10.2 MINIMUM-WEIGHT OPTIMIZATION OF HONEYCOMB-CORE SANDWICH PANELS SUBJECTED TO A UNIDIRECTIONAL COMPRESSIVE LOAD [1,2]

For this type of panel and load, there can be four buckling failure modes: overall buckling [Equations (9.26) through (9.35)], core-shear instability [Equation (9.36)], face-wrinkling instability [Equations (9.39) and (9.40)], and monocell buckling [Equation (9.41)]. For the same construction there are four dependent variables that the analyst/designer can specify, namely, the face thickness (t_f), the core depth (h_c), the core cell size (d), and the core-wall thickness (t_c). In addition, the load is related to the face stress by Equation (9.25). Therefore, for any panel size, boundary conditions, and given face and core material, unique values of each dependent variable are determined by the philosophy expressed earlier. In addition, it is found that there exists one and only one face stress σ_0, for a given load index (N_x/b), that will result in minimum weight.

It was shown that overall buckling and core-shear instability occur at the same stress under the conditions given by Equations (9.37) and (9.38), that is,

$$V_x = k_1 B_1 r$$

Equations (9.38), (9.36), (9.40), (9.42), and (9.25) yield the universal relationship for optimum honeycomb-sandwich panels with orthotropic facings and core under uniaxial compression,

$$\left(\frac{k_2}{k_3 k_1 r}\right)^{1/2} \left(\frac{N_x}{b}\right) = \frac{2(3)^{1/2}(1 - \nu_{xy}\nu_{yx}}{\pi} \left(\frac{G_c}{E_c}\right)^{1/2} \frac{\sigma_0^2}{E_{fx}^{3/4} E_{fy}^{1/4}} \tag{10.1}$$

It must be noted that, in Equation (10.1), σ_0 must be limited to some predetermined allowable stress (usually defined by the yield strength or ultimate strength divided by appropriate factors of safety for each such that the lower value of the two is the allowable stress); hence, there is an upper bound on the load that can be carried by the panel for a given material system. Therefore, when given a set of boundary conditions (denoted by k_1) and any type of honeycomb-core construction (denoted by k_2, k_3, and r), it is seen that the universal relationship relates the load index, N_x/b, to material properties only. It is independent of geometric variables and it establishes a unique optimum buckling stress, σ_0, for each value of the load index.

Explicit values of the unique optimum geometric variables as functions of the load index, N_x/b, are determined to be

$$h_c = b\left(\frac{2}{\pi}\right)^{3/4} \left(\frac{k_2}{3k_3}\right)^{1/8} (1 - \nu_{xy}\nu_{yx})^{1/4} (k_1 r)^{3/8} \left(\frac{E_c}{G_c}\right)^{1/8} \frac{E_{fy}^{1/16}}{E_{fx}^{5/16}} \left(\frac{N_x}{b}\right)^{1/4} \tag{10.2}$$

$$d = \frac{b2^{1/4}}{\pi^{3/4}} \left(\frac{3k_3 k_1 r}{k_2}\right)^{3/8} (1 - \nu_{xy}\nu_{yx})^{1/4} \left(\frac{G_c}{E_c}\right)^{3/8} \frac{E_{fx_T}^{1/4} E_{fy}^{1/16}}{E_{fx}^{9/16}} \left(\frac{N_x}{b}\right)^{1/4} \tag{10.3}$$

$$t_c = b\left[\frac{3}{2k_2 k_3} \frac{1}{G_c E_c} \sqrt{\frac{E_{fx_T}}{E_{fx}}}\right]^{1/2} \left(\frac{N_x}{b}\right) \tag{10.4}$$

$$t_f = b\left(\frac{3k_3 k_1 r}{k_2}\right)^{1/4} \left[\frac{1 - \nu_{xy}\nu_{yx}}{2\pi}\right]^{1/2} \left(\frac{G_c}{E_c}\right)^{1/4} \frac{(N_x/b)^{1/2}}{E_{fx}^{3/8} E_{fy}^{1/8}} \tag{10.5}$$

The explicit values of the optimum geometric variables in terms of the optimum critical stresses, σ_0, are written as

$$h_c = b\left(\frac{2}{\pi}\right)(k_1 r)^{1/2}\,(1 - \nu_{xy}\nu_{yx})\,\frac{\sigma_0^{1/2}}{E_{fx}^{1/2}} \tag{10.6}$$

$$d = \frac{b}{\pi}\left[\frac{6(1 - \nu_{xy}\nu_{yx})k_3 k_1 r}{k_2}\right]^{1/2}\left(\frac{G_c}{E_c}\right)^{1/2}\frac{E_{fx_T}^{1/4}\sigma_0^{1/2}}{E_{fx}^{3/4}} \tag{10.7}$$

$$t_c = \frac{b3\sqrt{2}}{\pi k_2}\,(k_1 r)^{1/2}\,(1 - \nu_{xy}\nu_{yx})\,\frac{E_{fx_T}^{1/4}\sigma_0^2}{E_c E_{fx} E_{fy}^{1/4}} \tag{10.8}$$

$$t_f = b\left(\frac{3k_3 k_1 r}{k_2}\right)^{1/2}\frac{(1 - \nu_{xy}\nu_{yx})}{\pi}\left(\frac{G_c}{E_c}\right)^{1/2}\frac{\sigma_0}{E_{fx}^{3/4}E_{fy}^{1/4}} \tag{10.9}$$

It should be noted that if a designer/analyst decides that face dimpling can be ignored because the monocell buckling does not necessarily mean panel failure, then equating the three remaining buckling stresses provides for the core depth [Equation (10.6)], for the face thickness [Equation (10.9)], and a ratio for (t_c/d) equal to Equation (10.8) divided by (10.7).

The expressions for the weight-per-unit planform area for the optimum panel as a function of σ_0 and N_x/b, can be found by substituting the above expressions into Equation (9.49). The results are

$$W = \left(\frac{3k_3 k_1 r}{k_2}\right)^{1/2}\frac{2(1 - \nu_{xy}\nu_{yx})b}{\pi} \times \left(\frac{G_c}{E_c}\right)^{1/2}\frac{\sigma_0}{E_{fx}^{3/4}E_{fy}^{1/4}}\left[\rho_f + \rho_c\frac{k_2}{k_3}\frac{\sigma_0}{G_c}\right] + W_{ad} \tag{10.10}$$

$$W = \left(\frac{3k_3 k_1 r}{k_2}\right)^{1/4}\left[\frac{2(1 - \nu_{xy}\nu_{yx})}{\pi}\right]^{1/2} \times \left(\frac{G_c}{E_c}\right)^{1/4}\frac{b(N_x/b)^{1/2}}{E_{fx}^{3/8}E_{fy}^{1/8}}\left[\rho_f + \rho_c\frac{k_2}{k_3}\frac{\sigma_0}{G_c}\right] + W_{ad} \tag{10.11}$$

Several very important and interesting conclusions can be drawn from Equation (10.11) about the weight of *optimized* honeycomb-sandwich panels under uniaxial compressive loads. It is seen that in the selection of the materials to use in these optimum panels, the core material with the highest ratio of G_c/ρ_c will result in minimum weight. For the selection of a facing material, the material that has the highest ratio of $E_{fx}^{3/8}E_{fy}^{1/8}/\rho_f\ (1 - \nu_{xy}\nu_{yx})$ for the particular load index N_x/b will result in the panel of lowest weight. For optimized honeycomb-sandwich panel construction, the ratio of core weight to face weight is

$$\frac{W_c}{W_f} = \frac{\rho_c}{\rho_f}\frac{k_2}{k_3}\frac{\sigma_0}{G_c} \tag{10.12}$$

Therefore, for a construction employing the same material in both faces and honeycomb core, the great majority of the weight is in the faces. Also, note that the ratio of the core weight to face weight for an optimum panel is independent of boundary conditions and the orthotropy factor, *r*. In Equation (10.11), it is seen that the panel weight varies as the one-fourth power of the boundary condition factor, k_1. For a panel that is simply supported on the unloaded edges, $k_1 = 1$. In a panel that is clamped on the unloaded edges, $k_1 = \frac{3}{4}$. Therefore, the ratio of the weight of an optimum panel simply supported on the unloaded edges to the weight of an optimum panel clamped along the unloaded edges is 1.0745. It can be concluded that, for a given load index, an optimum honeycomb-sandwich panel simply supported on the unloaded edges weighs no more than 1.0745 times the weight of the optimum panel clamped on the unloaded edges.

This result has major implications. Foremost, it implies that in almost all cases of sandwich panels subjected to uniaxial compression, the optimizations should be conducted for simply-supported boundary conditions on all edges. (Note that the boundary conditions on the loaded edges have no effect for this loading.) Also:

(1) Such an optimization will result in the panel weighting at most 7.45% over an optimum panel whose unloaded edges are clamped. Choosing a simply-supported panel is thus conservative as far as all dimensions selected, enabling the panel to have additional structural integrity even when the edges are clamped or partially clamped.

(2) In actual construction, the auxiliary structural elements required to make the unloaded edges clamped would possibly offset the potential saving of 7.45% in weight resulting from the clamped boundary condition. In the final analysis, the totally clamped panel assembly would weigh more.

(3) It is virtually impossible to insure a clamped edge; hence, most panel edge conditions are somewhere between the conditions of fully simply-supported and totally clamped. The choice of simple-support conditions for the optimization is therefore conservative as well as rational.

For a honeycomb core sandwich panel subjected to a uniaxial in-plane compressive load, independent of the boundary condition, the "best" face material is determined by

$$FM = \frac{E_x^{3/8} E_y^{1/8}}{\rho_f \,(1 - \nu_{xy}\nu_{yx})^{1/2}} \tag{10.13}$$

It is interesting to use this Factor of Merit (*FM*) to some of the material systems currently available. The values in Table 10.1 are for 70°F, and because the Poisson's ratios are rarely given tables of mechanical properties for the new materials, it is assumed below that $(1 - \nu_{xy}\nu_{yx}) \approx 1$. It is seen clearly that in this application composite materials are better than any all-metallic construction. It is also clear that graphite/epoxy and boron/aluminum composites clearly result in the minimum weight construction. Even boron/epoxy is quite competitive.

The above comparison does not indicate maximum loads that the panel may carry. To investigate that, knowledge of maximum allowable stresses must be available and Equation (9.25) must be used to determine the maximum load index, N_x/b, to which the optimum panel may be subjected. Figure 10.1 illustrates this point, where the maximum allowable stresses are taken to be the ultimate compressive strength and 2024 aluminum core is arbitrarily chosen for the honeycomb core. It is seen that T300/934 is the lowest weight material, but limited to a load index of $N_x/b = 719$ psi. For higher load levels, boron/aluminum is the most efficient material to a load index of 5840 psi and boron/epoxy the best to a load index of 8740 psi. Beyond that, other panel architectures are perhaps required.

In all of the above, it is assumed the materials are elastic to the ultimate stress. If ductility permits and the stress-strain curves are known for stresses higher than the yield point, a plasticity reduction factor can be employed. Finally, it should be remembered that in any laminated face construction, in which $B_{ij} = (\ )_{16} = (\ )_{26} = 0$, E_x and E_y can always be obtained by utilizing the appropriate stiffness-matrix values.

10.3 MINIMUM-WEIGHT OPTIMIZATION OF FOAM-CORE SANDWICH PANELS SUBJECTED TO A UNIDIRECTIONAL COMPRESSIVE LOAD

For this type of sandwich panel and loading, there are three buckling failure modes: overall buckling [Equations (9.26) through (9.35)], core-shear

TABLE 10.1. *Face Material Comparison, Based on Factor of Merit for an Optimized Honeycomb Core Sandwich Panel Subjected to an In-Plane Uniaxial Compressive Load.*

Material	Configuration	V_F, %	E_x,msi	E_y,msi	ρ, lb/in.^{3}	σ_{ult}, ksi[a]	FM[b]	Ref.
T300/934	Unidir.	60	23.69	1.7	0.0555	105	63.09	2.3
C/epoxy	Cross-ply	58	12.04	12.04	0.0555	55.1(T)	62.50	10.2
T300/5208	Unidir.	72	22.2	1.58	0.0555	110	61.01	2.3
T300/5208	Unidir.	60	21.9	1.53	0.0555	164	60.46	2.3
B/2024A1	—	49	33.93	22.04	0.0916	220(T)	60.24	10.2
B/6061A1	—	49	33.79	22.04	0.0916	513.3	60.14	10.2
$B(B_4C)/A1$	—	49	34.95	18.85	0.0918	210.3(T)	59.62	10.2
B/5052A1	—	49	33.35	21.03	0.0916	168.2(T)	59.50	10.2
B/110A1	—	49	33.35	19.72	0.0916	323.3	59.03	10.2
B/3002A1	—	49	32.48	20.74	0.0916	365.4	58.81	10.2
B/6061	Unidir.	50	32.0	20.00	0.0915	250.0	58.33	10.2
AS/3501	—	67	20.2	1.30	0.0555	209.9	57.47	2.3
AS1/3501-6	—	—	18.85	1.52	0.0555	246.0	57.10	10.2
B/A1	—	48	30.02	21.03	0.0918	221.9(T)	57.09	10.2
Borsic/A1	—	45	31.03	20.16	0.0926	190.0(T)	57.03	10.2
B/epoxy	—	67	31.18	3.05	0.0740	468.4	56.43	10.2
B/epoxy	—	67	30.3	2.80	0.0740	362.6	55.24	2.3
B/epoxy	Cross-ply	60	15.37	15.37	0.0721	55.1(T)	54.37	10.2
T300/2500	—	—	17.55	1.17	0.0555	—	53.80	10.2
$\alpha A1_2O_3/M_8$	Cont. fiber	50	30.02	15.08	0.1009	275.5	49.89	10.2
Kevlar49/epoxy	Cross-ply	60	5.80	5.80	0.0505	—	47.72	10.2
SiC/6061A1	—	48	31.63	18.06	0.112		46.74	10.2
S glass/xp-251		67	8.29	2.92	0.0555	170.0	45.53	2.3
$\gamma A1_2O_3/A1$	Cont. fiber	50	21.75	15.95	0.1045	203.0	42.91	10.2
$\gamma A1_2O_3/A1$-5Cu	Cont. fiber	50	21.75	14.50	0.1045	319.0	42.41	10.2
SiC/Ti	Woven, ISO	39.5	28.70	28.70	0.1432	228(T)	37.41	10.2
$\alpha A1_2O_2/A1$	Cont. fiber	50	31.9	20.16	0.1172	406	36.14	10.2
SiC/Ti-6A1-4V	Woven, ISO	35	27.0	27.0	0.1492	245(T)	34.83	10.2
2024 A1	ISO	—	10.88	10.88	0.0973	—	33.88	10.2
Titanium	ISO	—	17.4	17.4	0.1600	145	26.07	10.2
E glass/epoxy	Cross-ply	57	3.12	3.12	0.0710	82.0	24.87	10.2
Welton 80 steel	ISO	—	30.45	30.45	0.283	126.9	19.50	10.2

[a]Compressive strength: (T) denotes tensile properties available only.
[b]Units of *FM* are 10^3 in.^{2}/lb$^{1/2}$.

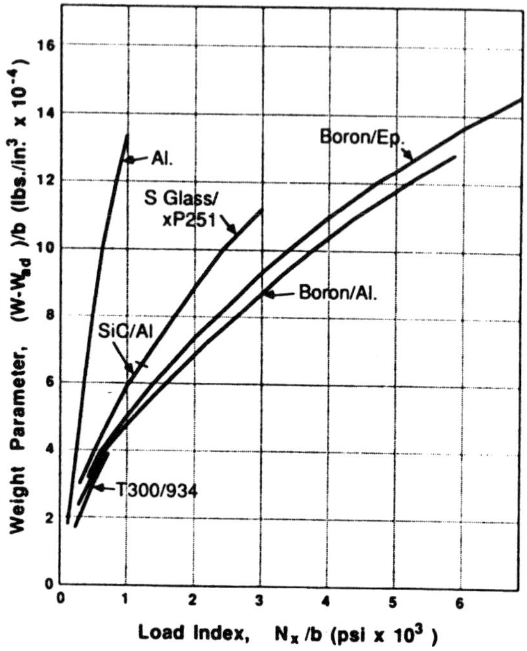

FIGURE 10.1. Materials comparison.

instability [Equation (9.51)] and face wrinkling [Equation (9.53)]. In this case of foam or solid core, one uses the actual core properties in Equation (9.31), namely

$$U_{xz} = G_{cx}h_c,\ U_{yz} = G_{cy}h_c \tag{10.14}$$

It is seen that overall panel buckling and core shear instability occur simultaneously when $V_x = k_1B_1r$, or more specifically, as in Equation (9.52),

$$\frac{\pi^2}{2(1 - \nu_{xy}\nu_{yx})}\frac{\overline{E}_{fx}}{G_{cx}}\frac{h_ct_f}{b^2} = k_1r \tag{10.15}$$

For the optimized construction the applied load-optimum stress relationship is

$$\frac{1}{(k_1r)^{1/2}}\left(\frac{N_x}{b}\right) = \frac{6(1 - \nu_{xy}\nu_{yx})^{3/2}\,\sigma_0^{7/2}}{\pi E_cE_{fx}E_{fy}{}^{1/2}} \tag{10.16}$$

The geometric variables and the core shear stiffness that result in minimum-weight design are as follows because of the linear relation discussed in Chapter 6 between core shear modulus and core density (G_{cx} is treated as a variable in the optimization):

$$t_f = \frac{3(1 - \nu_{xy}\nu_{yx})^{3/2}\, b\,(k_1 r)^{1/2}\,\sigma_0^{5/2}}{\pi E_c\,\overline{E}_{fx} E_{fy}^{1/2}} \tag{10.17}$$

$$h_c = \frac{2}{\pi}(1 - \nu_{xy}\nu_{yx})^{1/2}\, b\,(k_1 r)^{1/2}\,\frac{\sigma_0^{1/2}}{\overline{E}_{fx}^{1/2}} \tag{10.18}$$

$$G_{cx} = \frac{3(1 - \nu_{xy}\nu_{yx})\sigma_0^3}{E_c\sqrt{E_{fx}E_{fy}}} \tag{10.19}$$

The optimized panel weight in terms of applied load is

$$\frac{W - W_{ad}}{b} = \frac{6^{2/7}\,(k_1 r)^{1/7}\,(1 - \nu_{xy}\nu_{yx})^{3/7}\,(N_x/b)^{5/7}}{\pi^{2/7} E_c^{2/7}\,\overline{E}_{fx}^{2/7} E_{fy}^{1/7}} \times \left[\rho_f + \rho_c \frac{E_c \overline{E}_{fx}^{1/2} E_{fy}^{1/2}}{3(1 - \nu_{xy}\nu_{yx})\sigma_0^2}\right] \tag{10.20}$$

From Equation (10.20) it is seen that the best face material to use is the one with the largest value of

$$FM = \frac{\overline{E}_{fx}^{2/7} E_{fy}^{1/7}}{\rho_f} \tag{10.21}$$

Note that the modulus in the load direction is barred to indicate the use of a plasticity reduction factor if the stresses are above the proportional limit. For this loading, the face stress in the y-direction will never exceed the yield strength, hence E_{fy} is not barred.

10.4 MINIMUM-WEIGHT OPTIMIZATION OF TRUSS-CORE SANDWICH PANELS SUBJECTED TO A UNIDIRECTIONAL COMPRESSIVE LOAD

For the optimization in this case, there are three modes of buckling and four geometric variables discussed in Section 9.7. As a result, the approach is to solve for three of the variables in terms of the fourth variable, utilizing the three buckling equations and the load-stress equation. Then substitute these expressions into the weight equation such that the panel weight is a function of the core- and face-material properties, the load index, and the variable, θ. With five equations and six unknowns, a sixth equation is obtained by placing the weight relation in terms of one variable (namely θ) and setting the derivative of the weight with respect to θ equal to zero to obtain the value that will provide minimum weight. In the case of the faces and core made of the same orthotropic material, the weight equation in terms of the load index is written as

$$\frac{W - W_{ad}}{b} = \left[\frac{3(1 - \nu^2)}{K}\right]^{1/4} \left[\frac{N_x/b}{E}\right]^{1/2} \frac{\rho}{\pi} \frac{[1 + 4 \sin^2 \theta]^{3/4}}{\sin \theta\, (\cos \theta)^{1/2}} \qquad (10.22)$$

Setting the derivative of $(W - W_{ad})/b$ with respect to θ equal to zero, it is found that for the optimum construction,

$$\sin^2 \theta = 2/7 \quad \text{or} \quad \theta = 32.4° \qquad (10.23)$$

For core and faces of different materials the lengthy results are given in Reference [10]. The load index–optimum face stress relationship is

$$\frac{N_x}{b} = \frac{45}{2} \frac{(1 - \nu_{xy}\nu_{yx})^{1/2}\, \sigma_0^2}{\pi^2 K^{1/2} E_0^{1/2} (E_x E_y)^{1/4}} \qquad (10.24)$$

where E_0 is defined in Equation (9.70). The unique values for the geometric variables are

$$\frac{t_f}{b} = \frac{6(1 - \nu_{xy}\nu_{yx})^{1/2}\, \sigma_0}{\pi^2 K^{1/2} E_0^{1/2} (E_x E_y)^{1/4}} \qquad (10.25)$$

$$\frac{t_c}{b} = 3\left(\frac{7}{2}\right)^{1/2} \frac{(1 - \nu_{xy}\nu_{yx})^{1/2}\, \sigma_0}{E_0^{1/2} (E_x E_y)^{1/4} K^{1/2}} \qquad (10.26)$$

$$\frac{h_c}{b} = \frac{1}{(E_x E_y)^{1/4}} \left(\frac{15}{2} \frac{\sigma_0}{\pi^2 K}\right)^{1/2} \qquad (10.27)$$

The optimized panel weight is given by

$$\frac{W - W_{ad}}{b} = 3\left(\frac{5}{2}\right)^{1/2} \frac{(1 - \nu_{xy}\nu_{yx})^{1/4}\, \rho (N_x/b)^{1/2}}{K^{1/4} \pi E_0^{1/4} (E_x E_y)^{1/8}} \qquad (10.28)$$

from which it is seen that the factor of merit to use in materials selection is

$$FM = \frac{E_0^{1/4}\,(E_xE_y)^{1/8}}{\rho} \tag{10.29}$$

For the optimized case, using Figures 3 and 5 of Reference [11] to obtain the precise value of the coefficient K, use

$$\frac{E_c\bar{I}_c}{E_fI_f} = \frac{7}{24} \tag{10.30}$$

and

$$V = \frac{21(1 - \nu_{xy}\nu_{yx})\sigma_0}{(E_xE_y)^{1/2}\,K} \tag{10.31}$$

and perform a simple iteration.

If the truss-core sandwich panel is composed of the same isotropic face and core materials, then as noted before, the optimization relations hold also above the proportional limit, if a suitable plasticity reduction factor, η, is used, such that

$$\overline{E} = \eta E \tag{10.32}$$

All expressions will be written in the general form utilizing $\overline{E}$.

One then simplifies Equations (10.22) through (10.29) by letting $E_c = E_f = \overline{E}$, $\nu_c = \nu_f = \nu$, and $\sigma_c = \sigma_f = \sigma$. The buckling coefficient, K, is a slowly-varying function such that it can be considered constant, and can be determined later by a trivial iteration. The unknown variables are t_f, h_c, t_c, θ, σ, and $(W - W_{ad})$. For the case of the faces and core being the same isotropic material, the "universal relation," relating load index to a unique stress value for any set of material properties, which results in minimum weight is

$$\frac{N_x}{b} = \frac{45}{2}\,\frac{(1 - \nu^2)^{1/2}\,\sigma_0^2}{\pi^2\overline{E}K^{1/2}} \tag{10.33}$$

Unique values of all other variables are now found in terms of the applied load, N_x/b, and the material properties for the optimum, minimum weight isotropic construction.

$$\frac{t_f}{b} = 2\left(\frac{2}{5}\right)^{1/2}\frac{(1 - \nu^2)^{1/4}\,(N_x/b)^{1/2}}{\pi\overline{E}^{1/2}K^{1/4}} \tag{10.34}$$

$$\frac{t_c}{b} = \sqrt{\frac{7}{5}} \frac{(1 - \nu^2)^{1/4} (N_x/b)^{1/2}}{\pi \overline{E}^{1/2} K^{1/4}} \tag{10.35}$$

$$\frac{h_c}{b} = \left[\frac{5(N_x/b)}{2\overline{E}\pi^2(1 - \nu^2)^{1/2} K^{3/2}} \right]^{1/4} \tag{10.36}$$

$$\frac{W - W_{ad}}{b} = 3\left(\frac{5}{2}\right)^{1/2} \frac{(1 - \nu^2)^{1/4} \rho(N_x/b)^{1/2}}{\pi \overline{E}^{1/2} K^{1/4}} \tag{10.37}$$

Using Equation (10.33), analogous expressions can be written for each variable in terms of the optimum face stress, as was done earlier.

It is interesting to note that when the same core and face materials are used, the optimum angle, θ, is independent of load and materials used. Note also that in the optimum construction described above, for the isotropic case:

$$\frac{t_c}{t_f} = \sqrt{\frac{7}{8}} \tag{10.38}$$

and

$$\frac{W_c}{W_f} = \frac{7}{8} \tag{10.39}$$

where W_c and W_f are the weight of the core and face per unit plan form area, respectively.

The precise value of K is now determined, which appears in the weight equation to the 1/4 power. From Figures 3 and 5 of Reference [11], a trivial iteration can be performed utilizing the following for optimum isotropic construction:

$$\frac{\overline{E}_c \overline{I}_c}{\overline{E}_f \overline{I}_f} = \frac{7}{24} \tag{10.40}$$

$$V = \frac{21(1 - \nu^2)\sigma}{2\overline{E}K} \tag{10.41}$$

10.5 MINIMUM-WEIGHT OPTIMIZATION OF WEB-CORE SANDWICH PANELS SUBJECTED TO A UNIDIRECTIONAL COMPRESSIVE LOAD

Again, the philosophy of structural optimization is as follows: within the class of structures being studied and for each material system, a truly optimum structure is one that has a unique value for each geometric variable, that results in the minimum possible weight for a specified loading condition, and maintains its structural integrity. In this case, the optimum structure will have the characteristic that the panel will become unstable in all four buckling modes simultaneously.

The equations with which to optimize are: the four buckling equations [Equations (9.83), (9.85), (9.86) and (9.87)]; the load-stress relationship, [Equation (9.90)]; and the weight relationship [Equation (9.92)]. The known quantities for any optimization are N_x, a, b, and the material properties of both core and face material. The buckling coefficient K is a constant depending on the value of Equation (9.84) and is given by Equation (9.82). The unknown variables to be determined are t_f, h_c, t_c, d_f, θ, σ_f, and $(W - W_{ad})$. With six equations and seven unknown variables, a seventh equation is obtained by placing the weight equations in terms of one variable only and setting the derivative of weight with respect to that variable to zero as in the previous section. Thus, we find the value of that variable, and subsequently the value for every other variable, which results in minimum weight.

At the outset, independent of the material system, it is clear that in executing the above optimization philosophy, with Equations (9.85) and (9.86), that for optimum construction of the web-core sandwich,

$$\theta = 0^\circ \tag{10.42}$$

This is intuitively obvious because with $\theta = 0^\circ$, the upper and lower face plated components are of the same width. The result is that all other expressions used for optimization are simplified, and the construction shown in Figure 9.7, with $\theta = 0^\circ$, now assumes the familiar web-core configuration.

Proceeding with the optimization procedure for panels with faces and core of different orthotropic materials, the following expressions are obtained for the optimum minimum-weight construction. First, one obtains the "universal relationship" relating applied load index, N_x/b, to a unique value of face stress σ_{f_0}, as a function of given material properties for the optimum construction

$$\frac{N}{b} = \frac{(6)^{1/2}(1 - \nu_{xyf}\nu_{yxf})^{1/4}(1 - \nu_{xyc}\nu_{yxc})^{1/4}}{(E_{0f}E_{0c})^{1/4}\,(E_{fx}E_{fy})^{1/4}\,\pi^2 K^{1/2}} \left(\frac{E_{cx}}{E_{fx}}\right)^{7/4} \left(\frac{\rho_f}{\rho_c}\right) R^{3/2}\,\sigma_{f_0}^2 \tag{10.43}$$

Here, $R = 1 + 2(\rho_c/\rho_f)(E_{fx}/E_{cx})$, and K is determined from Figures 2 or 4 of Reference [11] in which $V \rightarrow \infty$ and for the optimum construction

$$\frac{E_c\bar{I}_c}{E_f I_c} = \frac{1}{6}\left(\frac{\rho_f}{\rho_c}\right)\left(\frac{E_{cx}E_{cy}}{E_{fx}E_{fy}}\right)^{1/2} \tag{10.44}$$

In a panel with all edges simply supported, K is given as follows, using the above expression for $E_c\bar{I}_c/E_f\bar{I}_f$

$$K = \frac{[(1/\beta) + \beta^2]}{1 - \nu_{xyf}\nu_{yxf} + 2(1 + \nu_{xyf})\beta^2} + \frac{E_c\bar{I}_c}{E_f\bar{I}_f\beta^2} \tag{10.45}$$

where$\beta = a/b$.

The significance of the universal relationship given by Equation (10.43) is that, given panel length, width, load, and materials, a panel designed for a face stress either higher or lower than that given by Equation (10.43), will be heavier than one designed for the optimum face stress σ_{f_0} given by Equation (10.43). Making use of Equation (10.43) then, the geometry of the panel for minimum-weight construction can be written as

$$\frac{h_c}{b} = \left[2\left(\frac{\rho_f}{\rho_c}\right)\left(\frac{E_{cx}}{E_{fx}}\right)\frac{R\sigma_{f_0}}{(E_{fx}E_{fy})^{1/2}\,\pi^2 K}\right]^{1/2} \tag{10.46}$$

$$\frac{h_c}{d_f} = \left[\left(\frac{\rho_f}{\rho_c}\right)^2\left(\frac{1 - \nu_{xyf}\nu_{yxf}}{1 - \nu_{xyc}\nu_{yxc}}\right)\left(\frac{E_{0c}}{E_{0f}}\right)\left(\frac{E_{fx}}{E_{cx}}\right)\right]^{1/4} \tag{10.47}$$

$$\frac{t_c}{b} = \frac{6^{1/2}}{\pi^2 K^{1/2}}\,\frac{(1 - \nu_{xyc}\nu_{yxc})^{1/2}}{E_{0c}^{1/2}(E_{fx}E_{fy})^{1/4}}\left(\frac{\rho_f}{\rho_c}\right)^{1/2}\left(\frac{E_{cx}}{E_{fx}}\right)R^{1/2}\sigma_{f_0} \tag{10.48}$$

$$\frac{h_c}{d_f} = (\rho_f/\rho_c)(t_f/t_c) \tag{10.49}$$

$$(W - W_{ad})/b = 3\rho_f(t_f/b) \tag{10.50}$$

Other alternative expressions for the optimum construction are given in Reference [10].

It can also be shown easily that even in this general material system, the ratio of the face weight W_f to the core weight W_c for the optimum construction is

$$W_f/W_c = 2 \tag{10.51}$$

Note that this is independent of applied load, material system, and panel boundary conditions.

The optimum construction for panels with faces and core of different isotropic materials can be obtained from Equations (10.43) through (10.51) by allowing $E_{ix} = E_{iy} = E_{0i} = E_i$ and $\nu_{xyi} = \nu_{yxi} = \nu_i$ where $i = c, f$.

The resulting expressions for the optimum construction for panels with faces and core of the same orthotropic material can be obtained from Equations (10.43) through (10.51) by letting each quantity $(\ )_f = (\ )_c = (\ )$. The results are

$$\frac{N_x}{b} = \frac{9(2)^{1/2}(1 - \nu_{xy}\nu_{yx})^{1/2}\sigma_0^2}{\pi^2 K^{1/2} E_0^{1/2}(E_x E_y)^{1/4}} \tag{10.52}$$

where K is determined from Figures 2 or 4 of Reference [12] or, for a simply supported panel, from Equation (10.45), in which

$$E_c\bar{I}_c/E_f\bar{I}_f = \frac{1}{6}$$

Making use of Equation (10.52) to obtain σ_0, the other geometric variables are obtained easily from the following:

$$h_c/b = \{6\sigma_0/[\pi^2 K(E_x E_y)^{1/2}]\}^{1/2} \tag{10.53}$$

$$d_f = h_c \tag{10.54}$$

$$\frac{t_c}{b} = \frac{3(2)^{1/2}(1 - \nu_{xy}\nu_{yx})^{1/2}\sigma_0}{\pi^2 K^{1/2}(E_x E_y)^{1/2} E_0^{1/2}} \tag{10.55}$$

$$t_f/t_c = 1 \tag{10.56}$$

$$\frac{W - W_{ad}}{b} = \frac{\rho}{\sigma_0}\left(\frac{N_x}{b}\right) = 3\rho\frac{t_f}{b} = 3\rho\frac{t_c}{b} \tag{10.57}$$

It is seen that for a panel in which the same orthotropic material is used in both face and core, the optimum geometry results in face and core of the same thickness ($t_c = t_f$ and square cells ($d_f = h_c$).

It is seen from Equation (10.52) and (10.57) that the factor of merit for selecting the best face material is

$$FM = \frac{E_0^{1/4}(E_x E_y)^{1/8}}{\rho}$$

Lastly, the expressions for optimum construction of panels with faces and core of the same isotopic material can be easily determined from Equations (10.52) through (10.57) by letting $(\ )_x = (\ )_y = (\ )$. With this material system, since the material is isotropic and the face and core are equally stressed, the expressions can be employed for loads resulting in stresses above the proportional limit by utilizing a suitably defined plasticity reduction factor η, such that $\overline{E} = \eta E$. Thus, in the following, $\overline{E}$ is used to denote that the expressions are valid in the range of inelastic deformations.

$$N_x/b = 9(2)^{1/2}\,(1 - \nu^2)^{1/2}\,\sigma_0^2/(\pi^2\overline{E}K^{1/2}) \tag{10.58}$$

$$h_c/b = d_f/b = (6\sigma_0/\pi^2 K\overline{E})^{1/2} \tag{10.59}$$

$$t_f/b = t_c/b = 3(2)^{1/2}\,(1 - \nu^2)^{1/2}\,\sigma_0/(\pi^2 K^{1/2}\overline{E}) \tag{10.60}$$

$$\frac{W - W_{ad}}{b} = \frac{3(2)^{1/4}\,\rho(1 - \nu^2)^{1/4}}{\pi K^{1/4}E^{1/2}}\left(\frac{N_x}{b}\right)^{1/2}$$

$$= \frac{\rho}{\sigma_0}\left(\frac{N_x}{b}\right) = 3\rho\,\frac{t_f}{b} = 3\rho\,\frac{t_c}{b} \tag{10.61}$$

Again, it is seen that the web and face have the same thickness and that the cells are square for the optimum construction. Note also that the isotropic material which has the highest ratio of $E^{1/2}/\rho$ is the material which will result in the lowest weight panel in the elastic range.

As an example of the use of the optimization procedures derived above, curves of the weight function $K^{1/2}\,(W - W_{ad})$ as a function of load index $(K^{1/2}\,N_x/b)$ are plotted in Figure 10.2 for the following materials:

(1) 7075-T6 Aluminum (clad)
(2) S994-181 HTS glass fabric, ERSB-0111 Resin
(3) 143 Glass fabric laminate with polyester resin (MIL-R-7575)
(4) 143 Glass fabric laminate with epoxy resin (MIL-R-9300)
(5) 181 Glass fabric laminate with epoxy resin (MIL-R-9300)
(6) Cross rolled beryllium
(7) Unidirectional boron fibers with epoxy resin
(8) Unidirectional Thornel-40 graphite fibers with epoxy resin
(9) AISI 4340 steel, 200,000 psi yield strength

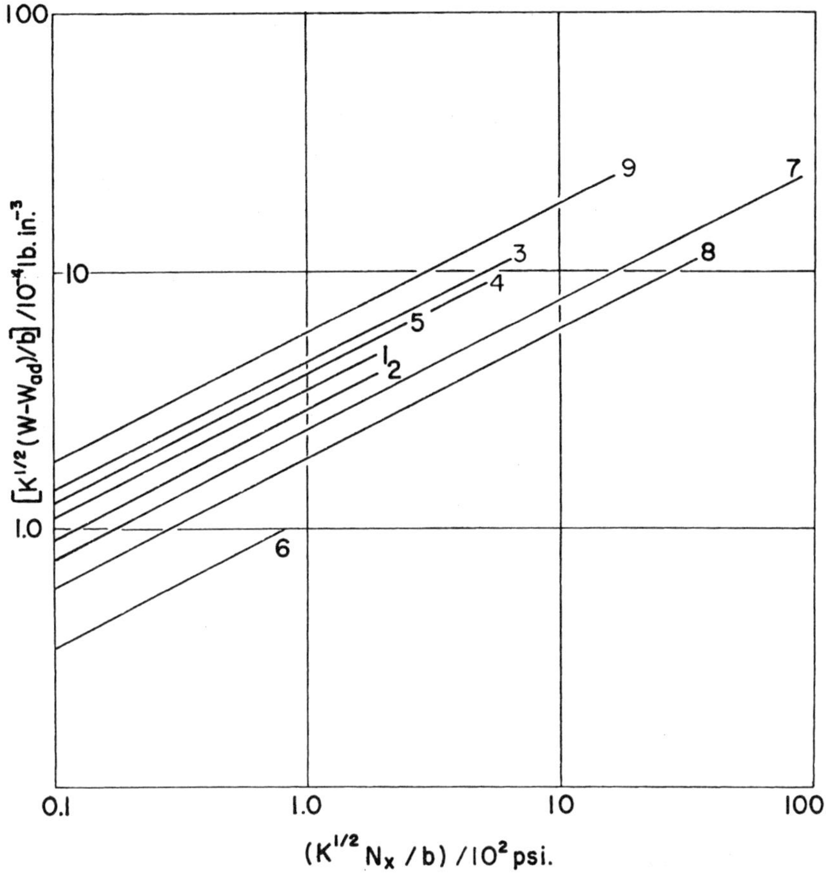

FIGURE 10.2. Weight parameter as a function of load index, web-core sandwich panel-unidirectional compressive load. (Reprinted from Reference [13].)

In each case, the load index is taken only to an optimum stress equal to the yield strength of the material system. The factor $K^{1/2}$, appearing as it does in both the ordinate and abscissa, permits the weight comparison for various material systems regardless of boundary conditions and aspect ratio, *a/b*.

It is clearly seen that the beryllium results in the lowest-weight construction, but the yield strength is reached at a low value of load index. It is also clearly seen that not only do boron-epoxy and graphite-epoxy materials provide lower weight construction than glass-reinforced plastic systems, steel, and aluminum, but of perhaps equal significance, higher loads can be carried by graphite-epoxy and boron-epoxy panels of given size and edge-support conditions than any other material system.

10.6 MINIMUM-WEIGHT OPTIMIZATION OF HONEYCOMB-CORE SANDWICH PANELS SUBJECTED TO IN-PLANE SHEAR LOADS

In this optimization, the faces must be sized such that the allowable shear stress is not exceeded. Also, the panel geometry must preclude overall buckling [Equations (9.94) through (9.97)], core shear instability [Equation (9.98)], face wrinkling [Equation (9.99)] and face dimpling [Equation (9.100)]. The optimization of the honeycomb core sandwich panel is then performed by equating the four buckling loads given by Equations (9.94) through (9.100). The results are as follows:

$$\frac{k_2}{k_3 k_1 S_{xi}}^{1/2}\left(\frac{N_{xy}}{b}\right) = \frac{2\sqrt{3}(1-\nu_{xy}\nu_{yx})}{\pi}\left(\frac{G_c}{E_c\overline{E}_{fx}\overline{E}_{fy}}\right)^{1/2}\sigma_0^2 \tag{10.62}$$

$$\frac{h_c}{b} = 2\left(\frac{k_3}{k_4}\right)^{1/4}(k_1 S_{xi})^{1/2}\left(\frac{1-\nu_{xy}\nu_{yx}}{\pi}\right)^{1/2}\frac{\sigma_0^{1/2}}{(E_{fx}E_{fy})^{1/4}} \tag{10.63}$$

$$\frac{d}{b} = \sqrt{6}\left(\frac{(1-\nu_{xy}\nu_{yx})}{\pi}\right)^{1/2}\left(\frac{G_c}{E_c}\right)^{1/2} \times \frac{(E_{fx_T}E_{fy_T})}{(\overline{E}_{fx}\overline{E}_{fx})^{1/2}}\sigma_0^{1/2}\left(\frac{k_3 k_1 S_{xi}}{k_2}\right)^{1/2} \tag{10.64}$$

$$\frac{t_c}{b} = 3\sqrt{2}\left(\left(\frac{k_3}{k_4}\right)^{1/4}\frac{(k_1 S_{xi})}{k_2}\right)^{1/2}\frac{(E_{fx_T}E_{fy_T})^{1/4}\sigma_0^2}{E_c(\overline{E}_{fx}\overline{E}_{fy})^{3/4}} \tag{10.65}$$

$$\frac{t_f}{b} = \left(\frac{3k_3 k_2 S_{xi}}{k_2}\right)^{1/2}\frac{(1-\nu_{xy}\nu_{yx})}{\pi}\left(\frac{G_c}{E_c}\right)^{1/2}\frac{\sigma_0}{(\overline{E}_{fx}\overline{E}_{fy})^{1/2}} \tag{10.66}$$

$$\left(\frac{W - W_{ad}}{b}\right) = \left(\frac{3k_3 k_1 S_{xi}}{k_2}\right)^{1/4}\left[\frac{2(1-\nu_{xy}\nu_{yx})}{\pi}\right]^{1/2}\left(\frac{G_c}{E_c}\right)^{1/4} \times \frac{(N_{xy}/b)^{1/2}}{(E_{fx}E_{fy})^{1/4}}\left[\rho_f + \frac{\rho_c k_2}{(k_3 k_4)^{1/2}}\frac{\sigma_0}{G_c}\right] \tag{10.67}$$

Therefore for the face material selection the factor of merit is

$$FM = \frac{(\overline{E}_{fx}\overline{E}_{fy})^{1/4}}{\rho} \tag{10.68}$$

TABLE 10.2. Face-Material Comparison, Based on the Factor of Merit, for an Optimized Honeycomb Core Sandwich Panel Subjected to In-Plane Shear Loads.

Material	Configuration	*FM* (10^3 in.^{2}/lb$^{1/2}$)
C/epoxy	Cross-ply	62.53
B/2024 al		57.09
B/6061 al		57.03
B/5052 al		56.18
B/3002 al		55.61
B/1100 al		55.29
B(B_4C)/al		55.18
B/al		54.60
B/6061 al		54.57
B/epoxy	cross-ply	54.37
Borsic/al		54.03
Kevlar/epoxy	cross-ply	47.68
αAl_2O_3/Mg	continuous fiber	45.72
T300/934	unidirectional	45.39
αAl_2O_3/Al-SCu	continuous fiber	44.04
T300/5208	unidirectional	43.85
SiC/6061 al		43.57
T300/SP286		43.35
αAl_2O_3/al	continuous fiber	42.97
E glass/epoxy		42.75
B/Epoxy		42.20
AS4/3501-6	unidirectional	41.88
AS1/3501-6		41.68
Boron/epoxy		41.62
HS graphite/ep.	unidirectional	41.56
αAl_2O_3/al	continuous fiber	41.30
B/EP	unidirectional	41.09
B/EP		41.02
Boron/EP		41.01
AS/3501		40.70
2024 al		35.79

With this factor of merit, a comparison of several materials is given in Table 10.2.

It can be shown [7] from Equation (10.68) that, for any laminated composite material used for the sandwich faces, the best stacking sequence to resist an in-plane shear load is a balanced cross-ply laminate.

10.7 MINIMUM-WEIGHT OPTIMIZATION OF SOLID- AND FOAM-CORE SANDWICH PANELS SUBJECTED TO IN-PLANE SHEAR LOADS

To minimize the weight, the three buckling critical stresses are equated, and the results for the minimum weight panel are

$$\left(\frac{1}{k_1 S_{xi}}\right)^{1/2}\left(\frac{N_{xy}}{b}\right) = \frac{6\sqrt{3}(1 - \nu_{xy}\nu_{yx})^2\, \sigma_0^5}{\pi E_c^{3/2} G_{cy}^{1/2} \overline{E}_{fx} \overline{E}_{fy}} \tag{10.69}$$

Again, because of the continuous linear relationship of core shear modulus to core density for foam cores, G_{cx} is treated as a variable; therefore,

$$G_{cx} = \frac{9(1 - \nu_{xy}\nu_{yx})^2\, \sigma_0^6}{G_{cy} E_c^2 \overline{E}_{fx} \overline{E}_{fy}} \tag{10.70}$$

$$t_f = \frac{\sqrt{3}(1 - \nu_{xy}\nu_{yx})^2\, b(k_1 S_{xi})^{1/2}\, \sigma_0^4}{\pi E_c^{3/2} G_{cy}^{1/2} \overline{E}_{fx} \overline{E}_{fy}} \tag{10.71}$$

$$h_c = \frac{2\sqrt{3}(1 - \nu_{xy}\nu_{yx}) b(k_1 S_{xi})^{1/2} \sigma_0^2}{\pi E_c^{1/2} G_{cy}^{1/2} \overline{E}_{fx}^{1/2} \overline{E}^{1/2}} \tag{10.72}$$

The optimum panel weight equation is

$$\left(\frac{W - W_{ad}}{b}\right) = \frac{2^{1/5} 3^{3/10} (1 - \nu_{xy}\nu_{yx})^{2/5}\, (k_1 S_{xi})^{1/10}\, (N_{xy}/b)^{4/5}}{\pi^{1/5} E_c^{3/10} \overline{E}_{fx}^{1/5} \overline{E}_{fy}^{1/5} G_{cy}^{1/10}} \times \left[\rho_f + \frac{\rho_c E_c (\overline{E}_{fx} \overline{E}_{fy})^{1/2}}{3(1 - \nu_{xy}\nu_{yx})\sigma_0^2}\right] \tag{10.73}$$

from which it is seen that the best face material can be selected from

$$FM = \frac{(\overline{E}_{fx} \overline{E}_{fy})^{1/5}}{\rho} \tag{10.74}$$

10.8 MINIMUM-WEIGHT OPTIMIZATION OF TRUSS-CORE SANDWICH PANELS SUBJECTED TO IN-PLANE SHEAR LOADS

Stated once again, the philosophy of the optimization for the truss-core sandwich panel with in-plane shear loads is as follows: a truly optimum structure is one which has a unique value for each dependent variable within the class of structure being studied (triangulated core sandwich panel, for example), for each set of materials for each set of boundary conditions and is the minimum possible weight for a specified set of design loads and will maintain its structural integrity (no mode of failure will occur at a load less than the optimum design load). In this case the optimum (minimum weight) structure will have the characteristic that the panel will become unstable in all three buckling modes simultaneously.

The governing equations pertaining to this construction to be used in the optimization are given by Equations (9.104), (9.106), (9.109), (9.111), (9.112), and (9.113). The known or specified quantities are N_{xy}, and the panel width b, that can be combined as the load index (N_{xy}/b), the material properties, and the panel boundary conditions. The buckling coefficients j, k_f and k_c are given in Figure 9.10 for any given set of variables, and hence are constants for the optimum construction being sought. The dependent variables with which to optimize the construction are the face thickness, t_f, the core depth, h_c, the web material thickness, t_c, the web angle, θ, the face stress, τ_f, the core stress, τ_c, and the weight, $W - W_{ad}$.

It is seen that there are six equations and seven unknowns. The seventh equation is obtained by placing the weight equation in terms of one convenient variable, taking the derivative of the weight equation with respect to this variable, and equating it to zero to obtain the unique value of that variable, which results in a minimum weight structure. Subsequently, one can determine the value of all other variables for the optimum construction. Manipulation of the equations listed above results in the weight equation involving only the dependent variable θ, as shown below

$$\frac{W - W_{ad}}{b} = \frac{3^{1/4}}{2}\frac{\rho_f}{(k_f j)^{1/4}}\frac{(N_{xy}/b)^{1/2}}{E_{sf}^{1/2}} \times \frac{4(\sin\theta)^{3/2} + \left(\dfrac{\rho_c}{\rho_f}\right)\left(\dfrac{G_{xyc}}{G_{xyf}}\right)^{1/2}\left(\dfrac{E_{sf}}{E_{sc}}\right)^{1/2}\left(\dfrac{k_f}{k_c}\right)^{1/2}}{(\sin\theta)(\cos\theta)^{1/2}} \tag{10.75}$$

where

$$E_{si} = [E_{iy}^3 E_{ix}]^{1/4}/(1 - \nu_{xyi}\nu_{yxi}), \ (i = c, f) \tag{10.76}$$

Taking the derivative of Equation (10.75) with respect to θ, and equating it to zero results in a value of θ in terms of material properties and buckling coefficients will result in minimum weight structure. This expression is

$$2(\sin\theta)^{3/2}(\cos\theta)^2 + 2(\sin\theta)^{7/2} - \left(\frac{\rho_c}{\rho_f}\right)\left(\frac{G_{xyc}}{G_{xyf}}\right)\left(\frac{E_{sf}}{E_{sc}}\right)^{1/2}\left(\frac{k_f}{k_c}\right)^{1/2}\left(\cos^2\theta - \frac{1}{2}\sin^2\theta\right) = 0 \qquad (10.77)$$

Note that the optimum web angle, defined by this equation is independent of the load and boundary conditions to which the panel is subjected.

A "universal relationship" may be obtained that relates N_{xy}/b to a unique value of face-shear stress, τ_{f_0}, for any set of material properties, that will result in a minimum weight panel. For a given N_{xy}/b, a panel designed to have a τ_f *higher* or *lower* than the value given by the following relationship will result in a panel that has a weight greater than can be achieved if this universal relationship is used,

$$\frac{N_{xy}}{b} = \frac{4\sqrt{3}\tan\theta}{k_f^{1/2}\bar{j}^{1/2}}\frac{\tau_{f_0}}{E_{sf}} \qquad (10.78)$$

where θ is obtained from Equation (10.77) and E_{sf} from Equation (10.76) and τ_{f_0} is the optimum face stress.

The remaining geometric variables t_f, t_c, and h_c, as well as the weight equation can now be expressed in terms of the optimum face stress τ_{f_0} obtained in Equation (10.78), and the optimum angle θ determined by Equation (10.77) above. For the case of faces and core of the same materials, these are

$$\frac{t_f}{b} = \frac{2/3\,(\tan\theta)\,\tau_{f_0}}{k_f^{1/2}\bar{j}^{1/2}E_s} \qquad (10.79)$$

$$\frac{h_c}{b} = \left(\frac{\tau_{f_0}}{E_s\bar{j}}\right)^{1/2} \qquad (10.80)$$

$$\frac{t_c}{b} = \frac{\sqrt{3}}{k_c^{1/2}\bar{j}^{1/2}}\frac{(\sin\theta)^{1/2}}{\cos\theta}\frac{\tau_{f_0}}{E_s} \qquad (10.81)$$

$$\frac{W - W_{ad}}{b} = \frac{\sqrt{3}\,\rho\tau_{f_0}}{k_f^{1/2}\bar{j}^{1/2}E_s}\,\frac{4(\sin\theta)^{3/2} + \left(\frac{k_f}{k_c}\right)^{1/2}}{\cos\theta\,(\sin\theta)^{1/2}} \qquad (10.82)$$

Employing Equation (10.78) the minimum weight panels for a given load index N_{xy}/b is given by

$$\frac{W - W_{ad}}{b} = \frac{3^{1/4}}{2} \frac{\rho}{k_f^{1/4}\bar{j}^{1/4}} \frac{(N_{xy}/b)^{1/2}}{E_s^{1/2}} \left[\frac{4(\sin\,\theta)^{3/2} + \left(\frac{k_f}{k_c}\right)^{1/2}}{\sin\,\theta(\cos\,\theta)^{1/2}} \right] \tag{10.83}$$

where E_s is given by Equation (10.76).

Two conclusions are drawn. First, the best composite material to use in such construction is the one with the highest ratio of

$$FM = \frac{E_s^{1/2}}{\rho} = \frac{[E_y^3 E_x]^{1/8}}{\rho(1 - \nu_{xy}\nu_{yx})^{1/2}} \tag{10.84}$$

Secondly, the ratio of face weights to core weight for optimum construction is

$$\frac{W_f}{W_c} = \frac{4(\sin\,\theta)^{3/2}}{(k_f/k_c)^{1/2}} \tag{10.85}$$

In the case of a panel with the same face and core materials, then $k_f = k_c$, $\theta = 28.4°$ and $W_f/W_c = 1.316$.

10.8.1 Certain Properties of the Buckling Coefficients $\bar{j}$, k_f and k_c

It is advantageous to discuss certain characteristics of the coefficients $\bar{j}$, k_f, and k_c that result in significant simplifications to the design procedures. It is shown in the references in Chapters 9 and 10 that in all calculations of optimum constructions

$$\bar{j} = j \text{ and } k_f = k_c \tag{10.86}$$

10.8.2 Conclusions

For truss-core sandwich panels in which both the faces and core are composed of the same isotropic or orthotropic materials, $\theta = 28.4°$ for minimum-weight construction whenever the edge restraint coefficients k_f and k_c are equal (the usual case). Under these conditions the weight ratio of face material to core material per unit planform area is 1.316.

For the case of core and face material being the same, the use of the orthotropic material which has the highest value of the following factor of merit will result in the least weight panel of specified geometry and loading. Since $(1 - \nu_{xy}\nu_{yx}) \approx 1$ for most composites, it is seen that a factor of merit (FM) can be defined as:

$$FM = \frac{E_y^{3/8} E_x^{1/8}}{\rho} \tag{10.87}$$

For unidirectional composites, the fiber direction for this type construction should be in the y-direction. This is the same factor of merit as that found for the faces of an optimum honeycomb core subjected to uniaxial compression [1,2,7]. In those references several dozen material systems were compared, and provide the wherewithal to select the best materials for the optimized structures of Figure 9.7.

However, to determine the limitations on N_{xy}/b to prevent overstressing, Equations (10.78) and (10.84) must be plotted to make the final comparison, as in Figure 10.3. There, typical composite materials given in an appendix of Reference [8] are plotted. The maximum values of N_{xy}/b correspond to the maximum shear strengths given in Reference [8]; and some fraction of those values as a factor of safety would terminate each curve proportionately for

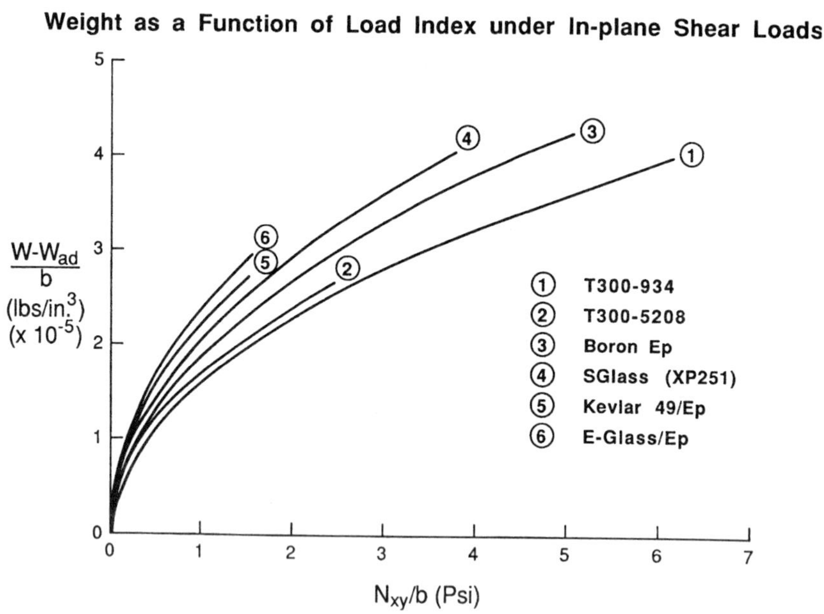

FIGURE 10.3. Weight as a function of load index under in-plane shear loads for a truss-core sandwich panel. (Reprinted from Reference [14].)

safety. It is seen that among these materials, T300-934 graphite-epoxy provides not only the lowest weight construction but is also usable to the highest load index.

10.9 MINIMUM-WEIGHT OPTIMIZATION OF WEB-CORE SANDWICH PANELS SUBJECTED TO IN-PLANE SHEAR LOADS

The equations to employ in the structural optimization of web-core sandwich panels subjected to in-plane shear loads are Equations (9.115), (9.127), (9.129), and (9.130). It should be noted that these equations can be used as is to design or analyze any nonoptimum web-core panel under these loads.

It can be shown that for minimum weight, the critical faces stress, τ_{fcr}, for overall instability and face-element instability must be equal, and this optimized value can be denoted τ_{f0}.

It is also seen that due to the geometry of the web-core construction there is no buckling criterion for the web element because with $\theta = 0$ no load can be introduced into the web plates where an in-plane shear load is applied to the panel. Therefore, the criteria for determining the web element thickness, t_c, is determined by strength alone.

Manipulating the equations enumerated above it is seen that

$$\frac{W - W_{ad}}{b} = \frac{\rho_f(N_{xy}/b)}{\tau_f} + \frac{\rho_c(1 - \nu_{xyf}\nu_{yxf})(t_c/b)2\sqrt{3}\,\tau_f^2}{k_f^{1/2}\bar{j}^{1/2}[E_{fy}^3 E_{fx}]^{1/4}(N_{xy}/b)} \tag{10.88}$$

where

$$\bar{j} = \frac{j}{1 + 4\left[\dfrac{K_m(V = 0)}{K_m} - 1\right]} \tag{10.89}$$

This weight equation is expressed in terms of the face stress, τ_f, and the core-wall thickness, t_c. As stated, t_c is not determined by buckling, and will temporarily be treated as a constant in Equation (10.88). Now setting the derivative of Equation (10.88) with respect to τ_f equal to zero, the following relationship provides the unique optimum value of τ_f for a minimum weight panel:

$$\frac{N_{xy}}{b} = \frac{2(3)^{1/4}}{k_f^{1/4}\bar{j}^{1/4}}\left(\frac{\rho_c}{\rho_f}\right)^{1/2}\frac{(1 - \nu_{xyf}\nu_{yxf})^{1/2}}{[E_{fy}^3 E_{fx}]^{1/8}}\left(\frac{t_c}{b}\right)^{1/2}\tau_{f0}^{3/2} \tag{10.90}$$

This is the "universal relationship" relating the applied load index to the optimum face stress for a specified t_c, as yet undetermined. Substituting Equation (10.90) into (10.88) results in

$$\frac{W - W_{ad}}{b} = \frac{(3)^{7/6}}{(2)^{1/3}} \frac{\rho_c^{1/3}\rho_f^{2/3}}{k_f^{1/6}\bar{j}^{1/6}} \frac{(1 - \nu_{xyf}\nu_{yxf})^{1/3}}{(E_{fy}^3 E_{fx})^{1/2}} (t_c/b)^{1/3}(N_{xy}/b)^{1/3} \tag{10.91}$$

From Equation (10.91) it is clear that the smaller the value of t_c, the lower the panel weight will be for given material systems, given panel geometry, and given load index. Likewise, from Equation (10.90), for given materials systems, panel geometry, and load index, as t_c decreases, τ_{f_0} increases. Therefore, it is concluded that minimum weight is obtained when τ_{f_0}, equals the materials' allowable shear stress, τ_{all}, i.e.,

$$\tau_{f_0} = \tau_{\text{all}} \tag{10.92}$$

It should be noted that this result differs from all of the previously presented sandwich panel optimizations, where the optimum face stress for minimum-weight construction was generally a unique stress below the maximum allowable value. That is, when the optimum face stress is the allowable stress, this is the upper bound for the applied load index.

Now substituting Equation (10.91) into (10.90) provides the optimum value of t_c for the minimum-weight construction:

$$\frac{t_c}{b} = \left(\frac{\rho_f}{\rho_c}\right) \frac{k_f^{1/2}\bar{j}^{1/2}E_{sf}(N_{xy}/b)^2}{4\sqrt{3}\tau_{\text{all}}} \tag{10.93}$$

where

$$E_{sf} = [E_{fy}^3 E_{fx}]^{1/4}/(1 - \nu_{xyf}\nu_{yxf})$$

The other variables for the minimum-weight construction are

$$(t_f/b) = (N_{xy}/b)/2\tau_{\text{all}} \tag{10.94}$$

$$(h_c/b) = (\tau_{\text{all}}/\bar{j}E_{sf})^{1/2} \tag{10.95}$$

$$(d_f/b) = \frac{k_f^{1/2}}{2\sqrt{3}} \frac{E_{sf}^{1/2}(N_{xy}/b)}{(\tau_{\text{all}})^{3/2}} \tag{10.96}$$

$$\frac{(W - W_{ad})}{b} = \frac{3\rho_f(N_{xy}/b)}{2\tau_{\text{all}}} \tag{10.97}$$

Other useful relations for the optimum construction are

$$(W_f/W_c) = 2 \tag{10.98}$$

where W_f and W_c are the weight of the face and core respectively, and

$$\frac{h_c}{d_f} = \left(\frac{\rho_f}{\rho_c}\right)\left(\frac{t_f}{t_c}\right) \tag{10.99}$$

Note that for the optimum construction, according to Equation (10.95), the core depth is practically independent of the applied load. Note also that the weight of the optimum construction is independent of all material properties except the allowable shear stress and the density of the face material, and the weight varies linearly with the load index. This differs markedly from the optimum constructions using other core geometries discussed earlier in Chapter 10.

Thus it is seen that to obtain a minimum-weight web-core panel subjected to in-plane shear loads, the best face material to utilize is the one having the highest ratio of τ_{all}/ρ_f. Thus, the factor of merit to determine the best materials system to use is

$$\text{Factor of Merit} = \frac{\tau_{\text{all}}}{\rho} \tag{10.100}$$

Also, it can be shown that $\bar{j}$ *and* k_f are constants for the optimum construction and no iteration is needed.

The Factor of Merit [Equation (10.100)] is used to compare various materials systems. Such a comparison is found in Table 10.3 for several unidirectional composites for which shear strengths were available. It is seen that among these systems graphite epoxy and boron aluminum look significantly better than the other materials.

10.10 OPTIMAL STACKING SEQUENCES FOR COMPOSITE MATERIAL LAMINATE FACES FOR VARIOUS SANDWICH PANELS SUBJECTED TO VARIOUS LOADS

If one is given a particular composite material laminate, with its associated material properties, then the methods presented in Sections 10.1 through 10.9

TABLE 10.3. Face-Material Comparison, Based on the Factor of Merit, for a Web-Core Sandwich Panel Subjected to In-Plane Shear Loads.

Rank	Material	V_f (%)	τ_u KSI	ρ lb/in.^{3}	FM 10^5in.
1	T300/934	?	14.8	0.0555	2.667
2	E Glass/Ep.	60	10.0	0.075	2.667
3	Boron/6061 Al	50	23.0	0.0915	2.515
4	AS/3501	67	13.5	0.0555	2.432
5	Boron/EP	?	18.0	0.0740	2.432
6	Kev 49/3501	62	11.7	0.050	2.340
7	Hi Str. Gr./Ep	60	12.0	0.057	2.105
8	Boron/Ep	67	15.2	0.074	2.054
9	Braided FP/Al	17	19.9	0.1001	1.988
10	T300/SP-286	60	10.5	0.0555	1.842
11	Kev 49/Ep	60	9.0	0.050	1.800

can be used to design a minimum-weight panel. However, if the designer has the freedom, this current section provides the means by which to select the particular stacking sequence to use with a given lamina of a filamentary composite in order to create a laminate for a minimum weight panel. The focus here is on the faces, as the majority of the weight of a sandwich panel is in the faces, because the faces take most of the in-plane loads.

It was seen earlier that the Factor of Merit for a honeycomb-core sandwich panel subjected to a uniaxial compressive load is given by Equation (10.13). Since the faces of the sandwich resist the applied compressive in-plane load prior to buckling, then for any face material, the best stacking sequence for this type of loading is one in which $A_{11}^3A_{22}$ is a maximum, because the Factor of Merit involves $E_x^{3/8}E_y^{1/8}$. Because the term $(1 - \nu_{xy}\nu_{yx})^{1/2}$ is approximately equal to one for most composites, it will not be included in this discussion. Therefore, to make $A_{11}^3A_{22}$ a maximum for a laminate involving 0° and 90° plies, let N equal the total number of plies, S equal the number of 0° plies, and N-S the number of 90° plies. If h_k is the uniform ply thickness then, from Equation (2.56):

$$A_{11} = \frac{SE_{11}h_k}{(1 - \nu_{12}\nu_{21})} + \frac{(N - S)E_{22}h_k}{(1 - \nu_{12}\nu_{21})} \tag{10.101}$$

$$A_{22} = \frac{SE_{22}h_k}{(1 - \nu_{12}\nu_{21})} + \frac{(N - S)E_{11}h_k}{(1 - \nu_{12}\nu_{21})} \tag{10.102}$$

So the Factor of Merit is proportional to

$$A_{11}^3 A_{22} = \{[SE_{11} + (N - S)E_{22}]^3 \times [SE_{22} + (N - S)E_{11}]\} \frac{h_k}{(1 - \nu_{21}\nu_{21})} \tag{10.103}$$

If one defines $\phi = E_{22}/E_{11}$, then the Factor of Merit (FM) is proportional to

$$[S + (N - S)\phi]^3 [S\phi + (N - S)] \tag{10.104}$$

To maximize Equation (10.104), placing the derivative of FM with respect to S equal to zero results in

$$(4S - N)\phi + 3N - 4S = 0 \tag{10.105}$$

or

$$S = \frac{1}{4}\frac{(3 - \phi)N}{(1 - \phi)} \qquad 0 \leq S \leq N \tag{10.106}$$

Equation (10.106) and Table 10.4 show that for $\phi = 0$, or E_{22} of essentially zero, the best stacking sequence is 75% 0° plies and 25% 90° plies. As ϕ increases, the percentage increases until when $\phi = 1/3$, $s = 1$. At $\phi \geq 1/3$, a unidirectional laminate is used to achieve minimum weight. Incidentally, that means for any metal matrix composite, a unidirectional composite is best when the loading is uniform axial compression, because in every practical case, $E_{22} > (E_{11}/3)$.

For the case of a solid or foam-core sandwich panel subjected to uniaxial compression, the factor merit is given by Equation (10.21).

TABLE 10.4. Optimum Stacking Sequence for the Laminated Face Material as a Function of ϕ for an Optimized Honeycomb Core Sandwich Panel Subjected to an Uniaxial Compressive Load.

$\phi = E_{22}/E_{11}$	S
0	0.75 N
0.1	0.806 N
0.2	0.875 N
0.3	0.96 N
0.333	N
>0.33	N

TABLE 10.5. Optimum Stacking Sequence for a Laminated Composite Face Material as a Function of ϕ for an Optimized Solid- or Foam-Core Sandwich Panel Subjected to an Uniaxially Compressive Load.

$\phi = E_{22}/E_{11}$	S
0	$0.667\ N$
0.1	$0.704\ N$
0.2	$0.75\ N$
0.3	$0.810\ N$
0.4	$0.889\ N$
0.5	N

As before, the Factor of Merit is proportional to

$$(E_x^2 E_y)^{1/7} \text{ or } (A_{11}^2 A_{22})^{1/7} \tag{10.107}$$

Utilizing Equation (10.101) and (10.102) the *FM* is proportional to

$$[S + (N - S)\phi]^2\,[S\phi + (N - S)] \tag{10.108}$$

Setting the derivative of the above with respect to S equal to zero results in the optimum stacking sequence of

$$S = \frac{1}{3}\frac{(2 - \phi)}{(1 - \phi)} N \tag{10.109}$$

For the foam- or solid-core sandwich panel subjected to a uniform compressive uniaxial load, it is seen from Equation (10.108) and shown by Table 10.5 that for $\phi = 0$, two thirds of the plies should be 0°, and only after $\phi = 1/2$ (i.e., $2E_2 \geq E_1$) should the laminate be unidirectional.

Examining at Equation (10.29) for the truss-core panel subjected to a uniaxial compressive load and at Equation (10.58) for the web-core sandwich panel, one sees that they are identical. Remembering that $2E_0 = (E_x E_y)^{1/2} + \nu_{yx}E_x + 2G_{xy}(1 - \nu_{xy}\nu_{yx})$, one sees that for the truss-core panel and for the web-core panel, the FM is proportional to $(A_{11}A_{22})^{1/4}$. Therefore, for minimum weight

$$S = N/2 \tag{10.110}$$

Thus, the optimum stacking sequence in each case is cross-ply with the same number of 0° plies and 90° plies. Incidentally, using the same procedures as above, it can be shown that an angle ply laminate is never better than a cross-ply or unidirectional laminate to prevent buckling for any of the four sandwich architectures when the panel is subjected to a uniaxial compressive load.

Turning now to the honeycomb core sandwich panel subjected to an in-plane shear load, it is seen that the factor of merit given by Equation (10.68) is proportional to $(E_xE_y)^{1/4}$. As a result the cross-ply laminate is best, as shown by Equation (10.110). The same holds true for the solid- or foam-core panel subjected to in-plane shear loading.

For the truss-core sandwich panel subjected to in-plane shear loads, from (10.84), it is seen that the FM is proportional to

$$(E_xE_y^3)^{1/8} \tag{10.111}$$

By the same process used before, it is seen that

$$S = \frac{1}{4}\frac{(1 - 3\phi)}{(1 - \phi)}N \tag{10.112}$$

with the tabular values given in Table 10.6. It is seen that if $\phi = 0$, 75% of the fibers should be perpendicular to the flutes, and for $\phi \geq 1/3$, the best stacking sequence for the truss-core sandwich subjected to in-plane shear is to have all fibers (i.e., unidirectional) perpendicular to the flutes.

Finally the web-core sandwich subjected to in-plane shear loads differs for all the rest in that the factor of merit is strictly strength (not stiffness) dependent. Repeating Equation (10.100),

$$FM = \tau_{all}/\rho$$

Again for any of the four panel architectures subjected to in-plane shear loads, an angle ply laminate is never the best choice to prevent buckling.

TABLE 10.6. Optimum Stacking Sequence for Laminated Composite Face Materials as a Function of ϕ for an Optimized Truss-Core Sandwich Panel Subjected to In-Plane Shear Loads.

$\phi = E_{22}/E_{11}$	S
0	0.25 N
0.1	0.194 N
0.2	0.125 N
0.3	0.036 N
0.333	0
>0.333	0

10.11 PROBLEMS

10.1 Consider a honeycomb sandwich panel 20 in. long and 12 in. wide subjected to an in-plane compressive load in the longer direction of $N_x = -2000$ lb/in. of width. The faces are made of boron/epoxy ($V_f =$ 67%) and the core is aluminum each with the following properties:

Boron/Epoxy	Aluminum Honeycomb Core
$E_x = 30.3 \times 10^6$ psi	$E_c = 10.5 \times 10^6$ psi
$E_y = 2.8 \times 10^6$ psi	$\nu = 0.348$
$\gamma_{xy} = 0.21$	$G_c = 3.9 \times 10^6$ psi
$G_{xy} = 0.93 \times 10^6$ psi	$\rho_m = 0.101$ lb/in.^{3}
$\rho_f = 0.074$ lb/in.^{3}	
$\sigma_{\text{all}} = 362{,}000$ psi	

If the panel is simply supported on all four edges:

a. What is the optimum face stress, σ_f?
b. Is the panel overstressed?
c. What are the optimum values of each dependent variable (in this case h_c, d, t_c, and t_f) for the honeycomb construction?
d. If the boron/epoxy ply thickness is 0.0055 in., how many laminates are necessary to approximate the optimum construction?
e. What is the weight of the panel per unit planform area?

10.2 Repeat Problem 10.1 for a foam-core sandwich, using the foam core of Problem 7.2.

10.3 Compare the results of Problems 10.1 and 10.2.

10.4 Consider a sandwich panel measuring 30″ × 24″ in planform area, composed of 5052 aluminum faces whose properties are: $E = 10 \times 10^6$ psi, $\nu = 0.3$, $\rho_w = 0.10$ lb/in.^{3}, $\sigma_{\text{all}} = 70{,}000$ psi. The uniaxial compressive load per unit width, in the longer direction is $N_x = 4000$ lb/in. The sandwich plate is simply supported on all four edges. What are the optimum t_f, h_c, t_c, d, and $(W\text{-}W_{ad})$ for a honeycomb core also made of 5052 aluminum?

10.5 If the sandwich panel of Problem 10.4 had a rigid Klegecell foam core, what are the optimum values of t_f, h_i, G_c, and $(W\text{-}W_{ad})$ using Figure 13.1 to determine the weight of the rigid Klegecell foam?

10.6 If the sandwich panels of Problem 10.4 involved a truss core also of 5052 aluminum, what are the optimum values of t_f, h_c, t_c, θ, and $(W\text{-}W_{ad})$ for this construction?

10.7 The sandwich panel of Problem 10.4 involved a web core also made of

5052 aluminum. What are the optimum values of t_f, h_c, t_c, d and (W-W_{ad}) for this construction?

10.8 For Problems 10.4 through 10.7, compare the various constructions as to the weight, the various total panel thickness, and the various face thicknesses?

10.12 REFERENCES

1. Vinson, J.R., "Minimum Weight Composite Material Honeycomb Sandwich Panels Under Uniaxial Compression," *Transactions of the First European Conference on Composite Materials*, September 1985.
2. Vinson, J. R., "Optimum Design of Composite Honeycomb Sandwich Panels Subjected to Uniaxial Compression," *AIAA Journal*, Vol. 24, 1986, pp. 1690–1696.
3. *Strength of Metal Aircraft Elements*, MIL-HDBK-5, March 1959, Armed Forces Supply Support Center, Washington, D.C. (newer addition is *Metallic Materials and Elements for Flight Vehicle Structures*, MIL-HDBK-5A, Feb. 1966, U.S. Government Printing Office, Washington, D.C.).
4. Vinson, J. R. and S. Shore, "Minimum Weight Web-Core Sandwich Panels Subjected to Uniaxial Compression," *Journal of Aircraft*, Vol. 8, No. 11, November 1971, pp. 843–847.
5. *Plastics for Flight Vehicles, Part 1, Reinforced Plastics*, MIL-HDBK-17, Change Notice 1, May 1964, Armed Forces Supply Support Center, Washington, D.C.
6. Beryllium in Aerospace Structures, Brush Beryllium Co., Cleveland, Ohio.
7. Handel, P. I. and J. R. Vinson, "Optimal Stacking Sequences of Composite Faces for Various Sandwich Panels and Loads to Attain Minimum Weight," *Proceedings of the 29th AIAA/ASME/ASCE/AHS/ASC Structures, Structural Dynamics and Materials Conference*, AIAA, New York, 1988, pp. 999–1006.
8. Vinson, J. R., "Minimum Weight Triangulated (Truss) Core Sandwich Panels Subjected to In-Plane Shear Loads," *Composites 86—Recent Advances in Japan and the United States*, Editors, K. Kawata, S. Umekawa and A. Kobayashi, Japan Society of Composite Materials, Publishers, Tokyo, 1986, pp. 663–671.
9. Vinson, J. R., "Minimum Weight Web-Core Composite Sandwich Panels Subjected to In-Plane Shear Loads," *Proceedings of the Army Symposium on Solid Mechanics—1986*, West Point, October 7–9, 1986, U.S. Army Materials Technology Laboratory Report MTL MS 86-3, Watertown, MA, 1986.
10. Vinson, J. R. and S. Shore, "Structural Optimization of Corrugated Core and Web Core Sandwich Panels Subjected to Uniaxial Compression," Naval Air Engineering Center Report NAEC-ASC-1109, May 1967.
11. Seide, P., "The Stability Under Longitudinal Compression of Flat Symmetric Corrugated-Core Sandwich Plates with Simply Supported Loaded Edges and Simply Supported or Clamped Unleaded Edges," NACA TN 2679, 1952.
12. Anderson, M. S., "Local Instability of the Elements of a Truss-Core Sandwich Plate," NASA TR R-30, 1959.
13. Vinson, J. R., "Minimum Weight Web-Core Composite Sandwich Panels Subjected to In-

Plane Compressive Loads," presented at the International Symposium on Composite Materials and Structures, Beijing, China, June 13–16, 1986.

14. Vinson, J. R., "Minimum Weight Triangulated (Truss) Core Composite Sandwich Panels Subjected to In-Plane Shear Loads," in *Composites 86—Recent Advances in Japan and the United States*, Eds. K. Kawata, S. Umekawa and A. Kobayashi, Proceedings of the Third Japan-U.S. Conference on Composite Materials, Japan Society of Composite Materials, Publishers, Tokyo, 1986, pp. 663–671.

CHAPTER 11

Sandwich Shells

11.1 INTRODUCTION

A shell is a thin-walled body (like beams and plates) whose middle surface is curved in at least one direction. For instance, a cylindrical shell and a conical shell have only one direction in which the middle surface is curved. On the other hand, in a spherical shell there is curvature in both directions. Such common items as a front fender of a car or an eggshell are examples of double curvature in shells.

Shell structures involve very complicated, lengthy equations that are further lengthened when the shell is composed of an anisotropic material such as any one of many fiber reinforced, composite materials. In addition, the effects of transverse shear deformation should be included for structures composed of composite materials in which the fibers are in the plane of the midsurface. However, for the most simple to the most involved form, the equations for shells involve certain characteristics that distinguish them from the equations and behavior of other structural components. In shells of any kind, the shell would like to behave as membrane with respect to stress and deformation. The membrane deformation solutions are the particular solutions of the partial differential governing equations and are usually easy to solve; membrane stresses are uniform across the shell wall thickness, and are usually found by a judicious use of free-body diagrams.

However, close to any structural discontinuity, such as a simply supported, clamped, or other restricted edge, adjacent to a hole, or at an abrupt change in wall thickness or material, there exists a bending boundary layer (BBL). In this BBL, the stresses can exceed the membrane stresses by factors of two, three, or more, due to the superposition of the bending stresses on the membrane stresses that exist throughout the shell; the deformations in the BBL are "wavy," and the curvatures are proportional to the bending stresses. Thus, within the BBL, reside the maximum stresses, the areas of engineering interest, and the sources of bad design in composite-sandwich shell structures and all shells in general.

Love [1] wrote the first paper on shell theory in 1888, and as recently as January 1997, an interesting if not quite accurate quote says that for shells "all you need is LOVE" [2]. Relatively little was done in advancing shell theory until its great emphasis during World War II, and with the advent of the missile and space ages. Many of this century's great researchers concentrated their efforts on shell structures during this period, including Timoshenko and Woinowski-Krieger [21], Reissner [3,4], Flügge [5,6], Goldenveizer [7], Vlasov [8], Naghdi [9], Hoff [10], Novozhilov [11], Kraus [12], Leissa [13], and Dym [14]. In fact the "golden era" of shell theory was in the 1940s–1970s. Nonlinear shell theory has been chronicled by Libai and Simmonds [15] and Palazotto and Dennis [16]. More recently, Bert [17] and his colleagues have been very prolific in the area of shell theory of composite materials. In addition to conventional shell theory, Bert has concentrated on the behavior of composites that have different properties in tension and compression; he calls them bimodulus composites. These are typical of some composites such as fiber-reinforced tires and some biological materials.

11.2 ANALYSIS OF CYLINDRICAL SANDWICH SHELLS UNDER AXIALLY SYMMETRIC LOADS

The simplest of all shell geometries is that of the circular cylindrical shell shown in Figure 11.1. The positive directions of the displacements u, v, and w are shown, as well as the positive directions of the coordinates x and ζ. The

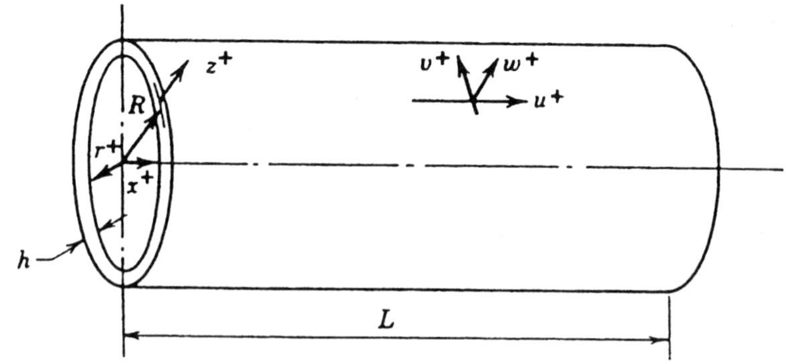

FIGURE 11.1. Circular cylindrical shell. (Reprinted from Reference [22].)

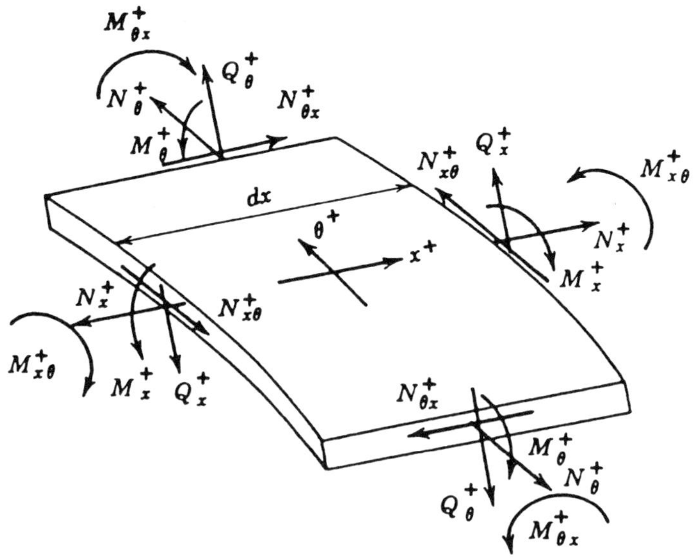

FIGURE 11.2. Positive directions of stress resultants and stress couples. (Reprinted from Reference [22].)

remaining coordinate is the circumferential coordinate θ. Here, the thickness, h, is the total thickness of the sandwich shell, i.e., $h = 2t_f + h_c$ for a sandwich with two equal face thicknesses. The positive value of all stress resultants and stress couples is shown in Figure 11.2.

In the classical shell theory discussed in this section, all of the assumptions used in classical plate theory of the previous chapters are utilized:

$$\sigma_\zeta = \varepsilon_\zeta = \varepsilon_{x\zeta} = \varepsilon_{\theta\zeta} = 0$$

$$u(x, \theta, \zeta) = u_0(x, \theta) + \zeta\beta_x(x, \theta) \tag{11.1}$$

$$v(x, \theta, \zeta) = v_0(x, \theta) + \zeta\beta_\theta(x, \theta)$$

where all of the terms have been explained in the previous chapters.

In addition, there is one other assumption known as Love's First Approximation, that is consistent with the neglect of transverse shear deformation:

$$h/R \ll 1 \tag{11.2}$$

It is true that the accurate analysis of shells of sandwich materials should include transverse shear deformation because the in-plane stiffnesses are dominated by the face properties, while the transverse shear modulus is dominated by the (usually much lower) core properties. However, for preliminary design and analysis one can neglect transverse shear deformation, with the resulting simplifications in the analytical complexity and time expended in design and analysis.

To derive the governing differential equations for cylindrical shells, one should begin with the elasticity equations in a curvilinear coordinate system, just as the elasticity equations in a Cartesian coordinate system were the starting point for the plate equation derivation. This is done in detail in the first two chapters of Reference [22]. Then one can proceed to develop the governing equations for a shell of general shape, and then specialize it to any particular configuration such as a circular cylindrical shell. This would require significant space, and would be too involved for the scope of this text. Therefore, the governing equations are presented below without derivation. The equilibrium equations are

$$\frac{\partial N_x}{\partial x} + \frac{1}{R}\frac{\partial N_{x\theta}}{\partial \theta} + q_x = 0 \tag{11.3}$$

$$\frac{\partial N_{x\theta}}{\partial x} + \frac{1}{R}\frac{\partial N_\theta}{\partial \theta} + \frac{Q_\theta}{R} + q_\theta = 0 \tag{11.4}$$

$$\frac{\partial Q_x}{\partial x} + \frac{1}{R}\frac{\partial Q_\theta}{\partial \theta} - \frac{1}{R}N_\theta + p(x, \theta) = 0 \tag{11.5}$$

$$\frac{\partial M_x}{\partial x} + \frac{1}{R}\frac{\partial M_{x\theta}}{\partial \theta} - (Q_x - m_x) = 0 \tag{11.6}$$

$$\frac{\partial M_{x\theta}}{\partial x} + \frac{1}{R}\frac{\partial M_\theta}{\partial \theta} - (Q_\theta - m_\theta) = 0 \tag{11.7}$$

where

$$q_x = \sigma_{\zeta x}(h/2) - \sigma_{\zeta x}(-h/2) = \tau_{1x} - \tau_{2x}$$

$$q_\theta = \sigma_{\zeta\theta}(h/2) - \sigma_{\zeta\theta}(-h/2) = \tau_{1\theta} - \tau_{2\theta}$$

$$m_x = \frac{h}{2}[\sigma_{\zeta x}(h/2) + \sigma_{\zeta x}(-h/2)] = \frac{h}{2}[\tau_{1x} + \tau_{2x}]$$

$$m_\theta = \frac{h}{2}[\sigma_{\zeta\theta}(h/2) + \sigma_{\zeta\theta}(-h/2)] = \frac{h}{2}[\tau_{1\theta} + \tau_{2\theta}]$$

These equilibrium equations are independent of the material system. They could also be written in terms of the arc distance, ds, where $ds = Rd\theta$. The q_x, q_θ, m_x, and m_θ quantities are functions of the surface-shear stresses on the outer surfaces of the sandwich faces.

For the case of no transverse-shear deformation, the relation between the rotations, β_x and β_θ, and the displacements is given by

$$\beta_x + \frac{\partial w}{\partial x} = 0 \tag{11.8}$$

$$\beta_\theta + \frac{1}{R}\frac{\partial w}{\partial \theta} - \frac{v_0}{R} = 0 \tag{11.9}$$

For simplicity, assume that the lamination stacking sequence is such that there is no bending-stretching coupling (that is $[B] = 0$), and that there are no other coupling terms (that is $(\ )_{16} = (\ )_{26} = 0$).

Also, in the following derivation, for simplicity first consider a shell of *one lamina only*. Subsequently, generalizations to the sandwich shell can be made. In this case, the stress-strain relations and the strain-displacement relations, utilizing the displacement assumptions of Equation (11.1) are:

$$\varepsilon_x = \frac{1}{E_x}(\sigma_x - \nu_{x\theta}\sigma_\theta) = \frac{\partial u_0}{\partial x} + \zeta\frac{\partial \beta_x}{\partial x} \tag{11.10}$$

$$\varepsilon_\theta = \frac{1}{E_\theta}[\sigma_\theta - \nu_{\theta x}\sigma_x] = \frac{1}{R}\left(\frac{\partial v_0}{\partial \theta} + w\right) + \frac{\zeta}{R}\frac{\partial \beta_\theta}{\partial \theta} \tag{11.11}$$

$$\varepsilon_{x\theta} = \frac{1}{2G_{x\theta}}\sigma_{x\theta} \tag{11.12}$$

The in-plane stiffnesses, K, and the flexural stiffnesses, D, are given as follows for the one lamina orthotropic shell:

$$\begin{gathered} K_x = \frac{E_x h}{(1 - \nu_{x\theta}\nu_{\theta x})},\ K_\theta = \frac{E_\theta h}{(1 - \nu_{x\theta}\nu_{\theta x})} \\ D_x = \frac{E_x h^3}{12(1 - \nu_{x\theta}\nu_{\theta x})},\ D_\theta = \frac{E_\theta h^3}{12(1 - \nu_{x\theta}\nu_{\theta x})} \end{gathered} \tag{11.13}$$

It is seen that the K values correspond exactly to A_{11} and A_{22}, and the D

values correspond exactly to the D_{11} and D_{22} values of Chapter 2 for a one-ply composite. The orthotropic relationship of course holds:

$$\frac{\nu_{x\theta}}{E_x} = \frac{\nu_{\theta x}}{E_\theta} \tag{11.14}$$

For this simplest of derivations, assume that the loads are axially symmetric; hence, $\partial(\;)/\partial\theta = 0$ for all quantities, since both the geometry and the loading are axially symmetric.

Doing the usual integration of Equations (11.10) and (11.11) by multiplying each by $d\zeta$, integrating from $-h/2$ and $h/2$, and utilizing the definitions of the stress resultants, results in

$$N_x = \nu_{x\theta}N_\theta = E_x h \frac{\partial u_0}{\partial x},\; N_\theta - \nu_{\theta x}N_x = E_\theta h \frac{w}{R}$$

Rearranging these, the following emerges:

$$N_x = K_x\left[\frac{\partial u_0}{\partial x} + \nu_{\theta x}\frac{w}{R}\right] \tag{11.15}$$

$$N_\theta = K_\theta\left[\nu_{x\theta}\frac{\partial u_0}{\partial x} + \frac{w}{R}\right] \tag{11.16}$$

Similarly, multiplying Equations (11.10) and (11.11) by $\zeta d\zeta$ and integrating from $-h/2$ to $h/2$ results in:

$$M_x - \nu_{x\theta}M_\theta = -\frac{E_x h^3}{12}\frac{\partial^2 w}{\partial x^2},\; M_\theta = \nu_{\theta x}M_x = 0$$

Rearranging these yields the following:

$$M_x = -D_x\frac{\partial^2 w}{\partial x^2} \tag{11.17}$$

$$M_\theta = \nu_{\theta x}M_x \tag{11.18}$$

Now, further assume for the following derivation that there are no surface-shear stresses; hence, $q_x = q_\theta = m_x = m_\theta = 0$. However, if they should be considered, they can be carried along remembering that for this derivation, they too are axially symmetric.

The above shell equations can be simplified to the following set for the case of axial symmetry in both geometry and loads:

$$\frac{dN_x}{dx} = 0, \text{ therefore } N_x = \text{constant} \tag{11.19}$$

$$\frac{dQ_x}{dx} - \frac{N_\theta}{R} + p(x) = 0 \tag{11.20}$$

$$\frac{dM_x}{dx} - Q_x = 0 \tag{11.21}$$

$$\beta_x + \frac{dw}{dx} = 0 \tag{11.22}$$

$$\beta_\theta = 0 \tag{11.23}$$

$$N_x = K_x\left[\frac{\partial u_0}{dx} + \frac{\nu_{\theta x} w}{R}\right] \tag{11.24}$$

$$N_\theta = K_\theta\left[\nu_{x\theta}\left[\nu_{x\theta}\frac{du_0}{dx} + \frac{w}{R}\right]\right] \tag{11.25}$$

$$M_x = -D_x\frac{d^2w}{dx^2} \tag{11.26}$$

$$M_\theta = \nu_{\theta x} M_x \tag{11.27}$$

$$Q_x = -D_x\frac{d^3w}{dx^3} \tag{11.28}$$

From Equations (11.19) and (11.24)

$$\frac{d^2u_0}{dx^2} + \frac{\nu_{\theta x}}{R}\frac{dw}{dx} = 0 \tag{11.29}$$

From Equations (11.28), (11.25) and (11.20), one obtains

$$-D_x\frac{d^4w}{dx^4} - \frac{K_\theta}{R}\left(\nu_{x\theta}\frac{du_0}{dx} + \frac{w}{R}\right) + p(x) = 0 \tag{11.30}$$

From Equations (11.24) and (11.30) the resulting governing equation is

$$\frac{d^4w}{dx^4} + \frac{K_\theta}{D_xR^2}(1 - \nu_{x\theta}\nu_{\theta x})w = \frac{1}{D_x}\left[p(x) - \nu_{\theta x}\frac{N_x}{R}\right] \tag{11.31}$$

For a composite laminate the coefficient of the second term can be written as

$$\frac{12(1 - \nu_{x\theta}\nu_{\theta x})}{h^2R^2}\frac{D_\theta}{D_x} \tag{11.32}$$

Now for the sandwich shell, ε^4 is defined to be

$$\varepsilon^4 = \frac{3(1 - \nu_{x\theta}\nu_{\theta x})}{h^2R^2}\frac{D_\theta}{D_x} \tag{11.33}$$

where here h is the total sandwich thickness,

$$h = h_c + 2t_f \tag{11.34}$$

Otherwise, in a composite laminate, h is the laminate thickness.

The governing differential equations for the lateral deflection, w, and the in-plane displacements, u_0, become

$$\frac{d^4w}{dx^4} + 4\varepsilon^4 w = \frac{1}{D_x}\left[p(x) - \frac{\nu_{\theta x}}{R}N_x\right] \tag{11.35}$$

$$\frac{d^2u_0}{dx^2} + \frac{\nu_{\theta x}}{R}\frac{dw}{dx} = 0 \tag{11.36}$$

The form of Equation (11.35) is desirable since it is uncoupled from the other governing equation (11.36). From Equation (11.19), N_x is a constant determined by boundary conditions. In fact, it is seen from Equation (11.35) that the presence of an axial in-plane force is equivalent to a constant laterally distributed load as far as the lateral displacement w is concerned. Upon determining w and u_0 from Equations (11.35) and (11.36), the stress resultants and stress couples can be obtained from Equations (11.24) through (11.28).

It is also seen that the governing differential equation, (11.35), has the same form as that of a beam on an elastic foundation, where D_x is replaced with the flexural stiffness of a beam, EI, and replacing $4\varepsilon^4D_x$ is replaced by k, the beam foundation modulus. Thus, the physical intuition of all of the solutions for beams on an elastic foundation can be utilized.

By standard methods, the roots of the fourth-order equation [Equation (11.35)] are obtained to be $\pm\varepsilon(1 \pm i)$ where $i = \sqrt{-1}$. The general solution can therefore be written as

$$w(x) = Ae^{-\varepsilon x}\cos\varepsilon x + Be^{-\varepsilon x}\sin\varepsilon x + Ce^{\varepsilon x}\cos\varepsilon x + Ee^{\varepsilon x}\sin\varepsilon x + w_p(x) \tag{11.37}$$

where A, B, C, and E are constants of integration determined by the boundary conditions, and $w_p(x)$ is the particular integral.

The in-plane displacement, $u_0(x)$, can be obtained through integrating Equation (11.24) as follows:

$$u_0(x) = \frac{N_x x}{K_x} - \frac{\nu_{\theta x}}{R}\int w\,dx + F \tag{11.38}$$

where N_x is constant and F is a constant of integration.

It is seen, therefore, that for circular cylindrical shells under axially-symmetric loads, there are six boundary conditions, three at each end. The natural boundary conditions for this case are

(1) Either u_x is prescribed or $N_x = 0$.
(2) Either dw/dx is prescribed or $M_x = 0$.
(3) Either w is prescribed or $Q_x = 0$.

In the above, the simply supported, clamped- and free edge boundary conditions are contained.

In solving Equation (11.35), another form may be utilized, other than that of Equation (11.37)—the so-called BBL solution. In that case, the solution to Equation (11.37) can be written as

$$w(x) = \frac{M_0}{2\varepsilon^2 D_x}e^{-\varepsilon x}(\sin\varepsilon x - \cos\varepsilon x) - \frac{Q_0}{2\varepsilon^3 D_x}e^{-\varepsilon x}\cos\varepsilon x + \frac{M_L}{2\varepsilon^2 D_x}e^{-\varepsilon(L-x)}[\sin\varepsilon(L-x) - \cos\varepsilon(L-x)] + \frac{Q_L}{2\varepsilon^3 D_x}e^{-\varepsilon(L-x)}\cos\varepsilon(L-x) + w_p(x) \tag{11.39}$$

where instead of A, B, C, and E of Equation (11.37) being the constants of integration, M_0, Q_0, M_L, and Q_L are the integration constants to be used.

The particular advantage of using Equation (11.39) rather than (11.37) is easily seen. Each term of the homogeneous solution contains trigonometric terms, which oscillate between ± 1, multiplied by an exponential term with a negative exponent, hence an exponential decay. If we say (somewhat arbitrarily) that any term is negligible when $e^{-\varepsilon x} \leq 0.006$ or $e^{-\varepsilon(L-x)} \leq 0.006$, then this occurs whenever εx or $\varepsilon(L - x)$ is greater than 5.15. That condition is met whenever

$$\begin{aligned} x &> 4\sqrt{Rh_c(D_x/D_\theta)^{1/2}} \\ L - x &> 4\sqrt{Rh_c(D_x/D_\theta)^{1/2}} \end{aligned} \tag{11.40}$$

This is most important because a distance greater than $4\sqrt{Rh_c(D_x/D_\theta)^{1/2}}$ away from either end of the shell, the entire homogeneous solution of (11.39) essentially goes to zero and can be neglected. In the case of continuous loads, p(x), the particular solution is one often leading only to membrane stresses and displacements. Because of this, the region $0 \leq x \leq 4\sqrt{Rh_c(D_x/D_\theta)^{1/2}}$ and $0 \leq L - x \leq 4\sqrt{Rh_c(D_x/D_\theta)^{1/2}}$ is called the "bending boundary layer," (BBL). It is seen that the length of the BBL is a function of the flexural stiffnesses D_x and D_θ.

Computationally, Equation (11.39) is very useful. For a shell longer than the BBL, that is, $L \geq 4\sqrt{Rh_c(D_x/D_\theta)^{1/2}}$ in the region $0 \leq x \leq 4\sqrt{Rh_c(D_x/D_\theta)^{1/2}}$, the terms involving M_L and Q_L are neglected; in the region $0 \leq (L - x) \leq 4\sqrt{Rh_c(D_x/D_\theta)^{1/2}}$ the terms involving M_0 and Q_0 can be ignored; and if the shell is longer than two BBLs in the region $4\sqrt{Rh_c(D_x/D_\theta)^{1/2}} \leq x \leq L - 4\sqrt{Rh_c(D_x/D_\theta)^{1/2}}$, the terms involving M_0, Q_0, M_L, Q_L can be ignored. Also, when the shell is longer than the BBL and when $d^2\, p(x)/dx^2 = d^3\, p(x)/dx^3 = 0$, then and only then,

$$\begin{aligned} M_0 &= M_x(0),\ M_L = M_x(L) \\ Q_0 &= Q_x(0),\ Q_L = Q_x(L) \end{aligned} \tag{11.41}$$

The implications of the BBL are most important and include:

(1) Bending and shear stresses only exist in the BBL; elsewhere the shell is in a state of membrane stress.

(2) It can be shown that the BBL also exists around such load discontinuities as concentrated load, ring load, and stringer loads. Beyond the BBL, the shell is in a state of membrane stress.

(3) Outside of the BBL in a laminated shell in which $d^2p/dx^2 = 0$, all interlaminar shear stresses are zero. Here, the implications are significant due to the expense and effort spent in NDE investigations over entire shell surfaces.

(4) The length of the BBL is so small that even a "tuna fish can" of length, L, is a long shell, i.e., $L > \text{(BBL)}$.

(5) In a shell structure composed of many segments, whenever a segment is longer than the BBL, then at $x = 0$, one can ignore the M_L and Q_L terms in Equation (11.39), and at $x = L$ one can ignore the M_0 and Q_0 terms in satisfying the boundary conditions at that edge. That is to say, that for many shells, the boundary conditions at each end are uncoupled from the boundary conditions at the other end, with great savings in an effort to obtain boundary value constants and overall solutions.

(6) The BBL for an anisotropic shell is dependent upon the material properties. If the axial flexural stiffness, D_{11}, is greater than the circumferential flexural stiffness D_{22}, then the BBL is longer than for an isotropic material or a shell with a "hoop" wrap, i.e., $\theta = 90°$, such that $D_{22} > D_{11}$.

The above behavior is unique to shells and to beams on an elastic foundation and does not exist in beams, plates, or rings.

Even though the above derivation through Equation (11.31) is for a single-layer orthotropic material, it can easily be extended to that of a laminated composite shell and to a sandwich shell if all couplings, i.e., $B_{ij} = (\ )_{16} = (\ )_{26} = 0$ are zero. In that case, the general solution can be written as in the following section.

11.3 A GENERAL SOLUTION FOR ORTHOTROPIC-SANDWICH CYLINDRICAL SHELLS UNDER AXIALLY SYMMETRIC LOADS

The following expressions are applicable for a wide variety of shells from a single-layer isotropic shell to an anisotropic-sandwich shell simply by using the appropriate stiffness values (See Section 2.9). In the following, the term, D_{11}, is used instead of D_x, and D_2 is used in place of D_θ in order to be consistent with Chapter 2 notation. $\theta = 0°$ means that the fiber direction is axial; $\theta = 90°$ indicates a hoop or circumferential wrap. The solution for the lateral displacement is therefore:

$$
\begin{aligned}
w(x) = {} & \frac{M_0}{2\varepsilon^2 D_{11}} e^{-\varepsilon x}(\sin \varepsilon x - \cos \varepsilon x) - \frac{Q_0}{2\varepsilon^3 D_{11}} e^{-\varepsilon x} \cos \varepsilon x \\
& + \frac{M_L}{2\varepsilon^2 D_{11}} e^{-\varepsilon(L-x)} [\sin \varepsilon(L-x) - \cos \varepsilon(L-x)] \\
& + \frac{Q_L}{2\varepsilon^3 D_{11}} e^{-\varepsilon(L-x)} \cos \varepsilon(L-x) + \frac{1}{4\varepsilon^4 D_{11}} \left[p(x) - \frac{A_{12}}{A_{11}} \frac{N_x}{R} \right]
\end{aligned}
\tag{11.42}
$$

where

$$\varepsilon^4 = \frac{3(1 - \nu_{x\theta}\nu_{\theta x})}{R^2 h^2} \frac{D_{22}}{D_{11}} \tag{11.43}$$

and $\nu_{\theta x}$ is replaced by A_{12}/A_{11} to reflect a laminated or sandwich construction.

$$\begin{aligned} \frac{dw}{dx} &= \frac{M_0}{\varepsilon D_{11}} e^{-\varepsilon x} \cos \varepsilon x + \frac{Q_0}{2\varepsilon^2 D_{11}} e^{-\varepsilon x} (\sin \varepsilon x + \cos \varepsilon x) \\ &+ \frac{M_L}{\varepsilon D_{11}} e^{-\varepsilon(L-x)} \cos \varepsilon(L - x) \\ &+ \frac{Q_L}{2\varepsilon^2 D_{11}} e^{-\varepsilon(L-x)} [\sin \varepsilon(L - x) + \cos \varepsilon(L - x)] \\ &+ \frac{1}{4\varepsilon^4 D_{11}} \frac{dp(x)}{dx} \end{aligned} \tag{11.44}$$

$$\begin{aligned} M_x &= -D_{11} \frac{d^2w}{dx^2} = M_0 e^{-\varepsilon x} (\sin \varepsilon x + \cos \varepsilon x) + \frac{Q_0}{\varepsilon} e^{-\varepsilon x} \sin \varepsilon x \\ &+ M_L e^{-\varepsilon(L-x)} [\sin \varepsilon(L - x) + \cos \varepsilon(L - x)] \\ &- \frac{Q_L}{\varepsilon} e^{-\varepsilon(L-x)} \sin \varepsilon(L - x) - \frac{1}{4\varepsilon^4} \frac{d^2p(x)}{dx^2} \end{aligned} \tag{11.45}$$

$$\begin{aligned} Q_x &= -D_{11} \frac{d^3w}{dx^3} = -2M_0 \varepsilon e^{-\varepsilon x} \sin \varepsilon x + Q_0 e^{-\varepsilon x} (\cos \varepsilon x - \sin \varepsilon x) \\ &+ 2M_L \varepsilon e^{-\varepsilon(L-x)} \sin \varepsilon(L - x) \\ &- Q_L e^{-\varepsilon(L-x)} [-\cos \varepsilon(L - x) + \sin \varepsilon(L - x)] - \frac{1}{4\varepsilon^4} \frac{d^3p(x)}{dx^3} \end{aligned} \tag{11.46}$$

$$N_x = \text{constant} \tag{11.47}$$

For the determination of stresses, if the shell wall under consideration is made of one lamina or a unidirectional laminate, the in-plane stresses are determined easily from

$$\sigma_i = \frac{N_i}{h} + \frac{M_i \zeta}{h^3/12} \qquad (i = x, \theta) \tag{11.48}$$

which is the same form for plate and beam structures. However, for any other laminated cylindrical shell, one must determine the in-plane stresses lamina by lamina and the stress values must be compared to the material strength allowable. In this case of axial symmetry,

$$\begin{Bmatrix} \sigma_x \\ \sigma_\theta \end{Bmatrix}_k = \begin{bmatrix} Q_{11} & Q_{12} \\ Q_{12} & Q_{22} \end{bmatrix}_k \begin{Bmatrix} \varepsilon_x^0 \\ \varepsilon_\theta^0 \end{Bmatrix} + \zeta \begin{bmatrix} Q_{11} & Q_{12} \\ Q_{12} & Q_{22} \end{bmatrix}_k \begin{Bmatrix} \kappa_x \\ 0 \end{Bmatrix} \tag{11.49}$$

where $\varepsilon_x^0 = \frac{du_0}{dx}$, $\varepsilon_\theta^0 = \frac{w}{R}$, $\kappa_x = -\frac{d^2w}{dx^2}$, and Q_{ij} stiffness quantities are given in Chapter 2.

Finally, the above equations can also be used to obtain the solutions for a sandwich shell, where for a symmetric sandwich, the face thickness is t_f and the core depth is h_c. In honeycomb- and foam-core sandwich construction, it is usually assumed all of the in-plane and bending loads are resisted by the faces alone. In that case the stresses are given by

$$\sigma_i = \frac{N_i}{2t_f} \pm \frac{M_i}{t_f h_c} \qquad (i = x,\ \theta) \tag{11.50}$$

In the Governing Equations (11.33) and (11.34) the flexural stiffness in the axial direction for the sandwich, D_x, is

$$D_{11} = \frac{1}{2} E_{fx} t_f h_c^2 \tag{11.51}$$

If the sandwich faces are laminates in which the stresses must be determined in each lamina, then the use of Equation (11.49) can be made wherein one assigns a k number to each face lamina as well as the core itself in a straightforward manner. In the case of axially symmetric loading, $\partial(\ )/\partial\theta = \kappa_\theta = 0$, the face stresses in the midplane symmetric-sandwich shell are given by

$$\begin{Bmatrix} \sigma_x \\ \sigma_\theta \end{Bmatrix} = \frac{1}{2t_f} \begin{bmatrix} A_{11} & A_{12} \\ A_{12} & A_{22} \end{bmatrix} \begin{Bmatrix} \varepsilon_x^0 \\ \varepsilon_\theta^0 \end{Bmatrix} \pm \frac{1}{t_f h_c} \begin{bmatrix} D_{11} & D_{12} \\ D_{12} & D_{22} \end{bmatrix} \begin{Bmatrix} \kappa_x \\ 0 \end{Bmatrix} \tag{11.52}$$

where $\varepsilon_x^0 = \frac{du_0}{dx}$, $\varepsilon_\theta^0 = \frac{w}{R}$, and $\kappa_x = -\frac{d^2w}{dx^2}$, and the stiffness quantities are given in Section 2.9. For a midplane symmetric-sandwich shell the midplane in-plane displacement, u_0, is found to be, analogous to Equation (11.38),

$$u_0(x) = \frac{N_x x}{A_{11}} - \frac{A_{12}}{A_{11}R}\int w dx + F \qquad (11.53)$$

where F is a constant of integration determined by where u_0 is specified.

11.4 SHELLS WITH MIDPLANE ASYMMETRY

With composite material laminates, the stacking sequences can be selected so that the stiffness matrix of Equation (2.64) will be asymmetric with respect to the shell wall midsurface. In that case the $B_{ij} \neq 0$. Moreover with resin transfer molding (RTM), SRIM, and other new manufacturing techniques it is easy to introduce preforms that are asymmetric if there is an advantage in doing so. In such cases, if the $[A]$, $[B]$, and $[D]$ matrix quantities are constant, the governing equations for the lateral deflection, $w(x)$, for axial symmetry in loading and geometry is found to be [18]:

$$\left(\frac{A_{11}D_{11} - B_{11}^2}{A_{11}}\right)\frac{d^4w}{dx^4} + \frac{2}{R}\left(\frac{A_{12}B_{11}}{A_{11}} - B_{12}\right)\frac{d^2w}{dx^2}$$

$$+ \frac{1}{R^2}\left(\frac{A_{11}A_{22} - A_{12}^2}{A_{11}}\right)w = p(x) - \frac{A_{12}}{A_{11}}\frac{N_x}{R} \qquad (11.54)$$

It is seen that the coefficient of the first term is the well known reduced flexural stiffness, D_e, often used in analyzing asymmetric beams. To simplify Equation (11.54), define

$$D_e = \frac{A_{11}D_{11} - B_{11}^2}{A_{11}}, \; \overline{p}(x) = p(x) - \frac{A_{12}}{A_{11}}\frac{N_x}{R} \qquad (11.55, 11.56)$$

$$4\varepsilon^4 = \frac{1}{D_e R^2}\left(\frac{A_{11}A_{22} - A_{12}^2}{A_{11}}\right) \qquad (11.57)$$

Using these, Equation (11.54) becomes much more simple, namely

$$\frac{d^4w}{dx^4} + \frac{2}{D_e R}\left(\frac{A_{12}B_{11}}{A_{11}} - B_{12}\right)\frac{d^2w}{dx^2} + 4\varepsilon^4 w = \frac{1}{D_e}\overline{p}(x) \qquad (11.58)$$

The second term appears only because of the midplane asymmetry, causing a nonzero $[B]$ matrix term B_{11} and B_{12}, where the shell symmetric, Equation (11.58), would have the same form as Equation (11.35).

Consider the coefficient of the second term to be small, i.e., less than unity. Vinson [18] has defined a midplane asymmetry parameter (MAP) as defined by

$$\phi = \frac{B_{11}}{\sqrt{D_{11}A_{11}}} \tag{11.59}$$

In all midplane asymmetric constructions, whether a sandwich or a reinforced laminate, this parameter has the properties of $-1 < \phi < +1$. Equation (11.58) is now written as

$$\frac{d^4w}{dx^4} + 2\left(\frac{A_{12}}{A_{11}} - \frac{B_{12}}{B_{11}}\right)\frac{\sqrt{D_{11}A_{11}}}{D_eR}\,\phi\,\frac{d^2w}{dx^2} + 4\varepsilon^4 w = \frac{1}{D_e}\,\overline{p}(x) \tag{11.60}$$

Consider the following form of solution, following [18],

$$w(x) = \sum_{n=0}^{\infty} w_n(x)\phi^n \tag{11.61}$$

Because of the property that 101 < 1, Equation (11.61) is a perturbation solution analogous to that used in Section 6.4. On substituting this into Equation (11.60) and collecting like powers of the perturbation parameter ϕ, it is seen that

$$\frac{d^4w_0}{dx^4} + 4\varepsilon^4 w_0 = \frac{1}{D_e}\,\overline{p}(x) \tag{11.62}$$

$$\frac{d^4w_n}{dx^4} + 4\varepsilon^4 w_n = -2\left(\frac{A_{12}}{A_{11}} - \frac{B_{12}}{B_{11}}\right)\frac{\sqrt{D_{11}A_{11}}}{D_eR}\,\frac{d^2w_{n-1}}{dx^2} \qquad n \geq 1 \tag{11.63}$$

Thus, a perturbation solution of the asymmetric shell can be made in terms of known solutions of the simpler midplane symmetric shell discussed in Section 11.3. It is also seen that a bending boundary layer (BBL) still exists, analogous to that for the midplane symmetric shell, given by Equation (11.41). It can be shown that for $\alpha < 1$, Equation (11.61) is another form of the exact solution for Equations (11.60) and (11.58).

11.5 OTHER CONSIDERATIONS

In sandwich shells, because of their initial curvature, bending moments (stress couples) give rise to transverse normal stresses in the core. Due to limited resistance (low stiffness) of the sandwich core to radial displacements, local thickness changes can occur; that could alter the structural stiffnesses locally. This phenomenon is similar to the Brazier effect [19]. Tensile normal stresses in the core may lead to tensile cracking in the core and/or debonding. Transverse compressive stresses could cause buckling of the honeycomb-core cell walls. This is discussed at some length by Starlinger and Rammerstorfer [20].

It must be remembered, however, that away from any load discontinuity, structural discontinuity, or material discontinuity, the stresses in a shell are membrane only, and there are no bending moments. Therefore, the Brazier effect discussed above could only happen in the bending boundary layer. There, because the bending stresses can be significant (or dominant) compared to the membrane stresses, the effect could be important.

11.6 PROBLEMS

11.1 Consider a circular cylindrical sandwich shell with faces composed of faces of four laminae, each 0.005 in. thick, of T300/5208 unidirectional graphite epoxy whose properties are given in Problem 4.1. The core is made of the same foam as in Problem 4.1, and the core depth $h_c = 0.8$ in. The shell radius $R = 14$ in.

   a. If the composite has all fibers in the axial direction, ($\theta = 0°$), what is the length of the BBL?
   b. If the composite face fibers are in the circumferential or hoop direction, ($\theta = 90°$), what is the length of the BBL?

11.2 Consider the sandwich shell of Problem 11.1 wherein the shell is clamped at each end, and the shell is subjected to an internal pressure of 100 psi, but no axial load ($N_x = 0$), such as in a very long pipeline. If the fibers are in the axial direction:

   a. What are the axial and circumferential face stresses in each face at each end (i.e., $x = 0$ or L)?
   b. What are the axial and circumferential face stresses in each face at the shell midlength (i.e., $x = L/2$) when the shell is considered to be long.

11.3 In the shell of Problem 11.1, with the fibers in the axial direction what are the membrane face stresses, σ_x, if the shell is subjected to a tensile axial load of 10^4/lb?

11.4 For the sandwich shell problem of Problem 11.1, if the shell is simply supported at one end, then the maximum face stress, σ_x, occurs at $x = \pi/4\varepsilon$. What are the maximum face stresses, σ_x? Compare the maximum face stress values, $\sigma_{x\max}$, for a simply-supported edge and a clamped edge.

11.5 Consider a portion of a deep-submergence instrumentation capsule to be a sandwich cylindrical shell composed of steel faces with a face thickness, $t_f = 0.1$ in., and a core depth, $h_c = 2$ in., where the core is the same used in Problem 11.1. At a depth of 1000 ft., the external uniform pressure is 450 psi. Assume an internal pressure of zero and that the shell is clamped at both ends. The steel properties are $E = 30 \times 10^6$ psi, $\nu = 0.3$, and the radius, $R = 50$ in., and the shell length, $L = 100$ in.

a. At each end, determine σ_x and σ_θ in both faces of the sandwich shell.
b. At the shell midlength, determine the σ_x and σ_θ in each face of the sandwich shell.

11.6 Consider a long composite sandwich cylindrical shell of radius, R, length, L, parameter, ε, and flexural stiffnesses, D_{ij}. The sandwich shell is midplane symmetric, specially orthotropic, subjected to a constant internal pressure, p_i, and no axial load, N_x. If the shell is simply supported at $x = 0$ and clamped at $x = L$, determine the constants M_0, Q_0, M_L, and Q_L, as discussed in Section 11.3.

11.7 Consider a sandwich circular-cylindrical shell subjected only to a constant lateral distributed load, $p(x) = p_0$. The axial load per unit circumference, N_x, is zero. The shell and the loading are axially symmetric. The shell is clamped at $x = 0$ and free at the end $x = L$. The shell is long, i.e., $L > 4[Rh(D_x/D_\theta)^{1/2}]^{1/2}$. Determine explicitly the constants M_0, M_L, Q_0, and Q_L given in Section 11.3.

11.7 REFERENCES

1. Love, A. E. H., "The Small Free Vibrations and Deformation of a Thin Elastic Shell," *Philosophical Transactions of the Royal Society of London*, Vol. 17, pp. 491–546, 1888.
2. Epstein, Marcello, verbal presentation at the Pan American Conference on Applied Mechanics—V, San Juan, Puerto Rico, January 2–4, 1997.
3. Reissner, E., "A New Derivation of the Equations for the Deformation of Elastic Shells," *American Journal of Mathematica*, 63, pp. 177–184, 1941.
4. Reissner, E., "On the Theory of Thin Elastic Shells," H. Reissner Anniversary Volume, J. W. Edwards, Ann Arbor, 1949, pp. 231–247.
5. Flügge, W., *Statik und Dynamik der Schalen*, Edwards Brothers, Inc., Ann Arbor, 1943.
6. Flügge, W., *Stresses in Shells*, Springer-Verlag, Berlin, 1962 (2nd Edition).

7. Goldenveizer, A.L., *Theory of Thin Shells*, Pergamon Press, New York, 1961.

8. Vlasov, V.S., "Basic Differential Equations in General Theory of Elastic Shells," NASA.TM 1241, 1951.

9. Naghdi, P.M., "A Survey of Recent Progress on the Theory of Elastic Shells," *Applied Mechanics Review*, Vol. 9, No. 9, 365–368, September 1956.

10. Hoff, N.J., "Buckling of a Thin Cylindrical Shell Under Hoop Stresses Varying in the Axial Direction," *Journal of Applied Mechanics*, Vol. 24, pp. 405–412, 1957.

11. Novozhilov, V.V., *Theory of Thin Shells*, E.P. Noordhoff, Ltd., Groningen, The Netherlands, 1959.

12. Kraus, H., *Thin Elastic Shells*, John Wiley and Sons, Inc., New York, 1967.

13. Leissa, A. W., *Vibrations in Shells*, NASA-SP-288, U.S. Government Printing Office, 1975.

14. Dym, C.L., *Introduction to the Theory of Shells*, Pergamon Press, Inc., Glasgow, 1974.

15. Libai, A. and J. G. Simmonds, *The Nonlinear Theory of Elastic Shells*, Academic Press, New York, 1988.

16. Palazotto, A. N. and S. T. Dennis, *Analysis of Nonlinear Shell Structures*, AIAA, New York, 1991.

17. Bert, C.W., "Structural Theory for Laminated Anisotropic Elastic Shells," *Journal of Composite Materials*, Vol. 1, pp. 414–423, 1, October, 1967.

18. Vinson, J.R., and A. T. Bee, "Use of Asymmetric Construction to Minimize Bending Stresses," Proceedings of the Fourth International Conference on Sandwich Constructions, Stockholm, June 1998, EMAS Publications, United Kingdom, 1998.

19. Brazier, L.G., "On the Flexure of Thin Cylindrical Shells and Other 'Thin' Sections," *Proceedings of the Royal Society, Series A, London*, Vol. 116, pp. 104–114, 1927.

20. Starlinger, A. and F.G. Rammerstorfer, "A Finite Element Formulation for Sandwich Shells Accounting for Local Failure Phenomena," in Sandwich Constructions 2—Proceedings of the Second International Conference on Sandwich Construction, Gainesville, Florida, Editors, D. Weissman-Berman and K-A. Olsson, EMAS Publications, United Kingdom, 1992, pp. 161–185.

21. Timoshenko, S. and A. Woinowski-Krieger, *Theory of Plates and Shells*, McGraw-Hill Book Co., Inc., New York, 2nd Edition, 1959.

22. Vinson, J. R., *The Behavior of Shells Composed of Isotropic and Composite Materials*, Kluwer Academic Publisher, Dordrecht, The Netherlands, 1993.

CHAPTER 12

Buckling of Sandwich Cylindrical Shells

12.1 BUCKLING OF A SOLID- OR FOAM-CORE SANDWICH CYLINDRICAL SHELL WITH ISOTROPIC FACES SUBJECTED TO AN AXIALLY-SYMMETRIC COMPRESSIVE END LOAD

The compressive stresses introduced into the faces of the sandwich subjected to a uniform axial compressive load, P, where $P = 2\pi RN_x$ are:

$$\sigma_{fx} = \frac{N_x}{2t_f} = \frac{P}{4\pi Rt_f} \tag{12.1}$$

for the case of the faces being of equal thickness and the same material, see Figure 11.1. Obviously, σ_{fx} must remain below some allowable stress of the material to avoid a failure due to overstressing. However, the sandwich shell can also undergo overall buckling. For a sandwich shell with isotropic facings, one widely used source for the buckling analysis of sandwich shells subjected to axial compression is a report by NASA [1]. In Section 4.4 of that report, the buckling of isotropic-sandwich cylinders is treated. One reason that this reference is used in sandwich-shell analyses and design is because it employs an empirical factor, γ, necessary to insure structural integrity. The buckling of shells in axial compression is very sensitive to initial imperfections such that shells buckle at a fraction of the load predicted by various theoretical analyses. To ignore this can produce unexpected, catastrophic failures. Since NASA meticulously studied this problem and incorporated all of the test data to that time to formulate the proper ϕ, it makes sense to utilize the NASA equations and the γ keyed to them.

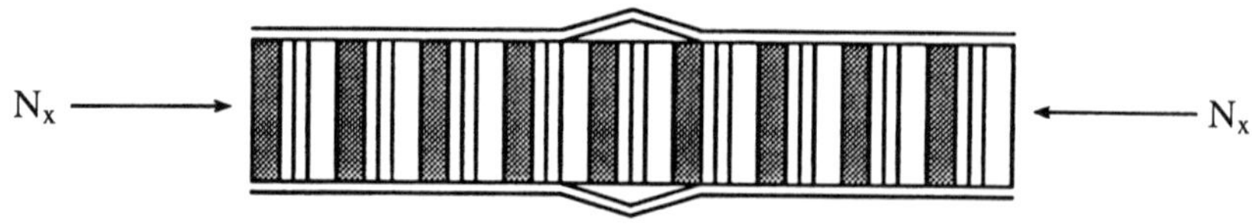

FIGURE 12.1. Face wrinkling. (Reprinted from Reference [17].)

From Reference [1], the critical value of the total compressive load per unit of circumference, N_{cr}, is

$$N_{x_{cr}} = \frac{2C_0 \, \gamma \, E_f}{(1 - \nu_f^2)^{1/2}} \frac{h_c t_f}{R} \tag{12.2}$$

where C_o is a factor determined by transverse shear deformation effects, as discussed in the NASA Report [1] and by Vinson [2]. In this equation, γ, the empirical factor related to initial imperfections, is given by

$$\gamma = 1 - 0.901 \, (1 - e^{-\phi}) \tag{12.3}$$

where

$$\phi = \frac{(2)^{1/2}}{29.8} \left(\frac{R}{h_c} \right)^{1/2} \tag{12.4}$$

In addition to overall buckling, a sandwich construction subjected to in-plane compressive loads can undergo several localized buckling instabilities. For the case of foam-core and honeycomb core construction, these consist of face wrinkling and core shear instability. Since face wrinkling and core shear instability (see Figures 12.1 and 12.2) are local failures, it is reasonable that these expressions, developed for flat panels, are equally usable for circular cylindrical shells in which the sandwich shell wall thickness to radius ratio is small compared to unity (Love's First Approximation, $h/R \ll 1$).

Core shear instability is given by Equation (12.5); this is the same equation used by Bert [3] for a cylindrical shell, and in numerous papers by Vinson

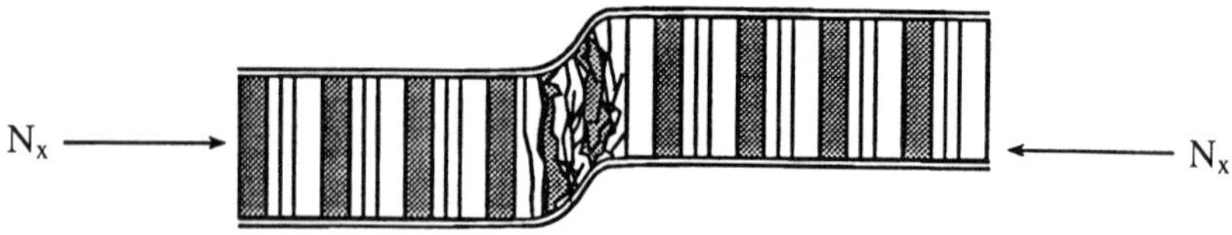

FIGURE 12.2. Core shear instability. (Reprinted from Reference [17].)

[11] for sandwich plates. Since it is a local instability, Equation (12.5) equally applies for beams, plates, and shells. Recently, Starlinger and Rammerstorfer [12] also stated that the influence of curvature on the local instability limit can be neglected if the shell radii of curvature are essentially larger than the wavelengths of the local instability.

$$(\sigma_{fx})_{cr} = \frac{G_{xz_c} h_c}{2t_f} \tag{12.5}$$

The most accurate equation for face wrinkling is developed by Heath [13] and given in Equation (12.6) where E_c is the modulus of elasticity of the foam or solid core in the radial or thickness direction.

$$(\sigma_{fx})_{cr} = \left[\frac{2}{3}\frac{t_f}{h_c}\frac{E_c E_f}{(1 - \nu^2)}\right]^{1/2} \tag{12.6}$$

Now, from Chapter 11 and Equation (12.1) one can analyze the stresses and deformations throughout a sandwich shell. By using Equations (12.2) through (12.4) for overall buckling and (12.5) and (12.6) for core shear instability and face wrinkling, one can analyze the buckling stresses to insure that buckling will not occur when a compressive load, P, is applied to the shell.

12.2 BUCKLING OF A SOLID- OR FOAM-CORE SANDWICH CYLINDRICAL SHELL WITH ORTHOTROPIC COMPOSITE FACES SUBJECTED TO AN AXIALLY-SYMMETRIC COMPRESSIVE LOAD

The stress in the faces in the axial direction, σ_{fx}, due to an axial load, P, is given by Equation (12.1). However, along with the σ_{fx}, there is also a circumferential face stress, $\sigma_{f\theta}$ in the shell, which is a Poisson's ratio effect, given in Reference [11] as,

$$\sigma_{f\theta} = -\nu_{\theta x}\,\sigma_{fx} \tag{12.7}$$

where $\nu_{\theta x}$ is a Poisson's ratio of the specially-orthotropic face material and which follows the orthotropic relationship

$$\frac{\nu_{x\theta}}{E_x} = \frac{\nu_{\theta x}}{E_\theta} \tag{12.8}$$

where x and θ are the axial and circumferential directions respectively.

In some unidirectional composites with the fibers in the axial direction, the shell structure can withstand a very high axial loading P, because the compressive failure stress in the axial direction is so large. Equation (12.7) is introduced here because the load carrying capability of the shell can be limited by the low circumferential tensile strength of the composite in the matrix-dominated direction, since the circumferential tensile stresses depicted by Equation (12.7) cannot be avoided.

Equations for the overall buckling of a specially-orthotropic sandwich shell subjected to axial compression are given in Reference [1] and by Bert et al. [3]. These equations are complicated, and extensive computation is needed to determine the axial and circumferential wave numbers m and n that result in a minimum critical load. However, Reese and Bert [4] provided a simple but accurate equation for overall buckling that is straightforward and amenable to easy computation while retaining accuracy.

$$(\sigma_{fx})_{cr} = \frac{\phi h_c}{R}\left(\frac{E_x E_\theta}{1 - \nu_{x\theta}\nu_{\theta x}}\right)^{1/2}\left\{1 - 1/2\left(\frac{E_x E_\theta}{1 - \nu_{x\theta}\nu_{\theta x}}\right)^{1/2}\left[\frac{t_f}{RG_{xz_c}}\right]\right\} \tag{12.9}$$

where

$$\phi = 1 \text{ or } \left\{\frac{2G_{x\theta}[1 + (\nu_{x\theta}\nu_{\theta x})]^{1/2}}{(E_x E_\theta)^{1/2}}\right\}^{1/2} \tag{12.10}$$

whichever is less.

In Equation (12.9), G_{xz_c} is the core-material shear modulus in the xz plane, where z is the radial coordinate. Equation (12.9) differs from the expression in Reference [4] only in that in Equation (12.9), it is assumed that $t_f \ll h_c$, which is the usual case in sandwich structures.

From the above, it is seen that for design and analysis, one can proceed to determine the applied face stresses in the axial direction and the critical face stresses for overall buckling, core shear instability, and face wrinkling. In a shell of given radius, R, length, L, core depth, h_c, face thickness, t_f, and specified core and face materials, one can calculate if the faces are overstressed and if any of the three buckling modes occur. For the design of a shell of given radius, R, length, L, and face and core materials, the designer can select h_c and t_f to insure that no failure occurs for a given axial compressive load, P.

12.3 BUCKLING OF A HONEYCOMB CORE SANDWICH CYLINDRICAL SHELL WITH COMPOSITE FACES SUBJECTED TO AN AXIALLY SYMMETRIC COMPRESSIVE END LOAD [5]

Analogous to the methods of Chapter 9, in this section, a prime is used to denote effective core properties for a honeycomb core, replacing all nonprimed quantities pertaining to the foam or solid core previously discussed in Section 12.2. These "effective" properties are functions of the core material properties (E_c and G_c), the core wall thickness t_c, and the cell size d. To truly optimize the structure for minimum weight, it is advantageous to relate these effective properties back to fundamental geometry and material properties. For hexagonal-cell honeycomb core having some double walls as shown in Figure 9.4, the properties are developed in Chapter 9 as Equations (9.43) through (9.47).

For the construction of Figure 9.4 as well as the square cell honeycomb, the values of k_2, k_3, and k_4 are given in Table 9.2, according to Kaechele [14] and Reference [15]. For the shell studied herein, x is taken as the axial direction, and y or θ is the circumferential direction. For other honeycomb configurations, the values of these constants can be easily derived using the methods of Kaechele [14].

Because the various governing equations used previously in Section 12.2 are herein modified only slightly to account for the effective core properties, they are simply presented below without comment.

For overall buckling,

$$(\sigma_{fx})_{cr} = \frac{\phi h_c}{R}\left(\frac{E_x E_\theta}{1 - \nu_{x\theta}\nu_{\theta x}}\right)^{1/2}\left\{1 - 1/2\left[\phi\left(\frac{E_x E_\theta}{1 - \nu_{x\theta}\nu_{\theta x}}\right)^{1/2}\frac{t_f}{RG'_{xz_c}}\right]\right\} \tag{12.11}$$

where ϕ is given by Equation (12.10).

For core shear instability, as before,

$$(\sigma_{fx})_{cr} = \frac{G'_{xz_c}\, h_c}{2t_f} \tag{12.12}$$

For face wrinkling, again,

$$(\sigma_{fx})_{cr} = \left[\frac{2}{3}\frac{t_f}{h_c}\frac{E'_c\,(E_x E_\theta)^{1/2}}{(1 - \nu_{x\theta}\nu_{\theta x})}\right]^{1/2} \tag{12.13}$$

where E'_c is the effective modulus of elasticity of the honeycomb core in the radial or thickness direction.

A fourth buckling mode can occur in a honeycomb sandwich construction—face dimpling or monocell buckling. The equation governing this failure is repeated here understood by [6,16]:

$$(\sigma_{fx})_{cr} = \frac{2\,(E_x E_\theta)^{1/2}}{(1 - \nu_{x\theta}\nu_{\theta x})}\left(\frac{t_f}{d}\right)^2 \tag{12.14}$$

In Equations (12.11) through (12.14) all material properties without the subscript c refer to the orthotropic face material properties. Overall buckling, core shear instability, and face wrinkling are synonymous with failure of the shell. Face dimpling can be tolerated in some cases.

From the above it is seen that for design and analysis, one can proceed to determine the applied face stresses in the axial direction, the resulting face stress in the circumferential direction, and the critical face stresses for overall buckling, core shear instability, face wrinkling and face dimpling. In a shell of given radius R, length L, core depth h_c, face thickness t_f, core wall thickness t_c, core cell size d, and specified core and face materials, one can determine if the faces are overstressed or if any of the four buckling modes occurs. For the design of a shell of given radius R, length L, and face and core materials, the designer can select h_c, t_f, t_c and d to insure that the shell does not fail for a given compressive load, P.

In any case a weight equation is desirable and is given by

$$(W - W_{ad}) = 2\, t_f \rho_f + \rho_c' h_c \tag{12.15}$$

12.4 OVERALL BUCKLING OF SANDWICH CYLINDRICAL SHELLS SUBJECTED TO AN OVERALL BENDING MOMENT [7]

Analogous to the approach of Section 12.1 for isotropic sandwich cylindrical shells, and Section 12.2 for composite sandwich cylindrical shells, Reference [1] recommends the use of Equations (12.2), with the correlation factor, γ, given by Equation (12.16) below, where ϕ is given by Equation (12.4).

$$\gamma = 1 - 0.731\,(1 - e^{-\phi}) \tag{12.16}$$

In the case of overall beam-type bending, the maximum compressive applied N_x due to a bending moment, M, is given by

$$N_x = \frac{M}{\pi R^2} \tag{12.17}$$

For a sandwich cylinder subjected to an overall bending moment M, the maximum compressive face stress will be as follows, unless the faces are laminates other than unidirectional composites. If that is the case the face stresses must be calculated lamina by lamina.

$$\sigma_f = \frac{MR}{\pi R^3\,(2t_f)} = \frac{M}{2\pi R^2 t_f} \tag{12.18}$$

Therefore, if this applied face stress exceeds the critical buckling stress for core shear instability, face wrinkling, or face dimpling, using the appropriate equations presented earlier, local buckling of the sandwich shell will occur.

12.5 BUCKLING OF A SANDWICH CYLINDRICAL SHELL DUE TO EXTERNAL PRESSURE

Limited information is available and is discussed on pp. 32–33 of Reference [1].

12.6 BUCKLING OF A SANDWICH CYLINDRICAL SHELL DUE TO TORSION

Limited information is available and is discussed on pp. 33–34 of Reference [1].

12.7 DYNAMIC BUCKLING

When dynamic loads cause buckling, the analysis becomes far more complicated because of inertia effects. It is beyond the scope of this text to treat the subject. However, Simitses [8–10] has written numerous papers and books on the subject, which are recommended as the starting point for further study.

12.8 PROBLEMS

12.1 Consider a cylindrical sandwich missile interstage structure wherein $L \geq 1.72\ Rh$, yet $L/R < 5$, composed of magnesium faces ($E = 6.5 \times 10^6$ psi, $\nu = 0.3$, $\sigma_{\text{all}} = \pm 40{,}000$ psi) and the foam core of Problem 4.1. If the shell is of $R = 30$ in., $t_f = 0.05$ in., $h_c = 1$ in.,

a. What is the compressive axial face stress?
b. What is the shell buckling stress?
c. Will the shell buckle or be overstressed first if the load increases?

12.2 A long cylindrical sandwich section of a missile structure of radius 40 in., face thickness = 0.2 in., and core depth 6 in. is subjected to both bending and axial compression during launch. If the maximum-intended bending moment is 30×10^6 in-lb, what axial load can be tolerated without buckling the aluminum-faced foam-core sandwich shell? For the aluminum $E = 10 \times 10^6$ psi, $\nu = 0.3$, $\sigma_{all} = \pm 30 \times 10^3$ psi; for the foam use properties for Problem 4.1.

12.3 A cylindrical-sandwich shell with steel faces and a foam core has the face properties of $E = 30 \times 10^6$ psi, $\nu = 0.3$, and the foam is given in Problem 4.1. This shell is of radius 25 in., is 125 in. long, and has a face thickness of 0.1 in., and a core depth of 2 in.

a. What is the total axial load, in pounds, that the shell can withstand before buckling?
b. What is the face stress at that buckling load?
c. What is the length of a BBL in this sandwich shell?
d. What is the maximum beam-type bending moment this shell can withstand before buckling?
e. With that critical bending moment, what is the maximum face stress in the shell?

12.4 Consider a circular cylindrical sandwich shell of a radius 30″, length 80″, clamped at each end, subjected to an axially-symmetric compressive load P (lb).

The shell has foam-core sandwich construction; the faces are made of 7075-T6 aluminum each 0.125″ thick; the core is Rigid Klegecell foam with a density of 15 lb/ft^3 and 2.5″ thick.

a. Determine the load carrying capacity P of this shell.
b. In what mode will the shell fail?
c. What is the overall shell weight?

12.9 REFERENCES

1. Anon., "Buckling of Thin Walled Circular Cylinders," NASA SP-8007 Revised, August 1968.
2. Vinson, J. R., "Analysis and Optimization of Composite and Metallic Sandwich Cylindrical Shells," in Sandwich Constructions 2—Proceedings of the Second International Conference on Sandwich Construction, Gainesville, Florida, Editors, D. Weissman-Berman and K-A. Olsson, EMAS Publications, United Kingdom, 1992, pp. 377–400.
3. Bert, C. W., W. C. Crisman, and G. M. Nordby, "Buckling of Cylindrical and Conical Sandwich Shells with Orthotropic Facings," *AIAA Journal*, Vol. 7, No. 2, pp. 250–257, February, 1969.
4. Reese, C. D. and C. W. Bert, "Simplified Design Equations for Buckling of Axially Compressed Sandwich Cylinders with Orthotropic Facings and Core," *AIAA Journal of Aircraft*, pp. 515–519, November–December 1969.
5. Vinson, J. R., "Analyses and Optimization of Composite Honeycomb Sandwich Cylindrical Shells," *Proceedings of the American Society for Composites Seventh Technical Conference*, Penn State, Technomic Publishing Co., Inc., Lancaster, PA, 1992, pp. 576–586.
6. Norris, C. B., "Short-Column Compressive Strength of Sandwich Constructions as Affected

by Size of Cells of Honeycomb Core Materials," Forest Products Laboratory Report 026, January 1964.

7. Vinson, J. R., "On the Optimization of Composite Cylindrical Sandwich Shells Subjected to Beam Type Bending," in Proceedings of the ASME Aerospace Division, Structures and Materials Committee, ASME-AD-Vol. 55, New York, 1997, pp. 387–399.
8. Huyan, Xiaozhi, and G. J. Simitses, "Dynamic Buckling of Imperfect Cylindrical Shells Under Axial Compression and Bending Moment," *AIAA Journal*, Vol. 35, No. 8, pp. 1404–1412, August 1997.
9. Simitses, G. J., "Instability of Dynamically-Loaded Structures," *Applied Mechanics Reviews*, Vol. 40, No. 10, pp. 1403–1408, 1987.
10. Simitses, G. J., "Dynamic Snap-Through Buckling of Low Arches and Shallow Spherical Caps," Ph.D. Dissertation Aeronautics and Astronautics, Stanford University, June 1965.
11. Vinson, J. R. and R. L. Sierakowski, *The Behavior of Structures Composed of Composite Materials*, Martinus-Nijhoff Publishers (now Kluwer Academic Publishers) Dordrecht, The Netherlands, 1986.
12. Starlinger, A. and F. G. Rammerstorfer, "A Finite Element Formulation for Sandwich Shells Accounting for Local Failure Phenomena," in Sandwich Constructions 2—Proceedings of the Second International Conference on Sandwich Construction, Gainesville, Florida, Editors, D. Weissman-Berman and K-A. Olsson, EMAS Publications, United Kingdom, 1992, pp. 161–185.
13. Heath, W. G., "Sandwich Construction, Part 2: The Optimum Design of Flat Sandwich Panels," *Aircraft Engineering*, Vol. 32, August 1960, pp. 230–235.
14. Kaechele, L. E., "Minimum Weight Design of Sandwich Panels," USAF. Project Rand Research Memorandum, RM 1895, AD-133011, March 1957.
15. Anon., "Materials, Properties and Design Criteria Part II, Sandwich Construction for Aircraft," *Military Handbook 23* (AMC-23), Department of the Air Force Research and Development Command; Department of the Navy, Bureau of Aeronautics; and the Department of Commerce, Civil Aeronautics Administration, 2nd Edition, 1955.
16. Vinson, J. R. and S. Shore, "Methods of Structural Optimization for Flat Sandwich Panels," Naval Air Engineering Center Report, NAEC-ASC-1083, April 15, 1965.
17. Vinson, J. R., "Optimum Design of Composite Honeycomb Sandwich Panels Subjected to Uniaxial Compression," *AIAA Journal*, Vol. 24, No. 10, October 1986, pp. 1690–1696.

CHAPTER 13

Minimum-Weight Optimization of Sandwich Cylindrical Shells

13.1 GENERAL DISCUSSION

In examining various foam-core materials used in sandwich construction, we see that the core-shear modulus, G_c, varies almost linearly with the foam density, ρ_c. Since there is a continuous spectrum of shear moduli commercially available within a sizable range of densities, G_c can be considered as a continuous variable and not limited to discrete values for a given material. For example, consider closed-cell PVC foams [1,2]. One can plot the properties as in Figure 13.1. For these materials, the relationship between G_c and ρ_c is almost linear. Regardless of this relationship, however, is the important conclusion that one can select any value of G_c as a variable in a foam-core sandwich construction just as one may select values of t_f and h_c. This has been utilized by Vinson [3–5] in the optimization of foam-core sandwich panels subjected to in-plane loads.

13.2 MINIMUM-WEIGHT OPTIMIZATION OF A SOLID FOAM-CORE SANDWICH CYLINDRICAL SHELL WITH ISOTROPIC FACINGS SUBJECTED TO AN AXIALLY COMPRESSIVE LOAD

In the following it is assumed that the two facings are identical. For the shell of length, L, radius, R, and an applied axially symmetric compressive load, P, the face stresses are given by Equations (12.1), while the overall buckling load per unit circumference is given by Equations (12.2) through (12.4). In addition, core shear instability and face wrinkling are characterized by Equations (12.5) and (12.6).

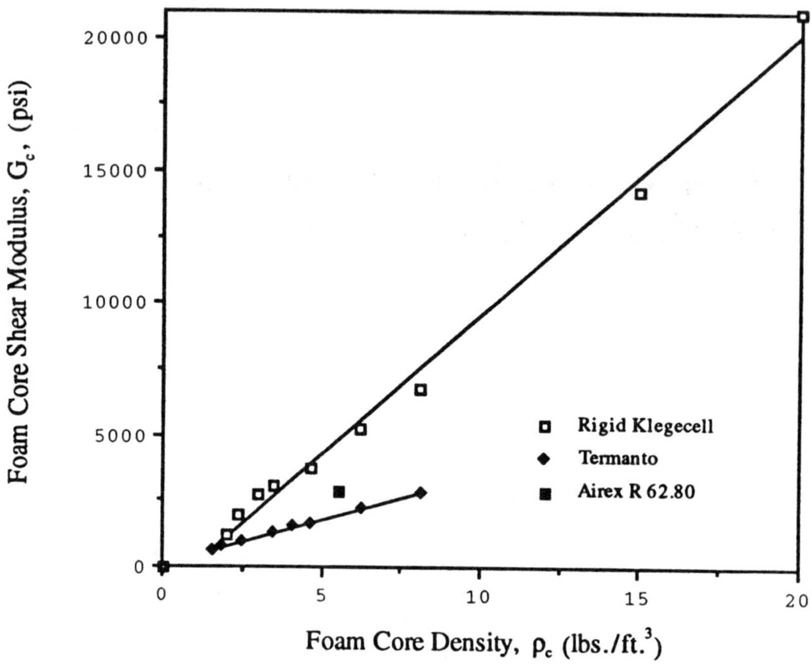

FIGURE 13.1. Typical foam core properties. (Reprinted from Reference [8].)

The equation to determine the sandwich shell weight per unit midsurface area, W, is straightforwardly given by

$$W = 2\rho_f t_f + \rho_c h_c + W_{ad}$$

where ρ_f and ρ_c are the weight densities of the face material and the foam or solid core respectively, and W_{ad} is the weight per unit area of the bonding or adhesive material between the core and face materials. W_{ad} is nonanalytic and is a function of the material used, the core, experience of the worker, and various other factors. Therefore, in what follows, the equation above shall be written as

$$(W - W_{ad}) = 2t_f\rho_f + \rho_c h_c \tag{13.1}$$

With these equations, any shell may be designed and/or analyzed, keeping in mind that the face stress must be limited to an allowable stress determined by the face material employed, and that the face stress cannot exceed any of the critical stresses given by Equations (12.2) through (12.6). Trial and error may be used to decide on materials and to select dimensions to search for minimum weight combinations. However, as shown in previous chapters, a much better procedure exists [6].

There are three critical stresses. These are overall instability, face wrinkling, and core shear instability. For a given set of materials, there are three variables, namely face thickness, t_f, core depth, h_c and the core shear modulus, G_c. Thus, there are three equations and three unknowns. This suggests a unique solution may be available for these dependent variables.

Since any one of the instabilities causes structural failure, it follows that the most efficient structure occurs when all three buckling stresses are equal. This follows the philosophy of the weakest link in the chain. Otherwise, material is being used inefficiently and can be shifted or eliminated to reduce weight and/or increase load-carrying ability. In carrying out the optimization, one can employ the following elasticity relationship for the isotropic foam core:

$$G_c = \frac{E_c}{2(1 + \nu_c)} \tag{13.2}$$

Because all three buckling stresses are equated in the optimized construction, these critical face stresses shall be combined and are simply labeled σ_o. It must be remembered that the optimized face stress σ_o must not exceed some allowable stress based on static strength, factors of safety, fatigue limits, etc.

As a result of considerable manipulation, the optimum values of the variables can be written in terms of the optimum face stress as follows:

$$\frac{h_c}{R} = \frac{(1 - \nu_f^2)^{1/2}\,\sigma_o}{\gamma E_f} \tag{13.3}$$

$$\frac{t_f}{R} = \frac{3^{1/2}\,(1 - \nu_f^2)\sigma_o^{3/2}}{2^{3/2}\,\gamma E_f^{3/2}\,(1 + \nu_c)^{1/2}} \tag{13.4}$$

$$G_c = \frac{3^{1/2}\,(1 - \nu_f^2)^{1/2}\,\sigma_o^{3/2}}{2^{1/2}\,(1 + \nu_c)^{1/2}\,E_f^{1/2}} \tag{13.5}$$

Equally important, a "universal relation" from the optimization is found equating the optimum face stress and the applied load index (N_x/R) for the face materials used in the optimum construction:

$$\frac{N_x}{R} = \frac{3^{1/2}\,(1 - \nu_f^2)\,\sigma_o^{5/2}}{2^{1/2}\,\gamma E_f^{3/2}\,(1 + \nu_c)^{1/2}} \tag{13.6}$$

Of course by using Equation (13.6), the optimum values of h_c, t_f, and G_c

given in Equations (13.3) through (13.5) can be written in longer expressions involving the load index (N_x/R).

The universal relation (13.6) means that for a given load index (N_x/R) and given face and core material there is one and only one face stress σ_o that will result in minimum-weight construction. If the face stress in a design is higher or lower than σ_o, the structure will weigh more than it would if the face stress is σ_o. As stated previously, this concept was found by NASA-Langley researchers for stiffened panels decades ago.

For the optimized construction, the weight equation can be written in terms of either the optimum face stress σ_o or in terms of the applied load per unit circumference, N_x. The result is

$$\frac{(W - W_{ad})}{R} = 2\rho_f\,(t_f/R) + \rho_c\,(h_c/R)$$

$$\frac{(W - W_{ad})}{R} = \frac{\rho_f\,3^{1/5}\,(1 - \nu_f^2)^{2/5}\,(N_x/R)^{3/5}}{2^{1/5}\,\gamma^{2/5}\,E_f^{3/5}\,(1 + \nu_c)^{1/5}} + \frac{\rho_c\,2^{1/5}\,(1 - \nu_f^2)^{1/10}\,(N_x/R)^{2/5}\,(1 + \nu_c)^{1/5}}{3^{1/5}\,\gamma^{3/5}\,E_f^{2/5}} \tag{13.7}$$

One significance of the weight Equation (13.7) is that in selecting various face materials, the isotropic or quasi-isotropic composite material that has the highest ratio of $E_f^{3/5}/\rho_f$ will result in the minimum weight structure. Hence the Factor of Merit (FM) for this structure is

$$FM = \frac{E_f^{3/5}/\rho_f}{(1 - \nu_f^2)^{2/5}} \approx E_f^{3/5}/\rho_f \tag{13.8}$$

In the optimized construction, the empirical factor, γ, is given by Equation (12.3) wherein ϕ, given in (12.4), can be written for the optimum construction as

$$\phi = \frac{1}{29.8}\sqrt{\frac{2\gamma E_f}{(1 - \nu_f^2)\sigma_o}} \tag{13.9}$$

It is seen that a simple iteration must be performed to obtain γ for the optimum construction.

As stated above, Equation (13.8) provides a way of selecting various materials to achieve a minimum-weight circular shell. Several typical materials are

TABLE 13.1. Factors of Merit for Various Materials for a Quasi-Isotropic Sandwich Cylindrical Shell with an Axial-Compressive Load.

Material	Conf.	V_f (%)	E(msi)	ρ(lb/in.^{3})	σ(ksi)[a]	FM (Eq 13.8)[b]
C/Epoxy	Cross-ply	58	12.04	0.0555	55.1 (T)	80.18
B/Epoxy	Cross-ply	60	15.37	0.0721	55.1 (T)	71.46
Kevlar 49/Ep	Cross-ply	60	5.8	0.0505	—	56.85
SiC/Ti	Woven, Iso.	39.5	28.7	0.1432	228 (T)	52.33
SiC/Ti-6Al-4V	Woven, Iso.	35	27.0	0.492	245 (T)	48.42
7075-T6 Alum	Iso		10.4	0.101	73.0 (Y)	40.36
Titanium	Iso		17.4	0.16	145	34.69
E glass/Ep	Cross-ply	57	3.12	0.0710	82.0	27.88
Welton Steel	Iso		30.45	0.283	126.9	27.43

[a]Compressive strength; (T) denotes tensile properties only are available. (Y) denotes yield.
[b]Here (E in msi)$^{3/5}$/(ρ in lb/in.^{3}).

analyzed and shown in Table 13.1. From Table 13.1, it is simple to select the materials to use from the *FM* column. However, cost, manufacturability, availability, environmental effects, etc., must also be considered; but the factor of merit is a starting point. Also, the *FM* does not portray what maximum load-carrying capability the sandwich shell may have for this configuration and load, because the *FM* does not provide an upper value of face stress, ie. $\sigma_o \leq \sigma_{all}$.

As examples, three materials are selected to indicate the use of the optimization method of Equations (13.3) through (13.9) above. They also point out some interesting conclusions that can be obtained from the procedures. The materials selected from Table 13.1 are carbon/epoxy, E-glass/epoxy, and 7075-T6 aluminum for face materials. In each of the optimizations the core properties given for rigid Klegecell is used for the core material, with the core properties given in Figure 13.1. The optimized constructions are given in Table 13.2 and are plotted in Figure 13.2 with the shell weight as a function of load index. One obvious result is that the use of C/Epoxy faces results in a cylindrical shell weighing less than if 7075-T6 faces were used, which in turn weigh less than if E-glass/epoxy faces were used. This was predicted by the $FM = E^{3/5}/\rho_f$ ratio in Table 13.1. However, only by the use of the universal relationship of Equation (13.6) can one find the upper limit of the load that each material system can resist. Although the carbon/epoxy cross-ply is the lightest of the three materials shown in this example, it is limited to a load index of N_x/R = 25.45. At that load index the maximum allowable face stress of 55,100 psi is reached. Likewise, 7075-T6 aluminum is limited to a load index, of 60.09. If

TABLE 13.2. Optimized Structures for Three Material Systems.

Face Material	σ_o (10^3 psi)	γ	N_x/R (psi)	$(W - W_{ad})/R$ (10^{-5} lb/in.^{3})	h_c/R 10^{-3}	t_f/R 10^{-4}	t_f/h_c	G_c (psi)	ρ_c (10^{-3} lb/in.^{3})
C/Epoxy	30	0.54	6.289	1.770	4.402	1.048	0.0238	1428	1.38
C/Epoxy	30	0.575	12.12	2.546	5.511	1.516	0.0275	2200	1.57
C/Epoxy	55.1	0.61	25.45	4.410	7.156	2.310	0.03227	3556	2.58
E-glass/Epoxy	10	0.53	3.116	2.512	5.769	1.558	0.027	540	0.521
E-glass/Epoxy	20	0.65	14.37	6.490	9.407	3.583	0.038	1556	1.50
E-glass/Epoxy	30	0.695	37.04	11.14	13.20	6.173	0.0468	2806	1.8
E-glass/Epoxy	82	0.79	102.4	59.44	31.73	24.54	0.0773	12730	7.75
7075-T6	30	0.56	7.554	3.271	4.914	1.259	0.0256	1537	1.48
7075-T6	50	0.62	24.46	6.594	7.397	2.447	0.0331	3307	2.22
7075-T6	73	0.65	60.09	12.42	10.3	4.116	0.0400	5834	3.99

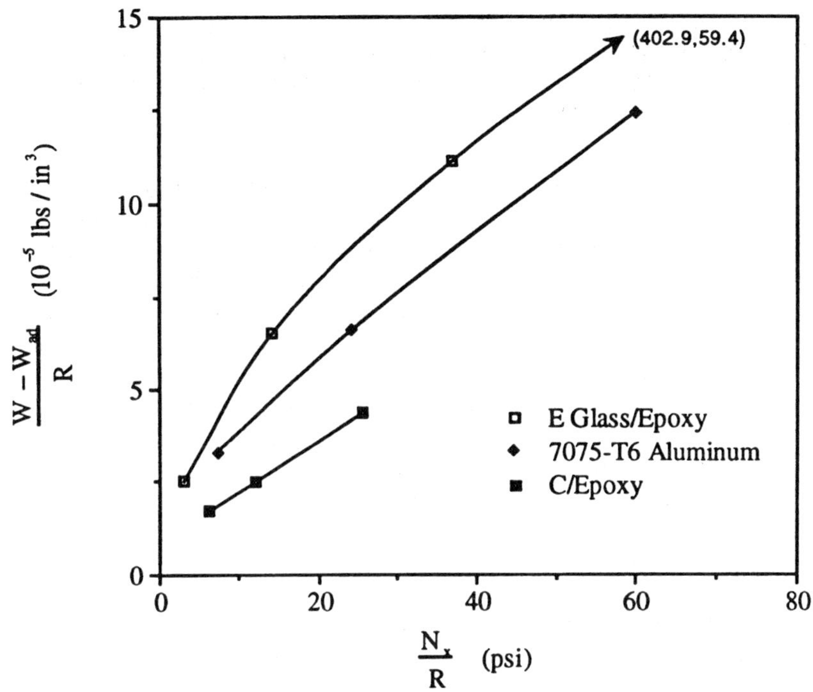

FIGURE 13.2. Comparison among optimum constructions for foam-core sandwich cylinders subjected to a uniform compressive axial load. (Reprinted from Reference [8].)

very high loads are imposed, of these three materials only the E glass/epoxy cross-ply can be used. This material can be used up to a load index of 402.9, over 6 times that of the 7075-T6 aluminum.

Comparisons of other variables are given in Reference [2]. However, several things are quite clear from the foregoing:

(1) In developing better quasi-isotropic or isotropic face materials, the parameters to improve are $FM = E_f^{3/5}/\rho_f$, as well as the allowable compressive strength in order to withstand higher load indices.

(2) For developing better core materials, increasing the slope of the G_c-ρ_c curve of Figure 13.1 is clearly the goal.

(3) From Reference [2] it is shown that straying from any of the optimized quantities results in a serious weight penalty or reduction in load-carrying ability or both.

13.3 MINIMUM-WEIGHT OPTIMIZATION OF A SOLID- OR FOAM-CORE SANDWICH CYLINDRICAL SHELL WITH ORTHOTROPIC COMPOSITE MATERIAL FACES SUBJECTED TO AN AXIALLY COMPRESSIVE LOAD

In Section 12.2, dealing with the analysis and design methods for this subject, three critical stresses are given pertaining to overall instability, face wrinkling, and core-shear instability, each of which would cause structural demise. For a given set of materials, there are three variables, namely, face thickness, t_f, core depth, h_c and core-shear modulus, G_c. Thus, there are three equations and three unknowns. This suggests a unique solution may be available for these dependent variables.

Since any one of the instabilities causes structural failure, it follows that the most efficient structure occurs when all three buckling stresses are equal. This follows the philosophy of the weakest link in the chain. Otherwise, material is being used inefficiently and can be shifted or eliminated to reduce weight and/or increase load-carrying ability. Vinson (and others) have used this technique, and have optimized honeycomb core, foam core, truss core, and web core sandwich panels subjected to in-plane compressive and shear loads, as discussed earlier in this text in Chapter 10.

In carrying out the optimization, one can again employ the following elasticity relationship for the isotropic foam core:

$$G_c = E_c/2\,(1 + \nu_c) \tag{13.10}$$

Because all three buckling stresses are equated in the optimized construction, these critical face stresses shall be combined and are simply labeled σ_o. It must be remembered that the optimized face stress, σ_o, must not exceed some allowable stress based upon static strength, factors of safety, fatigue limits, etc.

From equating (12.5) and (12.6)

$$G_c = \frac{6^{1/2}}{2}\frac{(1 - \nu_{x\theta}\nu_{\theta x})^{1/2}\,\sigma_o^{3/2}}{(1 + \nu_c)^{1/2}\,(E_x E_\theta)^{1/4}} \tag{13.11}$$

Utilizing Equations (12.5), (12.6), and (13.11), (12.9) can be written in the form of a perfect square in (h_c/R), with the result that for the optimum construction,

$$\frac{h_c}{R} = \frac{2(1 - \nu_{x\theta}\nu_{\theta x})^{1/2}\,\sigma_o}{\phi\,(E_x E_\theta)^{1/2}} \tag{13.12}$$

Proper substitution into one of several equations, results in

$$\frac{t_f}{R} = \frac{3^{1/2}}{2^{1/2}} \frac{(1 - \nu_{x\theta}\nu_{\theta x})\, \sigma_o^{3/2}}{\phi\,(1 + \nu_c)^{1/2}\,(E_x E_\theta)^{3/4}} \tag{13.13}$$

Substitution of Equation (13.13) into (12.1) provides the load index-optimum face stress relationship, referred to by previous authors as the "universal relationship":

$$\frac{N_x}{R} = \frac{6^{1/2}\,(1 - \nu_{x\theta}\nu_{\theta x})\, \sigma_o^{5/2}}{\phi\,(1 + \nu_c)^{1/2}\,(E_x E_\theta)^{3/4}} \tag{13.14}$$

This equation is very important because whenever σ_o is the allowable stress of the face material, then Equation (13.14) defines the upper limit of the applied load that the optimized structure can withstand. Any attempt to apply a larger load to this structure results in a prohibitive weight penalty.

Finally, the weight relationship (13.1) is written in terms of the applied load index for the optimized construction:

$$\left(\frac{W - W_{ad}}{R}\right) = \frac{6^{1/5}\,\rho_f\,(1 - \nu_{x\theta}\nu_{\theta x})^{2/5}\,(N_x/R)^{3/5}}{\phi^{2/5}\,(1 + \nu_c)^{1/5}\,(E_x E_\theta)^{3/10}}$$

$$+ \frac{2\,\rho_c\,(1 - \nu_{x\theta}\nu_{\theta x})^{1/10}\,(1 + \nu_c)^{1/5}\,(N_x/R)^{2/5}}{6^{1/5}\,\phi^{3/5}\,(E_x E_\theta)^{1/5}} \tag{13.15}$$

In this equation, the weight per unit area of the two faces is determined from the first term on the right-hand side, and the core weight per unit area is determined by the second term. Remember that in the second term, ρ_c, is uniquely related linearly to G_c, by an equation similar to Equation (13.16) below. This is used herein, because the optimum shear modulus is given by Equation (13.11) for the foam-core material.

$$G_c = A\rho_c \tag{13.16}$$

where G_c is in lb/in.^{2}, ρ_c is in lb/in.^{3} and A $= 1.728 \times 10^6$, from Figure 13.1.

Examining Equation (13.15) it is obvious that face material selection, the minimum weight foam-core sandwich orthotropic face material is given by a factor of merit (*FM*) defined by

$$FM = (E_x E_\theta)^{3/10}/\rho_f \tag{13.17}$$

where the moduli of elasticity are *compressive* moduli. Note that if the face

TABLE 13.3. Material Properties for Various Materials.

Matl.	Conf.	V_f (%)	E_1 (msi)	E_2 (msi)	G_{12} (msi)	υ^{12}	ρ (lb/in.^{3})	σ_1[a] (ksi)	σ_2[b] (ksi)
Boron/6061 Al	UNI	50	32	20	6	0.3	0.09145	250	20
T300 Gr/934 Epoxy	UNI	60	23.69	1.7	0.94	0.3	0.0555	105	?
Boron/Epoxy	UNI	?	30	3	1	0.3	0.074	400	12
Hi St Gr/EP	UNI	60	21	1.5	0.7	0.25	0.057	180	6
Glass/Epoxy	UNI	?	7.8	2.6	1.3	0.25	0.0555	150	4
7075-T6 Al	ISO	—	10.4	10.4	3.92	0.3	0.101	73.0[c]	73.0

[a]Compressive strength.
[b]Tensile strength.
[c]Compressive yield stress.

material were isotropic or quasi-isotropic Equation (13.17) is identical to Equation (13.8).

$$FM = E_f^{3/5}/\rho_f$$

In Table 13.3 are representative examples of several of the major fiber and matrix systems of interest. All quantities have been defined previously except V_f, which is the fiber volume fraction. It is interesting to note how often it occurs that one or more properties of a material are lacking, which prevents an optimization, design, or analysis from being complete. The value of ϕ, see Equation (12.10), and the factor of merit (*FM*), see Equation (13.17), for these materials are given in Table 13.4.

It is seen from Table 13.4 that the boron/aluminum metal-matrix composite has the highest Factor of Merit of the representative face-material systems. In fact, using this boron/aluminum metal-matrix composite results in faces that are (52.12/75.92) or 31.4% lighter than using faces of a boron/epoxy

TABLE 13.4. Factor of Merit of Face Materials of Table 13.3.

Material	φ	*FM*
Boron/6061 Al	0.766	75.97
T300 Gr/934 Ep	0.5657	54.60
Boron/Epoxy	0.4806	52.12
Hi St Gr/Ep	0.5159	49.39
Glass/Epoxy	0.8127	44.45
7075-T6 Al	1.0	40.36

composite. In fact, fourteen metal-matrix composites (including continuous fiber, short fiber, and 3-D braided) had factors of merit higher than the best polymer-matrix composite (a T300 graphite/934 epoxy unidirectional composite). Of course, the factor of merit does not identify how much load the shell can withstand before failure.

The upper limit of load-carrying capability is determined by the universal relationship Equation (13.14). Here, σ_o is the optimum stress, where in this construction it is an allowable compressive stress in the axial direction, or an allowable tensile stress in the circumferential directions. In the former, Equation (12.1), determines the compressive axial stress, bounded by the allowable compressive stress in the material; in the latter, Equation (12.7), determines the circumferential tensile stress, bounded by the allowable tensile stress. It is obvious that in using a specially orthotropic material like most composites, the 1-direction (i.e., the fiber direction) should be the axial direction and the 2-direction the circumferential direction, since in Equation (12.7) it is seen that the circumferential stress is a Poisson's ratio effect. Studying Table 13.3, it is seen that the circumferential tensile stress is the limit on both the unidirectional boron/aluminum MMC, and the unidirectional glass/epoxy composite. Therefore, in some cases, the 90° properties (i.e., properties normal to the fiber direction) limit the compressive, axial load-carrying capability of a shell composed of unidirectional composite.

In fact from Equation (13.17) and the research of Handel and Vinson [6], for a given fiber-matrix composite lamina system in this application, the best stacking sequence to attain minimum weight is a cross-ply laminate. However, it would probably be limited to lower maximum loads, because the cross-ply compressive strength in the axial direction would be less than that of the unidirectional composite of the same material.

Several optimizations are provided as examples utilizing the materials of Table 13.3, the optimized construction given by Equations (13.11) through (13.15), the values of ϕ in Equation (12.10), and the foam-core relationship given by Equation (13.16). The results are presented in Table 13.5.

From the table above, for representative materials involving popular fibers and matrix materials, it is interesting to note that the weight parameter–load index relationships are such that the materials fall naturally into two load categories—small loads and large loads. Even though the Factor of Merit values clearly show the relative merits of the materials, three of these (the B/6061 aluminum, the glass/epoxy, and the 7075-T6 aluminum faces) fall into a range of load index values below 135 psi, while the other two materials (the Boron/epoxy and the high strength graphite/epoxy) are capable of load indices nearly 35 to 100 times greater than this. Thus, from these example materials, one would select the unidirectional boron/epoxy or the unidirectional high-strength graphite/epoxy over the other materials to resist high load values

TABLE 13.5. Optimized Constructions for Table 13.3 Materials.

Face Matl.	σ_o (10^3 psi)	N_x/R (psi)	$\frac{W - W_{ad}}{R}$ (10^{-5} lb/in.^{3})	h_c/R (10^{-2})	t_f/R (10^{-4})	t_f/h_c	G_c (psi)	ρ_c (10^{-3} psi)
B/6061 Al	50	11.21	2.695	0.5015	1.121	0.0224	2235	1.293
	75	31.57	5.617	0.7523	2.060	0.0274	4106	2.376
	100	63.4	9.468	1.003	3.171	0.0316	6321	3.658
	106.7	74.5	10.70	1.070	3.494	0.0327	6964	4.030
Boron/Ep.	100	462	60.86	4.367	23.10	0.0529	10580	6.123
	200	2614	248.0	8.734	65.35	0.0748	29930	17.32
	300	7203	594.4	13.10	120.1	0.0917	59980	31.81
	400	14790	1129.9	17.47	184.9	0.1058	84660	49.0
Hi St G/Ep	50	167.9	31.41	3.446	16.79	0.0487	4873	2.820
	100	949.9	116.3	6.892	47.49	0.0689	13780	7.976
	150	2617.7	264.1	10.34	87.25	0.0844	25320	14.65
	180	4131	387.2	12.41	114.7	0.0924	33290	19.27
Glass/Ep	20	14.76	4.948	1.082	3.691	0.0341	1365	0.789
	30	40.69	9.878	1.622	6.781	0.0418	2508	1.451
	40	83.52	16.43	2.163	10.44	0.0483	3862	2.235
	48	131.8	22.86	2.596	13.73	0.0529	5078	2.939
7075-T6	25	5.551	2.542	0.4585	1.110	0.0242	1210	0.7002
	50	31.40	8.039	0.9170	3.140	0.0342	3422	1.980
	73	80.88	15.55	1.339	5.539	0.0414	6036	3.493

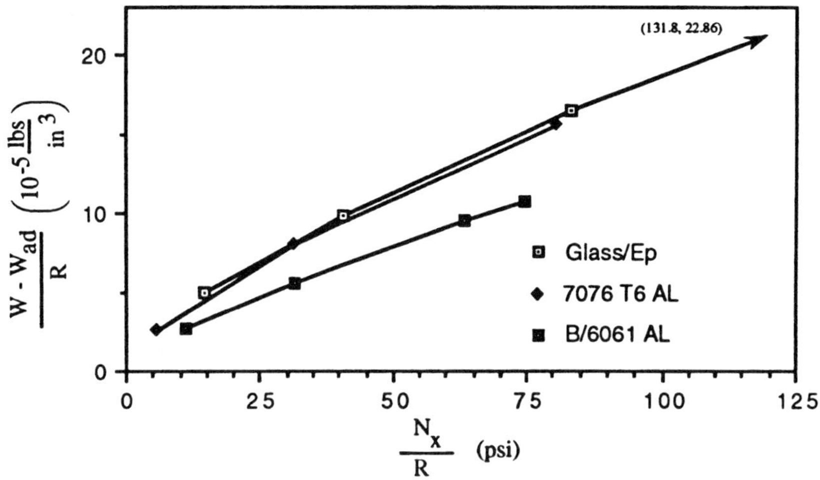

FIGURE 13.3. Weight parameter for various materials for optimized construction as a function of load index (low-load range).

for a given cylinder radius. In fact, for lowest weight, one would always select the boron/epoxy over the high-strength graphite/epoxy composite. This is shown in Figures 13.3 and 13.4.

Analytical methods to use in analyzing, designing, and optimizing foam-core (or solid-core) sandwich circular cylindrical shells subjected to an axially symmetric compressive load have been presented and discussed in this section and in Section 12.2. When using a specially orthotropic composite material the 1-direction is taken to be in the axial direction. For the optimized (minimum weight) construction Equations (13.11) through (13.15) are used. It has been shown in great detail [7] that any deviation from the optimized construction results in significant weight penalties.

In selection of face materials for a minimum-weight construction, Equation (13.17) is useful. For core materials, the core with the highest ratio of G_c/ρ_c is the best.

13.4 MINIMUM-WEIGHT OPTIMIZATION OF A HONEYCOMB CORE SANDWICH CYLINDRICAL SHELL WITH COMPOSITE MATERIAL FACINGS SUBJECTED TO AN AXIALLY-SYMMETRIC COMPRESSIVE LOAD

In Section 12.3, four critical stresses are given pertaining to overall instability, face wrinkling, core-shear instability, and face dimpling. For a given set

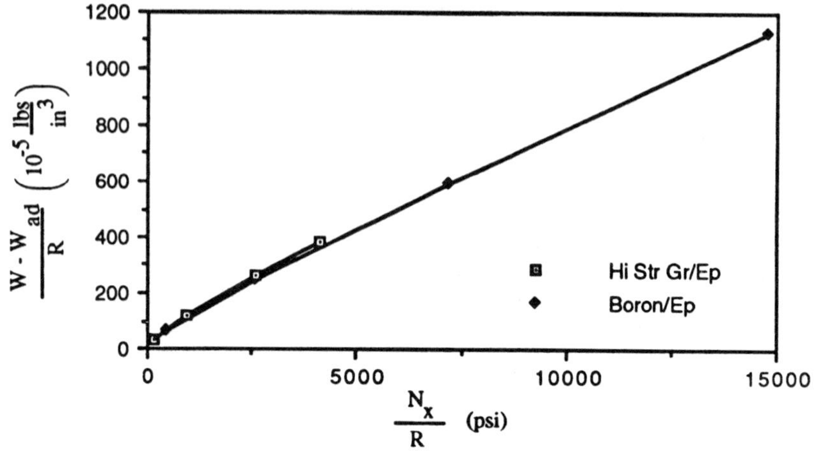

FIGURE 13.4. Weight parameter for various materials for optimized construction as a function of load index (high-load range).

of materials, there are four variables, namely, face thickness, t_f, core depth, h_c, cell wall thickness, t_c, and cell diameter, d. Thus, there are four equations and four unknowns. This suggests a unique solution may be available for these dependent variables.

Since any one of the instabilities can cause structural failure, it follows that the most efficient structure occurs when all four buckling stresses are equal. This follows the philosophy of the weakest link in the chain. Otherwise, material is being inefficiently used and can be shifted or eliminated to reduce weight and/or increase load-carrying ability. Therefore, in this process of optimization wherein all four buckling face stresses are equated, the optimum face stress, σ_{fx}, is denoted σ_o. Of course the upper bound on this face stress is an allowable material strength, either in the axial or the circumferential direction.

To optimize the honeycomb sandwich shell, one substitutes the effective core properties given by Equation (9.43) and (9.46) into the Buckling Equations (12.11) through (12.14) and proceeds to equate the four buckling face stresses (to be σ_o) in order to obtain values of three dependent variables such as t_f, t_c, and d in terms of the fourth variable, h_c. Substituting these geometric values into Equation (12.11), one finds h_c in terms of σ_o. The results for the optimum minimum-weight construction are found to be

$$\frac{h_c}{R} = \frac{2(1 - \nu_{x\theta}\nu_{\theta x})^{1/2}\,\sigma_o}{\phi\,(E_x E_\theta)^{1/2}} \tag{13.18}$$

$$\frac{t_f}{R} = \frac{3^{1/2}\,(k_3/k_2)^{1/2}\,(G_c/E_c)^{1/2}\,(1 - \nu_{x\theta}\nu_{\theta x})\,\sigma_o^{3/2}}{\phi\,(E_xE_\theta)^{3/4}} \tag{13.19}$$

$$\frac{d}{R} = \frac{6^{1/2}\,(k_3/k_2)^{1/2}\,(G_c/E_c)^{1/2}\,(1 - \nu_{x\theta}\nu_{\theta x})^{1/2}\,\sigma_o}{\phi\,(E_xE_\theta)^{1/2}} \tag{13.20}$$

$$\frac{t_c}{R} = \frac{3(2)^{1/2}\,(1 - \nu_{x\theta}\nu_{\theta x})\,\sigma_o^{5/2}}{\phi\,k_2\,E_c\,(E_xE_\theta)^{3/4}} \tag{13.21}$$

Substitution of Equation (13.19) into (12.1) provides the applied load index–optimum face stress relationship referred to by some previous NASA authors as the "universal relationship":

$$\frac{N_x}{R} = \frac{2(3)^{1/2}\,(k_3/k_2)^{1/2}\,(G_c/E_c)^{1/2}\,(1 - \nu_{x\theta}\nu_{\theta x})\,\sigma_o^{5/2}}{\phi\,(E_xE_\theta)^{3/4}} \tag{13.22}$$

As before, Equation (13.22) is very important because whenever σ_o is the allowable stress of the face material, then Equation (13.22) defines the upper limit of the applied load that the optimized structure can withstand. To attempt to apply a larger load to the honeycomb-core sandwich shell results in a prohibitive weight penalty, as shown in Reference [7].

Finally, the weight relationship (12.15) is written in terms of the applied load index for the optimized construction:

$$\left(\frac{W - W_{ad}}{R}\right) = \frac{2^{2/5}\,3^{1/5}\,(k_3/k_2)^{1/5}\,(G_c/E_c)^{1/5}\,\rho_f\,(1 - \nu_{x\theta}\nu_{\theta x})^{2/5}\,(N_x/R)^{3/5}}{\phi^{2/5}\,(E_xE_\theta)^{3/10}} + \frac{(k_2/k_3)\,\rho_c\,(N_x/R)}{G_c} \tag{13.23}$$

In this equation, the weight per unit area of the two faces is determined from the first term on the right-hand side, and the core weight per unit area is determined by the second term.

Studying Equation (13.23), it is obvious that for face-material selection, the lowest weight foam core sandwich face material is indicated by a Factor of Merit (*FM*) defined by

$$FM = (E_xE_\theta)^{3/10}/\rho_f \tag{13.24}$$

where the moduli of elasticity are *compressive* moduli. Equation (13.24) is identical to (13.17) for the foam-core sandwich. It is also seen that to minimize the core weight one should again utilize a material whose ratio of G_c/ρ_c is as high as possible.

The structural optimization of the honeycomb core composite face circular cylindrical sandwich shell subjected to axially compressive loads is now complete. It is seen that in the honeycomb core and the foam-core sandwich, the Factor of Merit for the facing material selection is the same. It is also interesting to note from Table 9.2 that in using the MIL-HDBK-23, values of k_2 and k_3, the maximum load-carrying capability (13.22) or weight (13.23) expressions for the optimum honeycomb sandwich cylindrical shell subjected to an axially symmetric compressive load, are unaffected by whether the honeycomb is hexagonal-cell or square cell.

It is not necessary to duplicate the comparisons of Section 13.3 for the foam-core sandwich because Equations (13.24) and (13.17) are identical. It is more relevant to compare the results of the foam-core optimized construction with the honeycomb core optimized construction for an example material—the Boron/Epoxy. In this comparison, an aluminum-honeycomb core is selected.

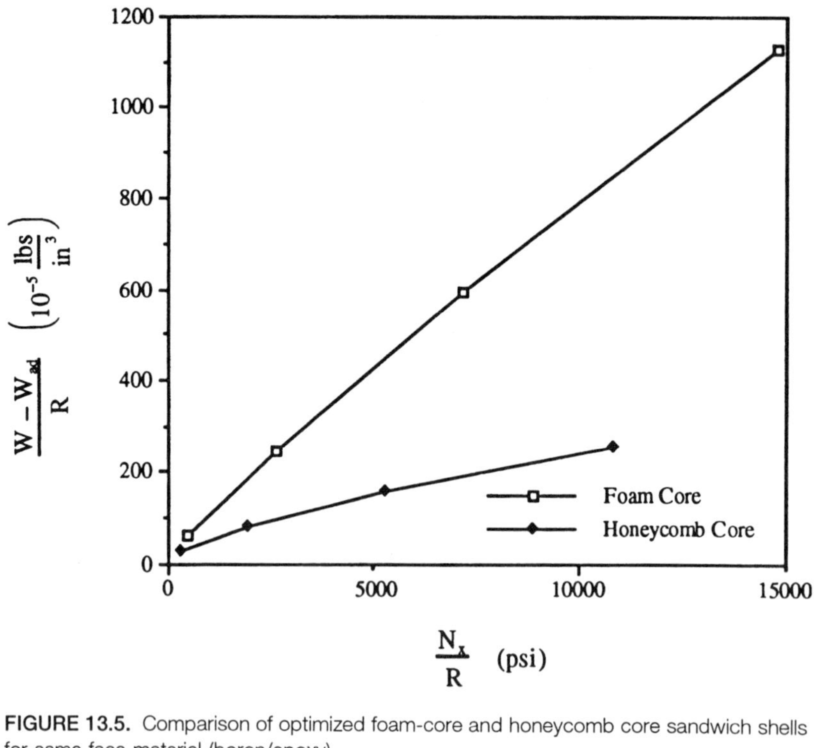

FIGURE 13.5. Comparison of optimized foam-core and honeycomb core sandwich shells for same face material (boron/epoxy).

The results are seen in Figure 13.5. It is seen that the maximum-load index for the optimum foam-core construction is 14,790 psi, while the maximum for the optimum honeycomb core shell is 10,849 psi (73.3% of the foam-core construction). Over the applicable load range, the honeycomb core construction with an aluminum-honeycomb core weighs less than the foam-core sandwich, with the optimized PVC foam. However, in many cases, the foam-core sandwich construction is less expensive than honeycomb core construction, and avoids moisture retention problems. In short, the selection of the proper core is an engineering decision, and the analyses presented herein assist in providing a rational quantitative comparison of weight as a function of load.

13.5 PROBLEMS

13.1 Using 7075-T6 aluminum facings and a proper foam core, determine the dimensions of an optimized foam-core sandwich to support the load of Problem 12.4.

a. What will this optimized shell weigh?
b. Choosing from among the materials of Tables 13.1 and 13.3, use the best face material and then determine the sandwich dimensions and shell weight for a foam-core sandwich to carry the load of (a) above.
c. Now optimize the sandwich construction for the best face material and an aluminum honeycomb-core and determine the sandwich dimensions and overall weight for the load of Problem 12.4a.
d. Lastly, consider an aluminum monocoque (7075-T6) shell to carry the load of Problem 12.4a, wherein the allowable strength is taken to be the yield strength (ie., F.S. = 1). What is the shell thickness h and the overall shell weight?
e. Compare the weights, the shell thicknesses, and the core thickness, for all of the above to lucidly see the relative advantages and disadvantages of the various constructions and materials.

13.6 REFERENCES

1. Technical Data Sheet, "Klegecell in Buoyancy," Klegecell Division of Hutchinson Corporation, Grapevine, TX.
2. "Termanti Rigid PVC Foam for Insulation Buoyancy and Sandwich Core Construction," Torin Inc., Ridgewood, NJ.
3. Vinson, J. R., "Comparison of Optimized Sandwich Panels of Various Constructions Subjected to In-Plane Loads," in *Sandwich Constructions I—Proceedings of the First International Conference on Sandwich Construction*, Stockholm, Sweden, Editors, K-A. Olsson and R. P. Reichard, publ., 1989, pp. 23–49.

4. Vinson, J. R., "Minimum Weight Solid Core Composite Sandwich Panels Subjected to In-Plane Shear Loads," American Society for Composites/University of Delaware CCM Joint Symposium on Composite Materials Science and Engineering, September 23–25, 1987.

5. Vinson, J. R., "Minimum Weight Solid Core Composite Sandwich Panels Subjected to In-Plane Compressive Loads," *Transactions of the International Conference on Composite Materials and Structures*, Madras, Jan. 6–9, 1988, EMAS Publications, United Kingdom, 1988.

6. Handel, P. I. and J. R. Vinson, "Optimal Stacking Sequences of Composite Faces for Various Sandwich Panels and Loads to Attain Minimum Weight," *Proceedings of the 29th AIAA/ASME/ASCE/AHS/ASC Structures, Structural Dynamics and Materials Conference*, pp. 999–1006, 1988.

7. Vinson, J. R., "Analysis and Optimization of Composite and Metallic Sandwich Cylindrical Shells," in *Sandwich Constructions 2—Proceedings of the Second International Conference on Sandwich Construction*, Gainesville, Florida, Editors, D, Weissman-Berman and K-A. Olsson, EMAS Publications, United Kingdom, 1992, pp. 377–400.

8. Vinson, J. R., *The Behavior of Shells Composed of Isotropic and Composite Materials*. Kluwer Academic Publishers, Dordrecht, The Netherlands, 1992.

APPENDIX 1

Core Materials

The following core material information is included to provide the student/engineer with an idea of what is commercially available. In no way is the information intended to be complete, nor is there any endorsement of any supplier intended. The L and W directions are explained in Figure A.1.

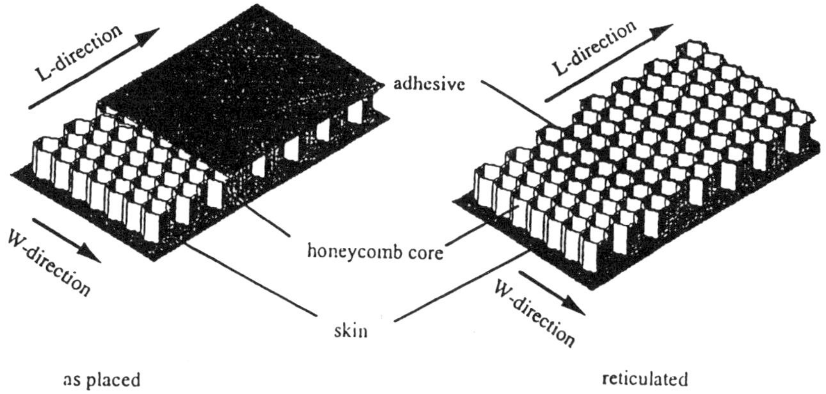

FIGURE A1. Sandwich-core nomenclature showing L and W directions.

ALUMINUM COMMERCIAL GRADE (ACG)

Cell Size (in.)	Density (lb/ft^3)	Shear Modulus	
		L (ksi)	*W* (ksi)
1/4	4.8	70	38
3/8	3.3	45	22
1/2	2.3	28	14
3/4	1.8	20	22
1	1.3	14	7

TPC FIBERGLASS-REINFORCED THERMOPLASTIC

Cell Size (in.)	Density (lb/ft^3)	Shear Modulus	
		L (ksi)	*W* (ksi)
3/16	4.0	28	7.5
3/16	5.5	32	12.5
1/8	3.0	19	6

ALUMINUM NAVAL GRADE (ANG)

Cell Size (in.)	Density (lb/ft^3)	Shear Modulus	
		L (ksi)	*W* (ksi)
3/8	4.2	65	29

HFT GLASS-REINFORCED PHENOLIC [1]

Cell Size (in.)	Density (lb/ft^3)	Shear Modulus	
		L (ksi)	*W* (ksi)
1/8	3.0	19	7.5
1/8	4.0	25	12
1/8	5.5	40	16
1/8	8.0	45	21.5
3/16	2.0	15	5
3/16	3.0	19	9
3/16	4.0	25	12
3/8	4.0	27	13

Composite Honeycomb Core [1]

HRP GLASS-REINFORCED PHENOLIC

		Shear Modulus	
Cell Size (in.)	Density (lb/ft^3)	*L* (ksi)	*W* (ksi)
3/16	4.0	13	6.5
3/16	5.5	19	11
3/16	7.0	30	14
3/16	8.0	33	19
3/16	12.0	48	28
1/4	3.5	10	5
1/4	4.5	15	8
1/4	5.0	20	10
1/4	6.5	25	13
3/8	2.2	6	3
3/8	3.2	12	5
3/8	4.5	14	8
3/8	6.0	25	12
3/8	8.0	27	18

NP GLASS-REINFORCED POLYESTER [1]

		Shear Modulus	
Cell Size (in.)	Density (lb/ft^3)	*L* (ksi)	*W* (ksi)
3/16	4.5	14	5.2
3/16	6.0	15	5.8
3/16	9.0	20	7.5
1/4	4.0	13	6
1/4	6.0	15	5.8
1/4	8.0	18	7
3/8	2.5	10	4
3/8	4.5	14	5.2

HFT-G GRAPHITE-REINFORCED HONEYCOMB [1]

Designation	Cell Size (in.)	Density (lb/ft^3)	Shear Modulus L (ksi)	Shear Modulus W (ksi)
HFT G	3/16	4.0	68	27
HFT-G-327	3/16	4.0	97	31
HFT-G-327	3/16	6.0	114	48
HFT-G-327	3/16	8.0	153	77
HFT-G-327	3/16	10.0	250	111
HDC-G-327	3/16	20.0	—	—
HFT-G-327	1/4	5.0	98	33
HFT-G-327	3/8	2.0	24	9

HRH-10 ARAMID FIBER-REINFORCED HONEYCOMB [1]

Cell Size (in.)	Density (lb/ft^3)	Shear Modulus L (ksi)	Shear Modulus W (ksi)
1/16	3/4	6	2.9
1/8	1.8	3.8	1.5
1/8	3.0	6.5	3.5
1/8	4.0	8.6	4.7
1/8	5.0	10.2	5.4
1/8	6.0	13.6	6.5
1/8	8.0	16	9.5
1/8	9.0	17.5	11
3/16	1.5	3	1.6
3/16	1.8	3.8	1.9
3/16	2.0	4.3	2.1

HRH-327 GLASS-REINFORCED POLYIMIDE HONEYCOMB [1]

		Shear Modulus	
Cell Size (in.)	Density (lb/ft^3)	*L* (ksi)	*W* (ksi)
	3.2		
1/8	19	3.4	
1/8	5.5	30	14.5
3/16	4.0	24	10
3/16	4.5	33	11
3/16	5.0	37	12
3/16	6.0	45	15
3/16	8.0	55	22
3/8	4.0	29	12
3/8	5.5	41	13
3/8	7.0	53	18
3/16	3.0	6.5	3.4
3/16	4.0	7.8	4.7
3/16	6.0	13	6.5
1/4	1.5	3	1.3
1/4	2.0	4.2	2
1/4	3.1	6.5	3
1/4	4.0	8	3.5
3/8	1.5	3	1.5
3/8	2.0	3.7	2.4
3/8	3.0	6.5	3

ECA-1 ARAMID FIBER (PHENOLIC RESIN COATED) [2]

Cell Size (in.)	Density (lb/ft^3)
1/4	1.5
1/8	3
3/16	1.8
3/8	2
5/32	1.8
1/2	2
1/2	2
3/4	1.5
1/4	1.5
1/8	3
3/16	1.8

Cell Size (in.)	Density (lb/ft^3)
3/4	1.5
3/8	2
5/32	1.8
3/16	1.8

ECP SPECIAL PHENOLIC FIBER H8 (POLYPROPYLENE [3]

Cell Size (in.)	Density (lb/ft^3)
3/16	3.1

FOAM CORE [3,5]

Core	Density (lb/ft^3)	Shear Strength (psi)	Shear Elongation (%)
Divinycell H80	5	135	15
Divinycell H100	6	190	14
Core-Cell A300	3	85	58
Core-Cell A400	4	105	61
Core-Cell A500	5	145	61
Core-Cell A600	6	190	60
Airex R63.50	3.75		
Airex R63.80	5.15	155	78
CK-100	9.5		
CK-150	15.5		
CK-57	6.5		
Airlite B-2.5	2.5		
Airlite B-3.75	3.75		
Airlite B-5.00	5		
Airlite B-6.25	6.25		
Airlite B-12.5	12.5		
Elfoam T200	2.0		
Elfoam T250	2.5		
Elfoam T300	3		

REFERENCES

1. Hexcel Honeycomb TSB 120 Mechanical Properties of Hexcel Honeycomb Materials, 1992.
2. *Professional Boat Builder,* "Marine Composites", 1997.
3. *ATC News,* February 1995.
4. Nida-Core Publication,
5. Christensen, R.M., Patent pending.

Face Materials

The mechanical properties of some of the materials that can be used for face materials in sandwich construction are found throughout the text. Rather than repeat them here, reference is made to their location.

Table/Problem	Page Number
Table 2.1	59
Table 2.2	60
Problem 4.1	99
Table 10.1	277
Table 10.3	298
Problem 10.1	302
Table 13.1	337
Table 13.3	342

American Society for Testing Materials (ASTM) Standards for Sandwich Structures and Materials

Listed below is a partial list of ASTM Standards presently in use:

	ASTM	MIL-STD
Edgewise Compressive Test	C364	401B
Long Beam Flex Test	C393	401B
Core Shear	C273	401B
Aging	C481	
Peel Test	D1781	401B
Impact Test	D3029	
Flatwise Tensile Test	C297	401B
Flatwise Compressive Test	C365	401B
NIST (NBS) Smoke Chamber Test	662	

INDEX